Studies in Systems, Decision and Control

Volume 86

About this Series

The series "Studies in Systems, Decision and Control" (SSDC) covers both new developments and advances, as well as the state of the art, in the various areas of broadly perceived systems, decision making and control- quickly, up to date and with a high quality. The intent is to cover the theory, applications, and perspectives on the state of the art and future developments relevant to systems, decision making, control, complex processes and related areas, as embedded in the fields of engineering, computer science, physics, economics, social and life sciences, as well as the paradigms and methodologies behind them. The series contains monographs, textbooks, lecture notes and edited volumes in systems, decision making and control spanning the areas of Cyber-Physical Systems, Autonomous Systems, Sensor Networks, Control Systems, Energy Systems, Automotive Systems, Biological Systems, Vehicular Networking and Connected Vehicles, Aerospace Systems, Automation, Manufacturing, Smart Grids, Nonlinear Systems, Power Systems, Robotics, Social Systems, Economic Systems and other. Of particular value to both the contributors and the readership are the short publication timeframe and the world-wide distribution and exposure which enable both a wide and rapid dissemination of research output.

More information about this series at http://www.springer.com/series/13304

Maurizio d'Amato · Tom Kauko
Editors

Advances in Automated Valuation Modeling

AVM After the Non-Agency Mortgage Crisis

Editors
Maurizio d'Amato
Property Valuation and Investment
DICATECh
Technical University Politecnico di Bari
Bari
Italy

Tom Kauko
School of the Built Environment
Oxford Brookes University
Headington, Oxford
UK

ISSN 2198-4182 ISSN 2198-4190 (electronic)
Studies in Systems, Decision and Control
ISBN 978-3-319-84231-8 ISBN 978-3-319-49746-4 (eBook)
DOI 10.1007/978-3-319-49746-4

Printed on acid-free paper

This Springer imprint is published by Springer Nature
The registered company is Springer International Publishing AG
The registered company address is: Gewerbestrasse 11, 6330 Cham, Switzerland

Foreword

The recent financial crisis has lead researchers to argue that existing valuation methods or the application of these techniques are inadequate to cope with the complexities of today's property market. This latest contribution from d'Amato, Kauko and guest authors presents new research examining current practice and providing examples of new methods and adaptations to improve the reliability of valuation and mass appraisal techniques. As such, it is a valuable addition to the literature on automated valuation and mass appraisal.

The book is divided into five parts with contributions from several experts in this field who focus on the different aspects associated with the application and development of real estate appraisal methods. Part I focuses on the emerging problems associated with property valuation. d'Amato and Kauko examine the theoretical background and application of property valuation and find that academic research still appears to focus on the integration of financial and property markets. The continued use of models based on the concept of a perpetually increasing income could, they argue, lead to a repeat of the errors which caused the 2008 global property crash. The role of automated valuation methods (AVM) in the 2008 financial crisis is discussed by Mooya who concludes that the use of AVMs are highly dependent on the both the assessor's understanding of a specific market and the inclusion of additional data to capture the market context.

In Part II, case studies from Italy, Germany and Turkey provide examples of AVM in practice and the impact of banking reform measures in emerging markets. Eilers and Kunert carry out an analysis of REITs in Turkey following the introduction Basel III. Analysing the performance of Turkish direct real estate investment compared with that of real estate investment companies (REICs) based on three assess classes (residential, office and retail), they found that REICs did not perform as well as direct real estate investments and had become a 'developer's vehicle' for construction companies and contractors.

The problems associated with calculating value when data is sparse is addressed by d'Amato, Cvorovich and Amoruso who explore the potential use of the Short Tab Market Comparison Approach as a method to statistically define the relationship between property prices and their characteristics where there is little data available. Moving away from the standard approach to AVM, Ciuna and Salvo propose an automatic procedure based on the Market Comparison Approach to define equations related to a specific market place rather than treating all markets as homogenous. Ciuna, De Ruggiero and Salvo address the use of the Income Approach, as recommended by the International Valuation Standards, in situations where there is a lack of comparable data. They propose calculating the capitalisation rate with an automated valuation model which is based on a real estate database built through a computerised geocoding automatic procedure rather than a capitalisation rate which is generally extracted from a different segment of the market.

Part III looks at the methodological challenges of using AVM. Del Giudice and De Paola undertake a spatial analysis of the residential rental market in central Naples with geoadditive models based on a penalised spline function in order to improve upon the usual Kriging techniques. They find that this approach is reliable, efficient and flexible and as such useful in modelling realistically complex situations.

Locational attributes are also the focus for the study by Curto, Fregonara and Semeraro who introduce a new approach to measure the relative improvement in price and asset liquidity prediction when the location is known. The use of different AVMs is examined by d'Amato and Amoruso in two case studies to explore first, the relationship between DCF inputs and outputs and second, the use of Locational Value Response Surface Modelling. More complex specifications of locational characteristics are investigated by Bidanset, Lombard, Davies, McCord and McCluskey who examine the impact of Kernel and Bandwidth specification of geographically weighted regression on the equity and uniformity of mass appraisal models.

The ability of AVM models to cope with diverse market conditions is explored by Kesken and Dunning who found that multilevel AVM is an ideal tool to calculate the effect of earthquakes on the housing market in Istanbul; a location with frequent seismic activity. Kesken and Dunning suggest that, appraisers working in segmented markets with natural disasters should consider the methodological advantages of using multilevel models to estimate value impacts in these locations. The appropriateness of using multilevel mass appraisal models approaches was also examined by Ciuna, Salvo and Simonotti in their appraisal of residential apartments in Palermo, Italy.

Part IV considers two different AVM approaches. First, the use of fuzzy logic is proposed by González to overcome the problems associated with defining market segments, where boundary lines for each location or sub-market are often blurred. Second, the main issue for many researchers and assessors is the lack or unavailability of data. D'Amato and Renigier-Bilozor found that where data is scarce the issue of calculating a single point estimate can be overcome using rough set theory.

In Part V, consideration is given to reducing inaccuracies in valuations. Although property valuation has been often been called an 'art' and not a 'science', this sentiment, arguably, reflects the assessors application of the available techniques rather than the valuation methods and underlying concepts themselves.

McCluskey and Borst focus on the way in which comparables are selected and weighted to reflect the subject property. They state that, with advances in research the subjectivity associated with comparable selection and the determination of variable weightings can be minimised and therefore the sales comparison approach can now be viewed as a more scientific approach, rather than one based on the knowledge and expertise of the assessor. Appraiser bias is also addressed by Lausberg and Dust who discuss the problems associated with anchoring heuristics, where appraisers anchor to reference points; such as a previous valuation and make adjustments to it to reach their estimate. They state that, while many studies have acknowledged the importance of the anchoring effect in appraisals, the accuracy of the valuation can be increased through the use of improved valuation software which includes a decision support tool.

The breadth of research presented here provides a sound basis for the next step in the evolution of AVM with exciting examples of new techniques to improve on the current valuation methods adopted by assessors.

Dr. Sally Sims
Oxford Brookes University

Preface

To extract key dimensions from a complex set of micro-level market data requires the use of high-quality data cross-sections and a robust modelling tool. Such tools have been developed within a realm known as mass appraisal: systematic economic valuation of groups of properties using standardised procedures largely based on the multiple regression analysis (MRA). Hitherto mass appraisal has been mostly restricted to taxation, although mortgage lending is fast becoming another widespread application area. In a more generic sense, mass appraisal offers an untapped possibility to link the property value with various characteristics of the building, plot and its vicinity, as well as social and functional features of the neighbourhood and local area. Ideally the data should cover differences in socio-economic aspect and differences in environmental aspects such as pollution. At present, valid property value data is easy to find in some countries and difficult (or even impossible) to find in others.

Following the experience of our edited book Mass appraisal methods—an international perspective for property valuers RICS Series, Blackwells, Oxford, 2008, a number of our colleagues who read it proposed a sequel focusing our attention on a concept known as automated valuation method/model (AVM, Automated Valuation Methodology). The present book picks on this request and poses some questions about AVM methodology. For this reason we have raised a number of issues: in particular, on the current methodological framework of AVM, about the main problems encountering AVM applications, and what we realistically could do to improve AVMs so as to make our financial—and by implication, social—world safer. This line of research seeks to contribute to the current debate on AVMs especially after the crisis of 2007–2008. After this extensive and tragic economic crisis we are entitled to have our doubts and we are also increasingly concerned about the social responsibility of AVM for the stability of our economies. As a consequence our field of research now has an opportunity to contribute, in an effective way, to improve the stability of our financial system. AVMs may be helpful in several fields. They can, for example, be used in the collateral estimation, in the valuation of real estate portfolios. According to Basel II agreement and EU Directive 2006/48/CE, banks should provide periodic automatic valuation to appraise properties for which acquisition has been financed in the mortgage lending process. In this valuation activity for mortgage lending purposes statistical and mathematical modelling may be used in combination with valuation.

When we examine strategic issues within mass appraisal AVMs are relevant due to their huge financial—and as a consequence also socio-economic—significance. How to avoid—or at least mitigate—a new financial crisis stemming from real estate market bubbles? So this is about socio-economic sustainability. The crisis showed that AVM can work in a normal situation with rising and stable prices, but not in a more abnormal one with falling prices. Since then a debate is emerging, but it is still not sufficiently developed in terms of conclusions between any connection between data, methods and the financial consequences.

To remind the initiated readers—and to demonstrate the point for the uninitiated ones—in our prior book on mass appraisal we followed a line of argumentation based in what we discovered was a contemporary problem—the difficulty of promoting development in the valuation paradigm. Since then, however times have changed, towards more favourable attitudes among the real estate research community, more people being involved, higher level of technical and methodological expertise, more and better datasets, greater R&D activity and data management responsibility of the private sector, the development of ICT and hardware, and not least, the new reality imposed on us by the massive global market meltdown (with consequences thereof) from 2007–2008 onwards. Because of these changes, the focus of our present book is rather different than what was the case documented in the prior book, almost a decade ago. In the present book each chapter makes a cut into the problem area we begun theorising in the previous book, rather than following a suggested line of argumentation—or vision—that would be common for all contributions. In the present compilation of papers the approach remains the same as in the prior one: we need to explore the unknown. This time we have not focussed on a competition amongst the results obtained applying different AVM methods as in the prior work. It is instead about an assemblance of different issues at stake, including best practices, real-life constrains, administrative procedures, software capabilities, expert competences, modelling frameworks, background theories and more.

When reading these books a detail in terminology is worth noting. Namely, in some instances the term computer-assisted mass appraisal (CAMA) is used instead of the term AVM. It is to observe that these two terms are not synonyms: AVM is about financial aims and mathematical procedures whereas CAMA pertains to any administrative end applications; however, plenty of overlap between these two realms exist as many methods can be used for both. At a technical level, the main difference between a CAMA estimate of value and the one produced by the AVM is the effective date of the appraised value estimate. CAMA systems value all properties in a jurisdiction as of a statutory valuation date such as January 1st of each year. On the other hand, AVMs usually are designed to produce a value estimate that coincides with the sale date of the property.

In USA the use of CAMA started in the 1970s and has since then spread around the world. During the last two decades CAMA has developed in an impressive way.

Here it is to note that, in 1999, the Appraisal Standard Board replaced the term 'estimate of value' with the 'opinion of value' in the USPAP. A clear distinction was made between two important and distinct definitions. The opinion of value regards the final results of an in person valuation and the estimate of value has been indicated as the final results of an AVM. It is worth noting that some institutions consider AVM assisted valuation more reliable than valuation in person.

Lastly, we would like to pay respect to the personal aspirations of all those colleagues, who have helped us develop our research agenda during the past 10–15 years period. To provide a brief summary, a group of academics with broadly similar interests (i.e. appraisal, valuation and market analysis) started working in two meeting organised by the OTB research Institute of Delft University of Technology, the Netherlands—this was in 2006 and 2007. For this reason we usually call this group the Delft group, even if frequent communications among many members of this group already existed a few years before that (the absolute starting point being the ERES meeting in Alicante in 2001). Then we continued with extending this network. Several authors joined our group after a large meeting arranged in Rome in 2010. While the list of authors in this book already gives an idea of this consistence, the whole group of people involved is too large to list here, and to mention only a few names would not be fair to those left out. Here is an exception, however: in this vein we have dedicated this book to the memory of Prof. Koloman Ivanicka Jr. of STU Bratislava, a passionate researcher and a joyful friend of ours.

Bari, Italy
Headington, Oxford, UK
July 2016

Maurizio d'Amato
Tom Kauko

Contents

Editors and Contributors

About the Editors

Maurizio d'Amato is Associate Professor at DICATECh, Technical University Politecnico di Bari, Italy, since 2006 where he teaches real estate investment and valuation. He completed his undergraduate work in economics at the University of Bari and worked for several banks (Bank of Rome, Bank of Salento, Micos—Mediobanca) in real estate finance before entering the doctoral programme in Planning, specialising in Valuation methods, at the Politecnico di Bari. After completing this programme, he served as a contract professor in Real Estate Valuation for several years. During this time he received research grants from the Italian Council of Research (CNR) for projects undertaken at the University of Florida in 1997, 1998 and at the University of Alicante Spain in september 2000. He received the faculty appointment of Researcher at the Politecnico di Bari in 1999. He has been Scientific Director of the Real Estate Center of Italian Association of Real Estate Counselor (AICI). He has been also professor of Real Estate Finance at University of Rome III, Real Estate Appraisal at SAA School of Business Administration University of Turin and Real Estate Appraisal at on line University UNINETTUNO. He was appointed Fellow Member of Royal Institution Chartered Surveyors in June 2004 and Recognised European Valuer in May 2012.

Tom Kauko has been Associate Professor in Urban Geography at the Norwegian University of Science and Technology (NTNU) since 2006. His research interests cover housing market analysis; evaluation of planning and urban regeneration; locational quality in housing consumption; institutional, evolutionary and behavioural property research, and alternative property valuation techniques. In 1994 he obtained his M.Sc. in Land Surveying at Helsinki University of Technology (HUT), with a major in Real Estate Economics and Valuation. During 1995–1996, he worked as a planner for the research department of National Land Survey of Finland in Helsinki (Maanmittauslaitos). During 1996–1997 he participated in a course in physical planning at the Centre for Urban and Regional Research of HUT (YTK). He subsequently moved to Utrecht University, Faculty of Geographical

Sciences, where he completed his dissertation in June 2002. From 2001 to 2006, he worked as a researcher at OTB, Delft University of Technology, where he still has visitor's affiliation.

Contributors

Kerem Yavuz Arslanlı is currently employed at Istanbul Technical University Institute of Social Sciences. He has been worked at ITU *Urban and Environmental Planning and Research Centre* 2002–2012. His dissertation is titled "Spatially Weighted House Price Index Model for Istanbul Metropolitan Area". He was visiting researcher at University of Alicante Institute of International Economics in 2010 and Cass Business School Department of Finance in 2011. He started lecturing, "Real Estate Finance" and "Urban Transformations and Real Estate Sector", at *İ.T.Ü. Real Estate Development Master Programme* since 2012. He has academic articles in international/national journals and several conference papers in international conferences. He has been a member of the board of directors of the *European Real Estate Society* (ERES), the leading real estate research and education organisation in Europe.

Paul E. Bidanset is a current Ph.D. candidate in Ulster University's School of the Built Environment, completing his dissertation under the supervision of Dr. Billy McCluskey and Dr. Peadar Davis. He is the Real Estate CAMA Modeler for the City of Norfolk, VA, Office of the Real Estate Assessor. He received his M.A. in Economics from Old Dominion University in 2013 and his B.S. in Economics from James Madison University in 2009. He is a member of the editorial review board for the Journal of Property Tax assessment and Administration, and serves on the International Association of Assessing Officers (IAAO) Research Subcommittee. On the subject of spatial automated valuation models and their impact on property tax equity, he has presented research at conferences around the world.

Richard A. Borst has been engaged in managing the design, development and implementation of computer-based real property information systems since 1973. He was president of North America's largest mass appraisal firm while at the same time maintaining his contributions to the technical aspects of mass appraisal systems. His technical background is evidenced in a number of published articles and conference presentations. He introduced artificial neural networks to the assessment community in 1990. He was appointed to a 3-year term in 1997 as a Visiting Research Fellow at the University of Ulster, Belfast, Northern Ireland. During this tenure he collaborated with members of the faculty at the University performing research in the fields of valuation modelling and the application of location effects in the model structuring and calibration process. He obtained a Doctor of Technology from the University of Ulster, Northern Ireland, a Master of Science in Industrial Engineering from the State University of New York at Buffalo and a Bachelor of Engineering Science, with honours, from the Cleveland State University.

Marina Ciuna is a Civil Engineer. 2002 and earned a Ph.D. on Transport Technique and Organisation. Since 2008, she has been Assistant Professor of Real Estate Appraisal—Faculty of Engineering, University of Palermo; and Contract Professor in Economy and real estate evaluation—Faculty of Engineering, University of Palermo. Since 2006, she has been Contract Professor in Real Estate Appraisal—Faculty of Architecture, University of Palermo; and a member of the Società Italiana di Estimo e Valutazione (SIEV). Since 2008, she has been Component of the Scientific Committee of the Independent Real Estate Valuers Association E-valuation (Istituto di Estimo e Valutazioni). She is the author of scientific contributions on real estate valuations and damages of environmental resources evaluations; she develops studies on methodologies of the real estate evaluation (Real estate market data monitoring); market-oriented procedures evaluation and quantitative analysis applications, and mass appraisal methodologies.

Yener Coskun have both government experience, as senior specialist, and academic background. He has been working for Capital Markets Board of Turkey since 1995 by specifically focusing on investment banking and capital market activities. At the academic side, his research areas are real estate, housing finance, mortgage markets, real estate appraisal, and capital markets. Yener spent 10 months at Wharton School, University of Pennsylvania, as a visiting scholar in 2002–2003 and awarded Ph.D. in 2013 at Ankara University (Turkey) The Graduate School of Natural and Applied Sciences, Real Estate Development Department. Dr. Coskun has MRICS designation since 2010 and various local professional designations related to capital markets and real estate appraisal. He has two published books and several journal articles on capital markets, real estate and housing finance. Yener is consultative member of RICS Sustainability Task Force Europe since 2012/July and has acted as the chair for European Real Estate Society (ERES) Ph.D. Student Committee in 2010/June–2013/July. As visiting lecturer, Dr. Coskun has been serving both University of Sarajevo (for Facility Management & Housing Market Courses in the M.Sc. in Applied Finance & Property programme at the School of Economics and Business) and also Izmir University of Economics (for the Real Estate Finance and Financial Markets Course) since 2012.

Rocco Curto is Director of the Department of Architecture and Design, Politecnico di Torino. Since 1999, he has been Full Professor in Property Valuation. He is Dean of the II Faculty of Architecture, Politecnico di Torino (2006–2012). He is Scientific Coordinator of "Territorio Italia", the scientific magazine of the Territorial Agency, Italian Ministry of Economy and Finance. In addition, he is Director of the Master's course in "Real Estate: Territorial Planning and Property Market" (1999–2012) and Director of the Master's course in "Management of Cultural and Environmental Heritage". He is responsible for the Turin Real Estate Observatory. He is President Second Level Specialisation Course in Architecture (restoration and enhancement) and Three-Year Graduate Course in History and Conservation of Architectural and Environmental Heritage, Politecnico di Torino (2000–2006). His research topics include property market; application of statistical techniques (forecasting and probability), mass appraisals; functional and territorial segmentation of the property market; economic-financial evaluation of projects and plans; enhancement and management of historical and architectural heritage; analysis of real estate investments; reform of the land register and revision of the cadastral rent.

Pierfrancesco De Paola is Engineer and Ph.D. in Conservation of Architectural and Environmental Heritage (*Mediterranean* University of Reggio Calabria, Curriculum in Economic Evaluation of Conservation Projects, S.S.D. ICAR22/Estimo), Adjunct Professor on Economics and Real Estate Appraisal in the Course of Study on Architecture and Building Engineering at the University of Basilicata, former Research Fellow in Quantity Surveying at the Faculty of Engineering of the University of Calabria, Collaborator and Teaching Assistant in the Professorship of Real Estate Appraisal Basic Sciences School for University of Naples *Federico II*, former Professor in the Master for the "Management of Community Programmes: Real Estate" at the School of Business Administration of the University of Turin, already Professor of Academic Course "Quantitative Methods and Statistics for Urban Regeneration" within the International Doctorate Program in "Urban Regeneration and Economic Development", sponsored by the European Program Marie Curie Action and established by Mediterranean University of Reggio Calabria, University of Rome La Sapienza, the Aalto University (Finland), the Salford University (UK), the Northeastern University of Boston (Massachusetts, USA) and San Diego State University (California, USA). Since 2005 he has held activities of educational cooperation in the various training courses related, first at the Faculty of Engineering of the University of Calabria and currently at the Polytechnic and Basic Sciences School of the University of Naples Federico II. He is author and co-author of research and scientific and professional publications in the sectors of Real Estate Appraisal and Economic Evaluation of Investment Projects.

Manuela De Ruggiero is a Civil Engineer and earned is Ph.D. in Environmental Technologies and Planning. Since 2007, she has assisted the teacher of *Real Estate Appraisal*—Department of Civil Engineering (DICI), University of Calabria. Since 2011, she has been a teacher of *Real Estate Appraisal*—Department of Mathematical Sciences and Physics, University of Calabria. She is author of scientific contributions on real estate valuations, and she does her studies on methodologies of the real estate evaluation and use of geographic information system in real estate applications.

Vincenzo Del Giudice has graduated in civil engineering at the University of Naples. He is Full Professor of the Chair of Economy and Estimate in the Polytechnic and Basic Sciences School at the University of Naples Federico II. He is Lecturer of the Course in Forensic Estimate at the Master in Forensic Engineering (University of Naples Federico II). He has taught specialisation course Advanced Finance. Ist. San Paolo di Torino—Institute for Research and Educational Activities (IPE). He was formerly Full Professor of the Chair of Economy and Estimate in the Faculty of Engineering of the University of Calabria. He has already been Professor of Estimate, Economics and Estimate Environmental and Economic Evaluation of Projects in the Faculty of Engineering at the University of Salerno, which was part of the Integrated Academic Senate. He previously taught the corse "Perfecting in Analysis, Evaluation and Planning of the Landscape" (Department of Territorial Planning—UNICAL, according to the Escuela Técnica Superior de Arquitectura del Vallés of Politécnica de Cataluña and in collaboration with the School of Landscape of Montreal and the University of Lisbon—Modern School of Architecture) and taught specialisation course "Design of Protective Operations by Hydrogeological Events"—Department of Ground's Defense, UNICAL. He is author of numerous scientific publications in the fields Estimate and Projects Economic Evaluation. He is Component of the Studies Center of Real Estate Appraisal and Land Economics (CeSET—Florence), Component in Italian Committee of the European Real Estate Society (ERES), Founding partner of Italian Society of Appraisal and Evaluation(SIEV), Component in the Technical and Scientific Committee "Confimprese", Board Member of Institute for research and educational activities (IPE, non-profit organisation under Presidential Decree no. 374/1981 on the proposal of the Ministry of Education), Board Member of F.O. I.S., Component of the Technical and Scientific Committee TEKNA—Environment, Energy and Technologies.

Richard Dunning is a Research Associate in the Department of Town and Regional Planning at the University of Sheffield. His principal research interest is applying behavioural analysis approaches to the study of housing and real estate markets. Prior to working as a researcher, Richard worked in the real estate industry. Richard has undertaken research for the EU, RICS, the Department for Communities and Local Government, the French Government, the Joseph Rowntree Foundation, Sheffield City Council and Rotherham City Council.

Anja Dust works as a Junior Consultant Valuation for an international real estate service firm in Frankfurt. In her first career she worked for a publishing house in Munich. In 2011 she started to study real estate at Nurtingen-Geislingen University, specialising in real estate valuation. She graduated in August 2014 with a with Bachelor of Science degree. Her bachelor thesis on the anchoring bias in property valuations was awarded the Aareon research prize.

Franz Eilers Head of Real Estate Research, joined vdpResearch, a subsidiary of the Association of German Pfandbrief Banks (vdp) in 2009. From 2000 to 2009, he was heading the real estate market research department in the the central division Valuation and Consulting at HypoVereinsbank, Germany's then second largest bank. He spent several years working at the institute for urban development and housing of the state of Brandenburg. From 1992 to 1996, he was employed at the research and consulting institute GEWOS in Hamburg. Prior to this position, he spent two years in the economics department of the Commerzbank in Frankfurt. He received his doctorate in Economics from the University of Hamburg.

Elena Fregonara has a degree in Architecture in 1989, Ph.D. in Urban Planning and the Real Estate Market. She is Associate Professor in Real Estate Appraisal and Economic Evaluation of Projects, at Architecture and Design Department, Politecnico di Torino. She is Vice-responsible of Turin Real Estate Market Observatory. She has been a lecturer of many courses, among others: Appraisal and Professional Practice, Economic Evaluation of Projects and Plans, Evaluation of Economic Sustainability of Projects. Member of the scientific committee at Master in "Urban Planning and The Real Estate Market" and at Master in "Management of Cultural and Environmental Heritage". Her research activity has been focused on: economic evaluation of assets and projects, in private and public context; economic and environmental sustainability; economic feasibility of capital investment in the real estate sector under risk and uncertainty; the real estate market analysis and monitoring of values and dynamics (statistical and econometric models); the Cadastral review, in particular about the definition of equal microzones for Turin municipality.

Marco Aurélio Stumpf González is Professor and Researcher at Civil Engineering Post Graduate Programme at Universidade do Vale do Rio dos Sinos (UNISINOS), where he has been teaching Real Estate Valuation since 1996. His research interests cover housing, sustainable buildings, economics of building renewal, market analysis, taxation of real estate, valuation and financial analysis of real estate, and alternative property valuation techniques (especially fuzzy rules and artificial neural networks models). In 1993 he obtained M.Sc. in Engineering at Universidade Federal do Rio Grande do Sul (UFRGS), with major in Real Estate Valuation. In 1998 he became a Ph.D. candidate at UFRGS (Civil Engineering), and completed his dissertation in 2002.

Berna Keskin is a University Teacher in the Department of Town and Regional Planning at the University of Sheffield. Her research interests focus on understanding the structure of urban housing market and specifically exploring the relative merits of different approaches to capturing neighbourhood segmentation. Following the completion of her Ph.D. in Housing Economics at the University of Sheffield, she has worked on funded projects for the INTERREG NWE Programme, Royal Institution of Chartered Surveyors, Department of Communities and Local Government, Department for Environment, Food & Rural Affairs and Investment Property Forum.

Andreas Kunert Senior Analyst, joined vdpResearch, a subsidiary of the Association of German Pfandbrief Banks (vdp) in 2009. From 2005 to 2009, he worked at HypoVereinsbank, Germany's then second largest bank, in Collateral Risk Management. Prior to this position, he was a researcher at the University of Essen and the North Rhine-Westphalia Institute for Economic Research (RWI). Andreas holds a Master's degree in Economics, specialising in econometrics and international economics. At vdpResearch, he focuses on collateral risk management, including measurement of price change in real estate markets, automated valuation models and classification of real estate market risks. He has developed and is responsible for the residential and commercial German real estate price indices, which are provided by vdp quarterly.

Carsten Lausberg is Professor of Real Estate Banking at Nurtingen-Geislingen University and head of the university's research institute for real estate information technology. He also works as a management consultant and valuer. He has taught courses at all levels at several universities and private institutions. In teaching, research, and consulting he has specialised on real estate finance, risk management, and portfolio management. Currently his research concentrates on decision support systems for the real estate industry. Dr. Lausberg studied business administration in Germany and the USA. He received a Master of Science (Finance) degree from Texas A&M University and a doctoral degree from Hohenheim University. From 1998 to 2005, he worked for a management consulting firm. In 2008 he became a full professor. He is a member of several academic associations and leads the working group on real estate risk management of the German chapter of the International Real Estate Society.

John R. Lombard joined the Department of Urban Studies and Public Administration at Old Dominion University in June 2002. He teaches graduate courses in research methods, urban and regional development, and urban resource allocation. Prior to his appointment in the department, he held a joint appointment as a research professor and as Vice President for Research and Information Services for the Hampton Roads Economic Development Alliance. Previously, Dr. Lombard was Vice President of Business Development for the Connecticut Economic Resource Center, the statewide marketing operation for economic development. He was responsible for all client-related activities in business recruitment and real estate relations. Prior to his tenure in Connecticut, Dr. Lombard was a consultant and head of research with the New York City firm of Moran, Stahl & Boyer. In addition to consulting assignments with Fortune 500 corporations, Dr. Lombard developed proprietary location measures such as the Labor Market Stress Index and the Underemployment Index. In a joint effort with Fortune Magazine, Dr. Lombard provided all the community research and analysis for the "Best Cities for Business" series. He received his Ph.D. and M.A. degrees from the State University at Buffalo.

William J. McCluskey is Reader in Real Estate and Valuation at the University of Ulster, where he received his Ph.D. in Real Estate Valuation in 1999. He has held various international positions including Visiting Professor of Real Estate at the University of Lodz, Poland, Professor of Property Studies at Lincoln University, Christchurch, New Zealand and is currently Visiting Professor in Real Estate at University of Technology, Malaysia. His main professional and academic interests are in the fields of real estate valuation, developing automated valuation methods and property tax policy. In addition, he has been an invited instructor in real estate at the African Tax Institute and the Lincoln Institute of Land Policy: China Programme. He is a faculty member of the Lincoln Institute of Land Policy and founding board member of the International Property Tax Institute.

Manya M. Mooya is a senior lecturer in Property Studies at the University of Cape Town's Department of Construction Economics and Management. He has previously taught in the Department of Land Economy of the Copperbelt University in Zambia (1997–2002) and the Department of Land Management at the Polytechnic of Namibia (2002–2006). He teaches property valuation on the undergraduate and postgraduate programmes in the department. His research on property valuation theory has been published in leading international journals and conferences. He is currently writing a research monograph on valuation theory. He holds a Ph.D. in Real Estate from the University of Pretoria, an M.Phil. from the University of Cambridge and a B.Sc. from the Copperbelt University.

Dilek Pekdemir earned her Ph.D. in Urban and Regional Planning and she is actually European Real Estate Society (ERES) Board member and fomer Vice chair of ERES 2015 22nd Annual Conference in Istanbul. She coordinates the research team of Cushman & Wakefield in Turkey, where she organises data collection, process and production of periodical market reports, market analysis and also client base market research, development and sales strategies, business strategy development reports. Dilek has an advisory and research experience of over 15 years. She has done various market research and consultancy reports for both national and international developers, investors, funds and finance companies. Formerly she worked at DTZ, Pamir & Soyuer Real Estate Advisory Services for eight years in the research department. She also gave lectures in real estate finance and real estate development and valuation in real estate master programmes at various universities. Currently she is a part-time lecturer at I.T.Ü. Real Estate Development Master Programme and gives real estate market analysis lectures.

Malgorzata Renigier (aka Renigier-Biłozor) has been Assistant Professor in the Department of Real Estate Management and Regional Development at the University of Warmia and Mazury, Olsztyn, Poland since 2005. Her major fields of research interest comprise systems of real estate management, value forecasting, nonlinear analysis in modelling of real estate value, influence analysis of stochastic factors on the real estate value, and application of artificial intelligence (AI) in real estate management. In 2000 she obtained her M.Sc. in the faculty of Geodesy and Space Management at the University of Warmia and Mazury. During 2001 she began her doctorate studies at the Department of Real Estate Management and Regional Development at the same university. In 2004 she obtained her Ph.D. in Geodesy and Cartography. In 2006 she received a prize from the Polish Minister for Building and Transport for her Ph.D. dissertation. Since 2006 she has been a member of the board of the Scientific Society of Real Estate. From 2004 to 2007 she has been a co-author of the programme concerning creation of a management system of real estate sources owned by local government, of the committee of scientific research.

Francesca Salvo is an engineer who earned his Ph.D. on Conservation of Architectural and Environmental Heritage. Since 2003, she has been University Researcher on Real Estate Appraisal—Department of Civil Engineering (DICI),

University of Calabria. Since 2000, she has been teaching Economy and Real Estate Evaluation and Real Estate Appraisal at University of Calabria. Since 2006, she has been Ordinary member of the Società Italiana di Estimo e Valutazione (SIEV) and Ordinary member of the Center Studies of Territorial Economy (CeSET) since 2013. She is author of scientific contributions on real estate valuations and damages of environmental resources evaluations; she develops her studies on methodologies of the real estate evaluation (real estate market data monitoring) and market-oriented procedures evaluation and quantitative analysis applications.

Patrizia Semeraro is Assistant Professor of Real Estate Appraisal and Economic Evaluation of Projects since 2011 at Architecture and Design Department, Politecnico di Torino. She is a graduate in Mathematics of University of Turin and holds a Ph.D. in Mathematics at University of Turin. She currently teaches *Real Estate Valuations: Theory and Methods* at Politecnico di Torino, *Market Segmentation* at graduate "Management of Cultural and Environmental Heritage" and *Mass Appraisal* at graduate "Urban Planning and The Real Estate Market" of Politecnico di Torino. She was a research assistant at Politecnico of Turin and at Dipartimento di Matematica Applicata "D. De Castro", University of Turin. She has taught Statistics at a graduate Master in Economics, Coripe Piemonte-Collegio Carlo Alberto. She has published in international finance and applied mathematics journals including *Journal of Theoretical and Applied Finance, Mathematics of Operations Research, Quantitative Finance* and *Journal of Computational and Applied Mathematics.*

Marco Simonotti is Full Professor on Real Estate Appraisal—Faculty of Engineering, University of Palermo. He develops studies on the general methodology in respect of the real estate, regarding mainly the experimental researches on the real estate market and the application of the statistics and appraisal analysis. He is author of an appraisal manual entitled "Bases of appraisal methodology" (Liguori, Naples, 1989), of a text book entitled "The appraisal of properties. With economy principles and appraisal applications" (UTET Libreria, Milan, 1997), of an appraising manual entitled "Manual of real estate appraising" (Geoval, Rome, 2005) compiled according to the international standards and of a book titled "Real Estate Appraising Methods" (Dario Flaccovio, Palerme, 2006). He has studied and introduced in Italy the market comparison approach (1997); he has formulated the critical capitalisation rate (1983), the parameters of the real estate segments (1998), the appraisal analysis standard of the real estate data (2003); he has applied the multiple regression analysis to the real estate properties (1988). He is author of the first one "Code of ethical behavior" of the technical professionals in the sector of the Italian appraising (2000).

Nikolaj Siniak (aka Mikalai) is Associate Professor at the Department of Economy and Management of Enterprises at the Belarusian State Technological University (BSTU), Minsk, where he has been working since 1998. His scientific interests cover economics, valuation and management of real estate, restructuring of enterprises, optimisation of production programmes, and simulation and formulation of economic problems on economy. At BSTU he obtained the following degrees: Diploma in Mechanical Engineering (1995), Ph.D. in Economical science (1998), and Diploma of Associate Professor in Economics (2005). He has developed grounding methods for a furniture factory production programme, the concept of real estate valuation for enterprises, and real estate market analysis. He has more than 70 publications to his credit. His teaching activities comprise the courses 'Economy of enterprise' and 'Methods of branch property valuation'.

Craig Watkins is Director of Research and Innovation for the Faculty of Social Sciences, Co-Director of the Sheffield Urban Institute and Professor of Planning and Housing at the University of Sheffield. Craig has written extensively on the economic structure and operation of housing and commercial property markets, urban development and on the relationship between public policy and markets. He has produced more than 130 research outputs including three books, 12 book chapters and more than 40 peer reviewed journal articles and has undertaken around 50 funded projects (half as Principal Investigator) including research for ESRC, Technology Strategy Board, EU, Joseph Rowntree Foundation, various Central and Local Government departments in the UK, RICS, Investment Property Forum and the Royal Town Planning Institute.

Acronyms

ANR	Annual Net Rent
AVM	Automated Valuation Model
BIST	Borsa İstanbul Stock Exchange
CAPM	Capital Asset Price Model
CBD	Central Business District
CBTR	Central Bank of Turkish Republic—Türkiye Cumhuriyeti Merkez Bankası
CMBT	Capital Markets Board—Sermaye Piyasaları Kurulu (SPK)
COD	Coefficient of Deviation
CREAS	Computerised Real Estate Appraisal System
DCFA or DCF	Discounted Cash Flow Analysis
DSS	Decision Support Systems
EPRA	European Public Real Estate Association
FTSE	Financial Times and Stock Exchange
GRM	Gross Rent Multiplier
HPM	Hedonic Price Model
IAAO	International Association of Assessing Officers
INREV	European Association for Investors in Non-Listed Real Estate Vehicles
IPD	Investment Property Databank
IPO	Initial Public Offering
IQR	Interquartile Range
IRR	Internal Rate of Return
ISE	İstanbul Stock Exchange—İstanbul Menkul Kıymetler Borsası (İMKB)
ITD	Investment Transaction Database (DTZ)
IVS	International Valuation Standards
LAF	Location Adjustment Factor
LVRS	Location Value Response Surface
MCA	Market Comparison Approach

MLV	Multilevel Models
MOR	Monthly Net Rent
MPT	Modern Portfolio Theory
NAREIT	National Association of Real Estate Investment Trusts
NCREIF	National Council of Real Estate Index Fiduciaries
NOI	Net Operate Income
OAR	Overall Capitalisation Rate
OMI	Osservatorio del Mercato Immobiliare—Italian Land Registry Observatory
PMI	Property Market Indicators
REICs	Real Estate Investment Companies—Gayrimenkul Yatırım Ortaklıkları (GYO)
REITs	Real Estate Investment Trusts
SAR	Spatial Autoregressive Model
SEM	Spatial Extension Method
TNHPI	Turkey New Housing Price Index
TRBOND	Turkish Government Bonds
VECM	Vector Error Correction Model
VIC	Value Influence Center
WACC	Weighted Average Cost of Capital
X100	ISE 100 index
XGMYO	REIC sector index

List of Figures

Automated Valuation Models for the Granting of Mortgage Loans in Germany

Emerging Markets Under Basel III: Can Moral Hazard Lead to Systematic Risk and Fragility? Analysis of REIT's in Turkey

An Application of Short Tab MCA to Podgorica

Spatial Analysis of Residential Real Estate Rental Market with Geoadditive Models

Spatial Analysis of Residential Real Estate Rental Market with Geoadditive Models

Location Value Response Surface Model as Automated Valuation Methodology a Case in Bari

Further Evaluating the Impact of Kernel and Bandwidth Specifications of Geographically Weighted Regression on the Equity and Uniformity of Mass Appraisal Models

Dealing with Spatial Modelling in Minsk

Using Multi Level Modeling Techniques as an AVM Tool: Isolating the Effects of Earthquake Risk from Other Price Determinants

The Multilevel Model in the Computer-Generated Appraisal: A Case in Palermo

List of Tables

An Application of Short Tab MCA to Podgorica

Spatial Analysis of Residential Real Estate Rental Market with Geoadditive Models

A Spatial Analysis for the Real Estate Market Applications

Location Value Response Surface Model as Automated Valuation Methodology a Case in Bari

Further Evaluating the Impact of Kernel and Bandwidth Specifications of Geographically Weighted Regression on the Equity and Uniformity of Mass Appraisal Models

Dealing with Spatial Modelling in Minsk

Using Multi Level Modeling Techniques as an AVM Tool: Isolating the Effects of Earthquake Risk from Other Price Determinants

Automated Procedures Based on Market Comparison Approach in Italy

Short Tab Market Comparison Approach. An Application to the Residential Real Estate Market in Bari

Part I
AVM, Valuation and Non-Agency Mortgage Crises: AVM, Mortgage Crises and Valuation in Person

A Brief Outline of AVM Models and Standards Evolutions

Maurizio d'Amato

Abstract The paper is a brief outline of the methodological evolution of Automated Valuation Modelling along the time. A further outlook is dedicated to the evolution of standards on Automated Valuation Methodology. Both the profiles may be useful to understand the increasing role of Automated Valuation Methodologies and the growth of their importance especially for the stability of the economic and social system and how they changed over the time.

Keywords Automated valuation methodology · Mass appraisal · Standards

1 Introduction

In this chapter is offered a general overview of the principal methodologies of AVM and their evolution along the time. The framework is not complete but may give an introduction on the methodological evolution of AVM and their standards. Before analysing Automated Valuation Methodologies it is important distinguish this kind of analysis from Mass Appraisal. Computer-assisted mass appraisal is normally applied to estimate using statistical modelling a large numbers of properties. Normally for this issue are used big quantity of data and the final result is normally used for tax purposes at a specific date. On the other side Automated Valuation Methodology using often the same methodological framework of mass appraisal, uses a statistical model and a large amount of property data to estimate the market value of an individual property or portfolio of properties. This valuation is normally delivered together with a confidence level to indicate how accurate the valuation is and they are provided essentially for lending purposes. In this case the date is not fixed and coincides with the appraisal date for mortgage lending purposes. Generally speaking automated valuation methodologies have a wide spectrum of applications. They may be used among the other for residential mortgage portfolio

M. d'Amato (✉)
DICATECh Technical University Politecnico Di Bari, Bari, Italy
e-mail: madamato@fastwebnet.it

M. d'Amato and T. Kauko (eds.), *Advances in Automated Valuation Modeling*,
Studies in Systems, Decision and Control 86, DOI 10.1007/978-3-319-49746-4_1

monitoring, capital requirement, loss given default modelling, securitization, residential mortgage covered bonds, whole loan portfolio trading, non performing loan management, quality control. The work is organized as follows in the next paragraph a brief methodological evolution of the AVM modelling is presented. A further paragraph is devoted to the evolution of AVM standards along the time and their increasing role in the economic system. Final remarks will be offered at the end.

2 AVM Models Evolution. A Brief Overview

Specification of a hedonic model has always been a critical issue. Hedonic model doesn't have a specified functional form by definition. It depends on assumptions arbitrarily chosen by a researcher (Cassel and Mendelson 1985). For this reason the debate on how to improve hedonic modelling is still opened with a fascinating and challenging evolution(d'Amato 2015; d'Amato 2013). The main issue is the same whether the approach is mass, or single, valuation: an accurate assessment of the value of many properties or a single property (McCluskey et al. 1997). Silverherz (1936) considers the reappraisal of St. Paul, Minnesota as the beginning of scientific mass appraisal. The introduction of computers accelerates the process. Several contributions highlighted the importance of mass appraisal, especially in the property market, analysing the linkage between property value, property characteristics, and urban social and economic features. Market behaviour is influenced by several characteristics like property prices, the high durability of this particular asset, geographic location (Robinson 1979; Harvey 1996). Hedonic price modeling has been proposed to define an econometric relationship between price and property characteristics. The application of hedonic price theory (Griliches 1971; Rosen 1974) is based on demand-side analysis in a static framework. Property valuation evolved from simple empirical judgments to automated valuation models and their applications have extended from single property to mass valuation (Clapp 2003). The dependent and independent variables are regressed using properties of known prices to determine the established relationships (coefficients) between the two types of variables (Adair and McGreal 1988). A first group relies on econometric modelling based on the definition of a cause effect relationship between property price and characteristics. The dependent and independent variables are regressed using properties of known prices to determine the established relationships (coefficients) between the two types of variables (Ciuna 2010; Ciuna 2011; Ciuna et al. 2015a; Ciuna and Simonotti 2014). Application of hedonic modelling are constructing constant-quality price indices for apartment buildings and vacant land in Geneva, Switzerland (Hoesli et al. 1997a); determining rental values of apartments in central Bordeaux, France (Hoesli et al. 1997b); explaining the housing market in Tel Aviv, Israel (Gat 1996); a confirmation of the rationality of condominium buyers and markets in Hong Kong (Mok et al. 1995). MRA calculates the coefficients with the least possible error (Benjamin et al. 2004) using the Ordinary Least

Squares (OLS), maximum likelihood (ML), or Weighted Least Squares (WLS) estimation Techniques. Among these OLS is he most popular (Ambrose 1990; Beach and MacKinnon 1978). The OLS method minimizes the sum of square of residuals or errors. The regression coefficients that are derived based on OLS shall be best linear unbiased estimator (BLUE) (Renigier-Bilozor and Bilozor 2016; Renigier-Bilozor et al. 2014; Renigier Bilozor et al. 2014b). In the hedonic modeling literature, locational proxies may be defined in various ways (cf., surveys by Ball 1973; Miller 1982; Lentz and Wang 1998). Another approach is *Comparable Sales Methods* (Cannaday 1989; Borst and McCluskey 2008a, b) This method is based on the selection of a number of comparable sales properties. Integration of adjustment grid methods and regression analysis was proposed by Colwell et al. (1983), in order to integrate the Ordinary-Least-Squares (OLS) estimation of adjustment factors to the standard method. Therefore using sales comparison approach a final adjusted sale price is determined for each comparable. At this point the estimates are weighted according to their similarity to the subject. At the end a final weighted sum of comparable estimate will give the single point estimate. The actual dissimilarity measure can be based on several measures as distance, differences in physical characteristics, date of sale, and the neighbourhood to which the comparable sale belongs.

In formal term we have

$$C_j = P_j + (V_{SK} - V_{Cj}) \tag{1}$$

C_j in this formula 1 is the estimated value of the subject based on the comparable j. P_j is the actual price of the comparable j while V_{SK} is the estimate selling of the subject and V_{Cj} is the estimate selling of the comparable C_j. However, there are some problems using MRA in property valuation normally correlated to spatial autocorrelation and heteroscedasticity, the two spatial effects inherent in property data (Mark and Goldberg 1988; Fletcher et al. 2000). The consequences are for example: the presence of excessive multicollinearity among attributes, spatial autocorrelation among residuals; diminishing the stability of regression coefficients (Dubin 1998; Anselin and Rey 1991; Des Rosiers and Thériault 1999). It is possible address the problem introducing in the model *neighbourhood binary variables* avoiding the dummy variable trap (Kauko and d'Amato 2011; d'Amato and Kauko 2012). Another opportunity is *market segmentation* based on specific characteristics as medium price or square meters (Borst 2015). Kang and Reichert (1991) constructed a locational-quality dummy based on levels of price per sqm. living-space. McCluskey and Anand (1999) used a solution, where the location was captured with a categorical 'ward'-variable comprising seven values based on mean transaction prices for a given area. For this reason neighbourhood factors should consider submarket specifics (Adair et al. 1996). A first attempt to deal with spatial heterogeneity is the *spatial expansion method* (Casetti 1972). Parameters in such models are themselves functions of location where the user determines the nature of the function (usually some linear polynomial). For example, if we take a simple model with two independent variables, x1 and x2, with three parameters a, b, and c, that are to be estimated, i.e.:

$$y = a + bx_1 + cx_2 \tag{2}$$

We then expand the parameters so that they are some linear function of the locations (ui, vi) of the observations. For example, if a 1st order linear polynomial is specified, then:

$$\begin{aligned} a_i &= \alpha_0 + \alpha_1 u_i + \alpha_2 v_i \\ b_i &= \beta_0 + \beta_1 u_i + \beta_2 v_i \\ c_i &= \gamma_0 + \gamma_1 u_i + \gamma_2 v_i \end{aligned} \tag{3}$$

A second way to deal with spatial effects are *location value response surface modelling*. These models has been provided (O'Connor 1982) for the first time for the appraisal of single family houses in Lucas County, representing a different approach to fixed neighbourhoods or composite submarkets analysis (Ward et al. 2002). The application of these models requires spatial interpolation of property prices or error term. This method has been applied in the U.S. (Eichenbaum 1989, 1995), in England (Gallimore et al. 1996), Northern Ireland (McCluskey et al. 2000) and Italy (d'Amato 2010) The impact of each value influence center on any property is determined using different possible measures of the distance from the property to the VIC (Eckert 1990; Eckert et al. 1993). The response surface is depending on the VIC positions and the adopted distance measure. The third approach is based on an interpolation grid, modelled to consider the influence on each property of the location ratio factors within its proximity. A further approach to deal with spatial analysis is *multilevel or hierarchical modelling* (Goodman et al. 1998; Salvo and De Ruggiero 2011; Salvo and De Ruggiero 2013; Salvo et al. 2013a; Salvo et al. 2013b). Spatial models can be computed at individual level or at cluster level (Francke 2008). Multi-level models are a variant on standard hedonic methods (Orford 1999; Leishman 2009) particularly useful when the observations being analysed are clustered and correlated (Ciuna et al. 2014a; Ciuna et al. 2014b; Ciuna et al. 2014c; Ciuna et al. 2015b). While the spatial modelling use coordinates x and y to define the observation at cluster level like neighbourhood. Problems that may be raised applying the cluster level can be undesirable discontinuities at the borders and a spatial structure that may be different from the selected clusters.

The use of coordinates previously seen in the spatial extension methods become increasing important with the development of spatial modelling. Pace et al. (1998) that showed the limits of real estate research using relatively spaceless tools. In fact real estate information may have spatial autocorrelation. It means that residuals are spatially correlated; off diagonal elements of the variance-covariance matrix of the estimated residuals deviate from zero indicating that the two observations that define the elements are spatially correlated. Spatially autocorrelation of residuals violate the assumption that OLS of residuals must be uncorrelated and normally distributed with zero mean and constant variance, i.e., e $\sim$ N $(0, \sigma^2 I)$. This makes the OLS estimated coefficients biased and unsuitable for inference. As a consequence predicted property prices are unreliable. The application of Spatial modelling is based on spatial stationarity. AVMs without spatial effect often produce inaccurate results because of the different behaviour of the market across geographic space (Berry and Bednarz 1975). According to Anselin (1988), there are two kinds of spatial effects: spatial autocorrelation and spatial heterogeneity. The former refers to a functional relationship between observations while the latter is connected to the lack of uniformity arising from space, leading to spatial heteroscedasticity and spatially varying parameters. It may occur that the observation showed spatial nonstationarity describing model relationships that are not constant across space. The process which underlies the relationship between variables changes with spatial context. In this case techniques as Moving Window Regression (Case et al. 2004) can be applied. In the application of this technique per each data point are selected a certain number of observation closer to the sample point and then included in the regression model. The peculiar aspect of Moving Window Regression is that each closer observation is included in the model having an equal weight.

In the application of Geographically Weighted Regression the observations closer to a specific sample point is weighted more than the other distant. In fact the value of the weight is a function of location, and go down as distance from the regression point increase. Bandwidth of the spatial kernel affect the final result of the valuation and it may be fixed of variable according to the nature of the data. In real estate market normally the bandwidth used is variable (Borst and McCluskey 2008a, b). It must be stressed that the partitioning of the area of interest in order to reach the condition of stationarity remains an alternative. In this case it will be possible the application of spatial modelling spatial autoregressive models or spatial panel data. Farber and Yeates (2006) showed that GWR is superior as accuracy and spatial bias than other spatial autocorrelated methods. The GWR model is applied using a Gaussian spatial kernel and a fixed bandwidth. Normally the Gaussian kernel with a fixed bandwidth achieves the lowest COD and is used in comparison against other spatial weighting functions tested (that is, bisquare kernel with adaptive bandwidth, bisquare kernel with fixed bandwidth, and Gaussian kernel with adaptive bandwidth). The Gaussian kernel have a distance decay function that gives a higher weight to the properties are closer to each point. In formal term Geographic Weighted Regression is indicated below:

$$P_j = X_j \beta_j + \varepsilon_j \tag{4}$$

where P is the vector of property prices, X is the matrix of the characteristics and β_i is the vector of marginal prices while the final is the error term. The j-annotation mean that the parameter estimates is specific to unit j. The weighted least squares method are used to estimate that is derived:

$$\beta_j = (X^T W_i X)^{-1} X^T W_i y \tag{5}$$

The matrix calculation (pseudo inverse Moore Penrose) is implemented with a Wi weighting matrix. It is a diagonal matrix whose diagonal reflects the weight of each unit with the respect of unit i. Spatial dependency in a regression model can be detected with either the spatial weight matrix or the direct specification of the covariance matrix (Dubin 1998). Spatial Auto Regressive (SAR) models, also called spatial models are a group of models that improves the accuracy of property price prediction of the MRA model taking into account spatial dependence of properties in the functional model. The spatial dependence parameters are estimated along with the regression coefficients and are based on the hypothesis of stationarity. SAR models incorporates a spatial weight matrices that are based on the concept of spatial neighbours. The most commonly used strategies to define spatial neighbours are Delaunay triangulation and k nearest neighbours when properties are represented by their centroids. In this way the properties that can be considered spatial neighbours to a subject property receive the value of one while those that are not spatial neighbours receive zero values. Normally an iteration of several various values of k until satisfactory results are obtained when the k nearest neighbours strategy. When the rows and columns of the weight matrix arranged such that the subject property is at the main diagonal, the weight matrices are usually sparse and banded. Literature regarding the application of these models in the property price valuation shows improvement in the property price prediction. The Spatial autoregressive models are: Spatial Lag Models and Spatial Error Models. The *Spatial Lag Model (SLM)* that models the dependency of property prices; the price of a property is dependent on the prices of its neighbouring properties. Normally this method is used to model a single submarket. The logic behind the autoregressive model is that there are 'spill over' effects in which the sales prices of nearby properties has an effect on the value of a given property more than those that are distant. Close-by locations are more correlated than values at locations that are far apart. Spatial lag model is appropriate when spatial structure is present in the variables of the model. This autoregressive model has the following simple formulation:

$$Y_i = \rho W_{ij} Y_j + \varepsilon \tag{6}$$

In the formula (lattice model) the term ρ is the coefficient of autocorrelation, W is a matrix of weights with the element of W which specify the strength between each point i (say observation) and the others j. The value of a single observation Y_i is the weighted average of its neighbouring Y_j observations. Can (1990, 1992) included the spatial lag model in his lattice version in the traditional hedonic models.

$$Y_i = \rho W_{ij} Y_j + XB + \varepsilon \tag{7}$$

In the formula the term Y is the vector of the observation on the dependent variable, the term ρ is the coefficient of autocorrelation W is the matrix of weights WY represent the spatial lagged dependent variable that specify the strength of the relationship between properties, X is a matrix of observations on explanatory variables ε is a vector of error term. W is a $n \times n$ matrix whose elements of represent the 'strength' of the connections between each property and all other properties. In the spatial lag models the weight matrix is used to model the spatial process directly.

Weight can be determined in several ways. It is possible to transform the weight in a dummy variable assuming 1 for the nearest neighbour and 0 otherwise or it is possible to define the inverse of distance for the nearest neighbour and 0 for the others (Wilhelmsson 2002). A further spatial autoregressive mode is the *Spatial Error Model (SEM)* that models the spatial dependence of the error terms; an error induced by a property may dependent on the error of nearby properties. In this model, the spatial influence is considered through the error term. Formally the model is

$$Y = XB + \varepsilon \tag{8}$$

In the formula Y is the vector of the observation on the dependent variables, X is a matrix of observations on explanatory variables ε is a vector of error term, assuming

$$\varepsilon = \lambda W \varepsilon + u \tag{9}$$

ε is a vector of spatially autocorrelated error terms, u a vector of errors λ. Spatial Durbin models take into account both the dependence of prices and errors of neighbouring properties, combining the SLM and SEM into one model.

$$\begin{aligned} Y &= \lambda W_1 Y + u \\ u &= \rho W_2 u + v \end{aligned} \tag{10}$$

Several contributions tried a comparison among different AVM techniques. Wilhelmsson (2002) compared the traditional multiple regression analysis using dichotomic variables with Spatial Autoregressive Models and Spatial Extension Method. In this case Spatial Autoregressive Model improved the accuracy of the multiple regression analysis with spatial dummies. Borst and McCluskey (2007) compared Geographic Weighted Regression and Comparable Sales Method. The latter method used in a spatial lag model diminish the Coefficient of Deviation more than Geographic Weighted Regression. Spatial Lag Model improves traditional multiple regression analysis performance considering spatial autocorrelation. Quintos (2013) use spatial lag models to create location adjustment factors to be included in traditional OLS. Further development is the use of spatio temporal modelling or spatial panel data are provided in order to include temporal variable as a further options in the spatial models previously indicated (Borst 2015). Despite the great number of applications, statistic data analysis has a theoretical weakness (Lentz and Wang 1998) and may not be efficient in those markets where uncertainty is high because of the unavailability, or the nature, of information and information sources. For this reason several contributions have analysed the role of alternative approaches to automated valuation. Neural networks (Borst 1992; McCluskey et al. 1997; Nguyen et al. 2001) is one of them. The application of these methods raised several criticisms

(Worzala et al. 1995) observing how outcomes vary with different architectures of neural network-based models. A further application is the self-organizing map (Kauko 2002; Kaklauskas et al. 2012). Another contribution to AVM comes from the application of fuzzy theory. Gonzales et al. (2002) have built a mass appraisal model, where fuzzy rules are extracted from the ANN. Another tool for valuation support is the analytic hierarchy process (AHP). The AHP has been applied by Kauko (2002) and Fischer (2003) also sees a more qualitative approach as an improvement. Using discrete mathematics Rough Set Theory may be applied to specific contexts with few data without a causal mathematical relationship. Originally based on several works d'Amato 2002, 2004 . A methodological improvement to RST has been developed by d'Amato (2004), Kauko and d'Amato (2011), to integrate the so-called valued tolerance relation (VTR). The final output of the model is an estimated of value based on an if-then rule instead of a continuous mathematical function.

3 AVM Standards Evolution. An Introductory Outline

Interestingly, The Appraisal of Real Estate created a difference between the *estimate* of value as the final output of an automated valuation model and the *opinion of value* considered the final output on an "in person" valuation. A difference that starts from the 1999. In 2003 the first standard on AVM was implemented by IAAO four years later the distinction indicated by the Appraisal Standard Two terms, different valuations. Obviously the distinction is not so evident. Since the first standard the general framework of the models has been proposed. The main phases of the construction of the models in Table 1.

After the definition of the sample and the stratification of the sample according to homogenous group of properties it is necessary the model specification phase or the selection of the appropriate model for the sample. In 2003 classification the use of spatial methods was not widely diffused. Therefore in the 2003 Standards are recurring traditional method of hedonic modeling. In the calibration phase the relevant variables will be selected. In a further phase the model is tested and applied to the sample. In AVM process it is growing the integration in several forms of in person valuation like Broker's Price Opinion. Computer Assisted Mass Appraisal

Table 1 The Phases of the Application of an AVM model (IAAO 2003)

Model specification	Meaning the definition of the model structure
	Comparable sales method, direct market approach (say hedonic), cost approach, income approach
Model calibration	Indication of the relevant variables for the model
Model testing	Use of specific ratio test to analyse the quality of AVM
Model application and value review	Application to the sample of properties and critical analysis

modelling (CAMA) increased in the last decades in a impressive way. Both the technical devices and the development of the model made the final output of AVM increasingly accurate. Since the first International Standards IAAO created and maintains standards that promote equity and fairness in real estate appraisals and assessments. For this reason in the testing phase have been selected different methods. Mainly among the other it is possible to recall the COD (coefficient of dispersion) as a measure of horizontal dispersion measuring the difference between price and appraised values.

$$\mathrm{COD} = \frac{100}{\mathrm{n}} \frac{\sum_{i=1}^{n} \left| \frac{\mathrm{EP}_t}{\mathrm{SP}_t} - \mathrm{Median}\left(\frac{\mathrm{EP}_t}{\mathrm{SP}_t}\right) \right|}{\mathrm{Median}\left(\frac{\mathrm{EP}_t}{\mathrm{SP}_t}\right)} \tag{11}$$

where EPi is the expected price from the application of the model to the of the ith property, and SPi is the sales price of the ith property. A further ratio is price-related differential (PRD) are two coefficients The Coefficient of Dispersion is indicated in the formula below:

$$\mathrm{PRD} = \frac{\mathrm{Mean}\left(\frac{\mathrm{EP}_t}{\mathrm{SP}_t}\right)}{\sum_{t=1}^{n} \frac{\mathrm{EP}_t}{\mathrm{SP}_t}} \tag{12}$$

The price-related differential is a score measuring vertical equity. According to the IAAO Standard on Automated Valuation Models, PRD values of less than 0.98 suggest evidence of progressivity, while PRD values of more than 1.03 suggest evidence of regressivity (IAAO 2003). In May 2005, the banking agencies delivered a guidance on the risks of home equity lines of credit and home equity loans. It cautioned financial institutions about credit risk management practices, pointing to interest-only features, low- or no-documentation loans, high loan-to-value and debt-to-income ratios, lower credit scores, greater use of automated valuation models (Bilozor and Renigier-Bilozor 2016), and the increase in transactions generated through a loan broker or other third party. While this guidance identified many of the problematic lending practices engaged in by bank lenders, it was limited to home equity loans. It did not apply to first mortgages. (Financial Crisis Inquiry Report, p. 200). In 2006 Appraisal Institute defined a differences between '*The output of an AVM is not, by itself, an appraisal. An AVM's output may become a basis for appraisal, appraisal review, or appraisal consulting opinions and conclusions if the appraiser believes the output to be credible for use in a specific assignment. However, the appropriate use of an AVM is, like any tool, dependent upon the skill of the user and the tool's suitability to the task at hand.*' (The Appraisal Foundation, 2006, Advisory Opinion 18: Use of an automated valuation model). In 2008 the Royal Institution of Chartered Surveyors AVM Standards Working Group proposed one of the first European definition of Automated valuation methodologies "*Automated Valuation Models use one or more mathematical techniques to provide an estimate of value of a specified property at a specified date, accompanied by a measure of confidence in the accuracy of the result, without human intervention post-initiation*" (RICS 2013, AVM Information Paper, p. 5). After the effect of the non agency mortgage crisis the 2008 HVCC (Chap. I Appraisal Independence Safeguards) stated "*...ordering, obtaining, using, or paying for a second or subsequent appraisal or automated valuation model (AVM) in connection with a mortgage financing transaction unless: (i) there is a reasonable basis to believe that the initial appraisal was flawed or tainted and such basis is clearly and appropriately noted in the loan file...*". In the same years a seminal paper offered and international overview of AVM application (Robson and Downie 2009). The work is based on a previous paper and a book of the authors that demonstrates the growing importance of AVM all over the world and the needs of specific information paper. After the non agency mortgage crisis in 2009 the European Mortgage Federation published a document addressing "*...AVM is an useful and efficient tool when used appropriately by an experienced operator. Lenders must be certain that the model used has been developed after rigorous testing using good quality data by experienced modelers...*". Furthermore the IAG 2010 indicates in the Appendix B: "*Institutions may employ AVMs for a variety of uses such as loan underwriting and portfolio monitoring. An institution may not rely solely on the results of an AVM to develop an evaluation unless the resulting evaluation is consistent with safe and sound banking practices and these Guidelines*" defining a limit in the integration between automated valuation and valuation in person. The RICS Information Paper Comparable Evidence in property valuation (2012) states in Sect. 4.3 that: "*Output from an AVM can be utilised as part of the evidence in support of a valuation. The valuer may consider such AVM outputs to have lesser or greater weight than other evidence that may be available. The validity of this, however, will depend greatly on the individual valuer's personal knowledge and experience of AVMs.*" On December 2013 Royal Institution of Chartered Surveyors published the first Information Paper on Automated Valuation Methodologies. The document is the first document produced in Europe on AVM. In fact in the paragraph 2.2.5 it is indicated as a further AVM model artificial network like in the IAAO standards of 2003. In the paragraph 2.3, unless the standards of IAAO it is officially provided a list or property attribute to be considered in implementing an AVM model. Interestingly this is one of the first official list indicated by an international standards and is reported in Table 2.

The phase of Model Stratification, Specification and Calibration (RICS 2013, AVM Information Paper, p. 8) are also indicated in continuity with the IAAO standards of 2003. The final output provides also confidence level determination in order to give more information about the quality of result (Salvo et al. 2014; Salvo et al. 2015; Salvo et al. 2015; Salvo et al. 2015; Simonotti et al. 2015). This is very important for the use of Automated Valuation Methodology as a risk analysis tool. In the paragraph 3 RICS Information Paper provides a classification of possible uses of AVM in Table 3.

Table 2 Main Variables to be included in an Automated Valuation Model (RICS 2013)

Description of main variables
Architectural style
Property type
Age (or year built)
Location (e.g. geographic co-ordinates or allocation of a 'locality' identifier)
Floor area
Number of rooms
Number of bedrooms
Number of bathrooms
Plot size
Conservatory
Outbuildings
Parking/garaging
Quality (i.e., specification of the property)
Condition

Table 3 Uses of AVM (RICS 2013)

By lenders for the loan origination process or subsequent revaluation for credit decision purposes	Lenders can now obtain an AVM that can sometimes be used prior to processing a case, to see if the proposed figures are likely to be adequate, without going to the cost of a full valuation by a human valuer. It introduces a level of risk that the lender necessarily accepts. Alternatively, an AVM can be used part-way through a mortgage term to check how a property value may be changing
In-arrears assessment and planning	Lenders may wish to check the value of a security to establish, where the customer is in arrears, if the loan is still likely to be secured by the value of the property and what scope there is in arrears planning. The AVM offers a cheap and quick indication
In an audit of valuations	Lenders, and those who audit valuations, will sometimes obtain a second valuation from an AVM supplier to serve as an audit of the original valuation. This can occasionally be applied across a range of properties, as well as in individual cases
For mass appraisal, such as for local taxation purposes	A model sometimes called computer aided mass appraisal (CAMA) is a sophisticated AVM capable of providing valuation estimates for thousands or millions of properties, cost-effectively

(continued)

Table 3 (continued)

For the provision of valuation estimates for individual capital tax purposes	An AVM can be used in these instances for individual properties or for a portfolio, to provide a quick estimate of likely tax implications to aid with tax planning
For the identification of fraudulent activity	An AVM applied to a range of properties and their valuations can distinguish activity that does not follow the normal market trends
For the provision of valuation estimates for large-scale asset valuations, for example, a portfolio of local-authority-owned residential properties, or the sale of a mortgage book in respect of 'securitisation'	Here an AVM can be used, potentially cost-effectively, to provide sample valuations for portfolio purposes or for the valuation of a whole portfolio
For estimating compensation payments to owners of residential property due to the effect of the use of new public works, for example, road schemes or airport expansion, and so on	Public authorities may find an AVM a quick and cost-effective way of estimating the likely cost of compensation as part of a total 'scheme' cost. An appropriate model can provide a pre-scheme value estimate of affected properties and keep this up to date as necessary
For cost/benefit analyses for potential public expenditure	Similar to 8 above, when estimating the cost or benefit of public works, an AVM can rapidly and economically provide a total cost estimate of the residential value of properties likely to be affected to aid a cost benefit exercise
For lending (capital adequacy purposes)	AVMs can be applied to a portfolio of property subject to mortgage finance, to obtain an indication of how well they are performing. This can assist the lender and their accountants in determining capital adequacy ratios

In the final Sect. 4.3 (RICS 2013) a ratio provided by Fitch rating is considered and proposed as a standard to connect property valuation with automated valuation result. This linkage is indicated only in this standard and is reported in the formula below:

$$PE = \frac{\text{AVMValue}}{\text{Surveyor Value} - \text{AVMValue}} \quad (13)$$

In the Formula (13) a percentage error (PE) is calculated dividing the final result of an AVM valuation per the difference between the surveyor value and the final result of AVM valuation. The formula proposes an adjustments that may be taken into account when a surveyor value is no longer available like in the selling of a pool of loans with AVM values. In this case the surveyor values will not be available, so the percentage adjustment that has been made will be based on the AVM value that was received. Consequently, when determining the adjustments, the differential between the AVM and the surveyor value should be consider as a percentage of the

Table 4 Main Documents Published by Rating Agencies on AVM

Guidelines for the Use of Automated Valuation Models for U.K. RMBS Transactions”—Standard and Poor’s, 20th February 2004
Guidelines for the Use of Automated Valuation Models for U.K. RMBS Transactions”—Standard and Poor’s, 26th September 2005
Criteria for Automated Valuation Models in the UK—Fitch, 22nd May 2007
Moody’s Approach to Automated Valuation Models in Rating UK RMBS—Moody’s, 21st August 2008
Criteria for Automated Valuation Models in EMEA RMBS; Addendum—Criteria for Automated Valuation Models in EMEA RMBS”—Fitch, 20th November 2008
Criteria for Automated Valuation Models in EMEA RMBS; “Automated Valuation Models in EMEA RMBS—Special Report”—Fitch, 12 August 2011

AVM value. This is an interesting case of contiguity between the rating agency standards and international standards on AVM. Since 2004 rating agencies provide an impressive and growing production of professional standards on AVM. Table 4 indicates the main documents published by rating agencies on AVM.

There is an increasing need of AVM even if still the vast majority of revaluations of residential properties is carried out by using national indices available in each of the EU 28 jurisdictions such as those provided by the respective national offices of statistics. Previously has been highlighted the role of the confidence level as a useful information provided, the use of indexed valuation do not give any information about confidence level and lacks as predictive measure of accuracy of valuation. The growth of the AVM in the mortgage lending process has also seen a strict integration with opinion of value and valuation in persons. In same cases replacement of the activity of appraising has been indicated (d’Amato 2013). It is possible a rapid classification of the integration between automated valuation methodology and valuation activity. In the Table 5 below it is possible to indicate the main form of this kind of integration namely called Hybrid Valuation.

The relationship between the two methods is variable according to the specific needs of the final user. It is quite evident that methods, standards and even the application is a challenging sector of analysis. In the last decades an impressive growth of the quality of models, the introduction of spatio temporal panel data and the growth of the awareness about risk in the lending process created a particular interest in this sector.

Table 5 Different form of hybrid valuation (EMF-EAA)

AAAVM	Analyst assisted automated valuation methodology an automatic valuation assisted by the expertise of a professional who may not be a qualified surveyor
SAAVM	Surveyor assisted automated valuation methodology or an automatic valuation assisted by a qualified surveyor to validate and supplement the output of AVM
AVMAA	Automated valuation assisted appraisal an automated valuation carried on without physical inspection, integrated with a professional valuer expertise to reach a legally compliant valuation

4 Conclusion

The chapter provides an introductory view of the methodological evolution of both methods and standards adopted for Automated Valuation Methodology purposes. The idea is to provide a general trend of what is happening in the AVM world in order to analyse and understand the change and the challenges of this sector. It is quite evident the rising role of the AVM in our system being a field of analysis at the heart of the economic process with the mission to provide equilibrium. In particular it is important to stress the change in the standards and the enrichment of the standard provided by the professional world (rate agency). At the moment the impact of the sophisticated methodologies is limited by the use of rough method as indexed valuation but seems quite evident the needs of more professional and advanced methodological solution to address the risk of a lending process.

References

Adair, A., & McGreal, S. (1988). The application of multiple regression analysis in property valuation. *Journal of Property Valuation and Investment, 6*, 57–67.

Adair, A. S., Berry, J. N., & McGreal, S. W. (1996). Hedonic modelling, housing submarkets and residential valuation. *Journal of Property Research, 13*(1), 67–83.

Ambrose, B. (1990). An analysis of the factors affecting light industrial property valuation. *Journal of Real Estate Research, 5*, 355–370.

Anselin, L. (1988). *Spatial econometrics: Methods and models*. Dordrecht: Kluwer Academic Publishers.

Anselin, L., & Rey, S. (1991). Properties of tests for spatial dependence in linear regression models. *Geographical Analysis, 23*(2), 112–131.

Appraisal Institute. (2006). *Advisory Opinion 18 Use of an automated valuation method*. Chicago, IL.

Ball, M. J. (1973). Recent empirical work on determinants of relative house prices. *Urban Studies*, 10, 213–223.

Beach, C. M., & MacKinnon, J. G. (1978). A maximum likelihood procedure for regression with autocorrelatederrors. *Econometrica, 46*, January, 51–58.

Benjamin, J. D., Guttery, R. S., & Sirmans, C. F. (2004). Mass appraisal: An introduction to multiple regression analysis for real estate valuation. *Journal of Real Estate Practice and Education, 7*, 65–77.

Berry, B. J., & Bednarz, R. S. (1975). A hedonic model of prices and assessments for single-family homes: Does the assessor follow the market or the market follow the assessor? *Land Economics, 51*(1), 21–40.

Biłozor, A., Renigier-Biłozor, M. (2016) The procedure of assessing usefulness of the land in the process of optimal investment location for multi-family housing function. *World Multidisciplinary Civil Engineering-Architecture-Urban Planning Symposium*. WMCAUS 2016 Praga.

Borst, R. A. (1992). Artificial neural networks: The next modelling/calibration technology for the assessment community. *Artificial Neural Networks, 1992*, P64–P94.

Borst, R. (2015) *Improving mass appraisal valuation models using spatio temporal methods*, International Property Tax Institute.

Borst, R. A., & McCluskey W. J. (2007) Comparative evaluation of the comparable sales method with geostatistical valuation models. *Pacifc Rim Property Research Journal 13*(1), 106–129.
Borst, R. A., & McCluskey, W. J. (2008). *The modified comparable sales method as the basis for a property tax valuations system and its relationship and comparison to spatially autoregressive valuation models Chapter 3*. In Kauko, T & d'Amato, M. (eds) Mass Appraisal Methods. An international perspectives for property valuers, Wiley Blackwell, London
Borst, R. A., & McCluskey, W. J. (2008b). Using geographically weighted regression to detect housing submarkets: Modeling large-scale spatial variations in value. *Journal of Property Tax Assessment and Administration, 5*(1), 21–51.
Can, A. (1990). The measurement of neighborhood dynamics in urban house prices. *Economic Geography, 66*, 254–272.
Can, A. (1992). Specification and estimation of hedonic housing price models. *Regional Science and Urban Economics, 22*, 453–474.
Cannaday, R. E. (1989). How should you estimate and provide market support for adjustments in single family appraisals. *Real Estate Appraiser and Analyst, 55*(4), 43–54.
Case, B., Clapp, J., Dubin, R., & Rodriguez, M. (2004). Modeling spatial and temporal house price patterns: A comparison of four models. *Journal of Real Estate Finance and Economics, 29*(2), 167–191.
Casetti, E. (1972). Generating models by the expansion method: Applications to geographic research. *Geographical Analysis, 4*, 81–91.
Cassel, E., & Mendelson, R. (1985). The choice of functional forms for hedonic price equations: Comment. *Journal of Urban Economics, 18*, 135–142.
Ciuna, M. (2010). L'Allocation method per la stima delle aree edificabili. *AESTIMUM, 57*, 171–184.
Ciuna, M. (2011). *The valuation error in the compound values*. AESTIMUM [S.l.], pp. 569–583, Aug. 2013. ISSN:1724-2118.
Ciuna, M., Salvo F., & Simonotti M. (2015a). Appraisal Value and Assessed Value in Italy. International journal of economics and statistics. Pages: 24–31. ISSN: 2309-0685.
Ciuna, M., & Simonotti, M. (2014). Real estate surfaces appraisal. AESTIMUM 64. *Giugno, 2014*, 1–13.
Ciuna, M., Salvo, F., & Simonotti, M. (2014a). The expertise in the real estate appraisal in Italy. Recent advances in civil engineering and mechanics. In *Proceedings of the 5th European Conference of Civil Engineering (ECCIE '14)* (pp. 120–129). Florence, Italy. November 22–24, 2014. North Atlantic University Union. Series: Mathematics and Computers in Science and Engineering Series, 36. ISBN:978-960-474-403-9. ISSN:2227-4.
Ciuna, M., Salvo, F., & De Ruggiero, M. (2014b). *Property prices index numbers and derived indices*. Property management (Vol. 32. Issue. 2, pp. 139–153). doi:(Permanent URL):10.1108/PM-03-2013-0021.
Ciuna, M., Salvo, F., & Simonotti, M. (2014c). Multilevel methodology approach for the construction of real estate monthly index numbers. *Journal of Real Estate Literature: 2014, 22* (2), 281–302.
Ciuna M., Salvo F., Simonotti M. (2015b). Compensation appraisal processes for the realization of hydraulic works in an agricultural area. In *Proceedings of XLIV INCONTRO DI STUDI Ce.S.E.T. Il danno. Elementi giuridici, urbanistici e economico-estimativi. Bologna, Italy*, (pp. 69–82). November 27–28, 2014. ISBN:978-88-99459-21-5.
Ciuna, M., Salvo, F., & Simonotti, M. (2015c). Parametric measurement of partial damage in building. In *Proceedings of XLIV INCONTRO DI STUDI Ce.S.E.T. Il danno. Elementi giuridici, urbanistici e economico-estimativi* (pp. 171–188). Bologna, Italy. November 27–28, 2014. ISBN:978-88-99459-21-5.
Clapp, J. M. (2003). A semiparametric method for valuing residential locations: Application to automated valuation. *The Journal of Real Estate Finance and Economics, 27*(Nov. 2003), 303–320.

Colwell, P. F., Cannady R. E., & Wu C. (1983). The analytical foundations of adjutment grid methods. *Journal of the American Real Estate and Urban Economics Association, 11*, 11–29.
d'Amato, M. (2002). Appraising property with rough set theory. *Journal of Property Investment and Finance, 20*(4), 406–418.
d'Amato, M. (2004). A comparison between RST and MRA for mass appraisal purposes. *A Case in Bari, International Journal of Strategic Property Management, 8*, 205–217.
d'Amato, M. (2010). A location value response surface model for mass appraising: An "iterative" location adjustment factor in Bari, Italy. *International Journal of Strategic Property Management, 14*, 231–244.
d'Amato, M. (2013). *Man Vs. Machine, Appraisal Buzz*, August 13.
d'Amato, M. (2015). Income approach and property market cycle. *International Journal of Strategic Property Management, 29*(3), 207–219.
d'Amato, M., & Kauko, T. (2012). Sustainability and risk premium estimation in property valuation and assessment of worth. *Building Research and Information, 40*(2), 174–185 (March–April 2012).
Des Rosiers, F., & Thériault, M. (1999) House prices and spatial dependence: Towards an integrated procedure to model neighborhood dynamics, Working Papers, Laval—Faculte des sciences de administration.
Dubin, R. A. (1998). Predicting house prices using multiple listings data. *Journal of Real Estate Finance and Economics, 17*(1), 35–59.
Eckert, j. (Ed.) (1990). Property appraisal and assessment administration. *International Association of Assessing Officers*, Chicago, IL.
Eckert, j., O'Connor, P., & Chamberlain, C. (1993). Computer-Assisted real estate appraisal: A california savings and loan case study. *The Appraisal Journal*, 524–532.
Eichenbaum, J. (1989). Incorporating location into computer-assisted valuation. *Property TaxJournal, 8*(2), 151–169.
Eichenbaum, J. (1995). The location variable in world class cities: Lessons from cama valuation in New York city. *Journal of Property Tax Assessment and Administration, 1*(3), 46–60.
EMF EAA. (2016). *Joint paper on the use of automated valuation methodology in Europe*. www.europeanavmalliance.org, last contact 07.07.2016.
Farber, S., & Yeates, M. (2006). A comparison of localized regression models in a hedonic house price context. *Canadian Journal of Regional Science, 29*(3), 405–420.
Federal Reserve Second Monetary Policy Report for 2008. (2008). Hearing before the Committee on Banking, Housing and Urban Affairs United States Senate One Hundred Tenth Congress Second Session on Oversight on the Monetary Policy Report to Congress Pursuant to the Full Employment and Balanced Growth Act of 1978.
Fischer, D. (2003). Multi-criteria analysis of ranking preferences on residential traits. In *10th ERES Conference*, Helsinki, Finland, 10–13 June.
Fitch. (2007). Criteria for automated valuation models in the UK, 22nd May
Fitch. (2008). Criteria for automated valuation models in EMEA RMBS; Addendum criteria for automated valuation models in EMEA RMBS, 20th November
Fitch. (2011). Criteria for automated valuation models in EMEA RMBS; Automated valuation models in EMEA RMBS, Special Report, 12th August
Fletcher, M., Gallimore, P., & Mangan, J. (2000). Heteroscedasticity in hedonic house price models. *Journalof Property Research, 17*, 93–108.
Francke, M. (2008). The hierarchical trend model. In Kauko, T., & d'Amato, M. (eds) Mass appraisal an international perspective for property valuers. Wiley Blackwell, London.
Gallimore, P., Fletcher, M., & Carter, M. (1996). Modelling the influence of location on value. *Journal of Property Valuation and Investment, 14*(1).
Gat D. (1996). A compact hedonic model of the greater tel aviv housing market. *4*(2) 162–172.
Gonzalez. M. A. S., Soiberman, L., & Formoso, C.T. (2002). Explaining results in a neural-mass appraisal model. In *9th European Real Estate Society Conference (ERES)*.
Goodman, A. C., & Thibodeau, T. G. (1998). Housing market segmentation. *Journal of Housing Economics, 7*, 121–143.

Griliches, Z. (1971). Hedonic prices indexes revisited, in prices indexes and quality change, federal reserve board, Massachuttes, pp. 3–15.
Harvey J. (1996). Urban land economics: The economics of real property markets, London, Macmillan.
Hoesli, M., Giacotto, C., & Favarger, P. (1997a). Three new real estate price for geneva. *Switzerland Journal of Real Estate Finance and Economics, 15*(1), 93–109.
Hoesli, M., Thion, B., & Watkins, C. (1997b). A hedonic investigation of the rental value of apartments in central bordeaux. *Journal of Property Research,14*, 15–26.
Home Valuation Code of Conduct. (2008). Chapter I Appraisal Indipendence Safeguards.
IAAO. (2003). Automatic valuation Standards.
Kaklauskas, A., Zavadskas, E. K., Kazokaitis, P., Bivainis, J., Galiniene, B., d'Amato, M., et al. (2012). Crisis Management Model and Recommended System for Construction and Real Estate. In R. Katarzyniak & G.-S. Jo (Eds.), *Ngoc Thanh Nguyen, Bogdan Trawi'nski* (pp. 333–343). Advanced Methods for Computational Collective Intelligence in Studies in Computational Intelligence Series edited by Janusz Kacprzyk: Springer Verlag Berlin.
Kang, H.-B., & Reichert, A. (1991). An empirical analysis of hedonic regression and grid-adjustment techniques in real estate appraisal. *AREUEA Journal, 19*(1), 70–91.
Kauko, J. T. (2002). *Modelling the locational determinants of house prices: Neural network and value tree approaches*. Utrecht, The Netherland: Labor Graphimedia.
Kauko, T., & d'Amato, M. (2011). *Neighbourhood effect, International encyclopedia of housing and home*, Edited by Elsevier Publisher.
Leishman, C. (2009). Spatial change and the structure of urban housing sub-markets. *Housing Studies, 24*(5), 563–585.
Lentz, G. H., & Wang, K. (1998). Residential appraisal and the lending process: A survey of issues. *Journal of Real Estate Research, 15*, 11–39.
Malpezzi, S. (2003). Hedonic pricing models: A selective and applied review. In T. O'Sullivan & K. Gibb (Eds.), *Housing Economics and Public Policy* (pp. 67–89). Oxford: Blackwell.
Mark, J., & Goldberg, M. (1988). Multiple regression analysis and mass assessment: A review of the issues. *Appraisal Journal, 56*, 89–109.
McCluskey, W., Deddis, W., McBurney, R. D., Mannis, A., & Borst, R. (1997). Interactive application of computer assisted mass appraisal and geographic information systems. *Journal of Property Valuation and Investment, 15*, 448–465.
McCluskey W.J., & Anand, S. (1999). The application of intelligent hybrid techniques for the mass appraisal of residential properties. *Journal of Property Investment and Finance, 17*(3), 218–238.
McCluskey W.J., Deddis, W.G., Lamont, I., & Borst R.A. (2000). The Application of surface generate interpolate models for the prediction of residential property values. *Journal of Property Investment and Finance, 18*(2), 162–176.
Miller, N.G. (1982). Residential property hedonic price models: A Review. In Sirmans C.F. (ed), *Urban Housing Markets and Property Valuation*. Research in Real Estate, Vol. 2, (pp. 31–56). Jai Press Inc. Greenwich, CT.
Mok H.M.K., Chan K.P.P., & Yin-Sun Cho. (1995). A hedonic price model for private properties in hong kong. *Journal of Real Estate Finance and Economics, 10*(1), 37–48.
Moody's. (2008). Moody's approach to automated valuation models in rating UK RMBS, 21st August
Nguyen, N., & Cripps, Al. (2001). Predicting housing value: A comparison of multiple regression analysis and artificial neural network. *Journal of Real Estate Research, 22* (3).
O'Connor, P. (1982). Locational valuation derived directly from the real estate market with the assistance of response surface techniques. *Lincoln Institute of Land Policy*.
Orford, S. (1999). *Valuing the built environment: GIS and house price analysis*. Aldershot: Ashgate.

Pace, R.K., & Gilley O.W. (1998). Generalizing the OLS and grid estimators. *Real Estate Economics*, *26*(2), 331–347.
Quintos, C. (2013). Spatial weight matrices and their use as baseline values and location—adjustment factors in property assessment models. *Cityscape*, *15*(3), 295–306.
Renigier-Biłozor, M., & Biłozor, A. (2016). Proximity and propinquity of residential market area —Polish and Italian case study. In *16th International Multidisciplinary Scientific GeoConferences SGEM*. Bułgaria. (web of science).
Renigier-Biłozor, M., & Biłozor, A. (2016). The use of geoinformation in the process of shaping a safe space. In *16th International Multidisciplinary Scientific GeoConferences SGEM Bułgaria*.
Renigier-Biłozor, M., Wiśniewski, R., Biłozor, A., Kaklauskas, A. (2014). Rating methodology for real estate markets—Poland case study. Pub. *International Journal of Strategic Property Management*, *18*(2), 198–212. ISNN:1648-715X.
Renigier-Biłozor, M., Dawidowicz, A., & Radzewicz, A. (2014b). An algorithm for the purposes of determining the real estate markets efficiency in Land Administration System. *Public Survey Review*, *46*(336), 189–204.
RICS. (2012). *Comparable Evidence in Property Valuation*. RICS Information Paper.
RICS. (2013). *Automated valuation models* (1st Ed.), RICS Information Paper.
Robson, G., & Downie, M. L. (2009). *Integrating valuation models with valuation services to meet the need of borrowers lenders and valuers, findings in built and rural environments*, November, (pp. 3–10).
Robinson, R. (1979). Housing economy and public policy. London, MacMillan.
Rosen S. (1974). Hedonic prices and implicit markets: Product differentiation in pure competition. *The Journal of Political Economy*, *82*, 34–55.
Salvo, F., & De Ruggiero, M. (2011). Misure di similarità negli adjustment grid methods. *AESTIMUM*, *58*, 47–58, ISSN:1592-6117.
Salvo, F., & De Ruggiero, M. (2013). Market comparison approach between tradition and innovation. A simplifying approach. *AESTIMUM*, *62*, 585–594, ISSN:1592-6117.
Salvo, F., Ciuna, M., & d'Amato, M. (2013a). Appraising building area's index numbers using repeat values model. A case study in Paternò (CT). In *Dynamics of land values and agricultural policies* (pp. 63–71), Bologna: Medimond International Proceedings, Editografica. ISBN:978-88-7587-690-6, Palermo, 22–23/11/2012.
Salvo, F., Ciuna, M., & d'Amato, M. (2013b). The appraisal smoothing in the real estate indices. In *(a cura di): Maria Crescimanno, Leonardo Casini and Antonino Galati, Dynamisc of land values agricultural policies* (pp. 99–111), Bologna: Medimond Monduzzi Editore International Proceeding Division. ISBN:978-88-7587-690-6, Palermo, 22–23/11/2012.
Salvo, F., De Ruggiero, M., Ciuna, M. (2014). Property prices index numbers and derived indices. *Property Management*, *32*, 139–153. ISSN:0263-7472, doi:10.1108/PM-03-2013-0021.
Salvo, F., De Ruggiero, M., & Zupi, M. (2015). The valorization of public real estate. A first outcome of the experiences in progress and a methodological proposal. In *XLIII INCONTRO DI STUDI Ce.S.E.T. Sviluppo economico e nuovi rapporti tra agricoltura, territorio e ambiente* (Vol. 67, pp. 135–146). AESTIMUM, FIRENZE: Firenze University Press, ISSN:1592-6117, Verona, 21/23 November 2013.
Salvo, F., Ciuna, M., & Simonotti, M. (2015). Compensation appraisal processes for completion of hydraulic works in an agricultural area. In *(a cura di) Alessandra Castellini Lucia Devenuto, Proceedings of XLIV INCONTRO DI STUDI Ce.S.E.T. Il danno. Elementi giuridici, urbanistici e economico-estimativi* (pp. 69–82), Mantova: Universitas Studiorum S.r.l., Bologna, 27/28 November 2014.
Salvo, F., Simonotti, M., & Ciuna, M. (2015). Multilevel methodology approach for the construction of real estate monthly index numbers. *Journal of Real Estate Literature*, *22*, 281–302. ISSN:0927-7544.
Silverherz, J. D. (1936). *The assessment of real property in the United States*. Albany: J.B. Lyon Co. Printers.
Simonotti, M., Salvo, F., & Ciuna, M. (2015). Appraisal value and assessed value in Italy. *International Journal of Economics and Statistics*, *3*, 24–31. ISSN:2309-0685.

Standard and Poor's (2004). Guidelines for the use of automated valuation models for U.K. RMBS transactions, 20th February.

Standard and Poor's (2005). Guidelines for the use of automated valuation models for U.K. RMBS transactions, 26th September.

Ward, R. D., Guilford, J., Jones, B., Pratt, D., & German, J. C. (2002). Piecing together location: Three studies by the Lucas County research and development staff. *Assessment Journal, 9*(5), 15–48.

Wilhelmsson, M. (2002). Spatial models in real estate economics. *Housing Theory and Society, 19*, 92–101.

Worzala, E., Lenk, M., & Silva, A. (1995). An exploration of neural networks and its application to real estate valuation. *The Journal of Real Estate Research, 10*(2), 185–201.

Appraisal Methods and the Non-Agency Mortgage Crisis

Maurizio d'Amato and Tom Kauko

Abstract Since the global economic crisis of 2007–08 an increasing amount of attention has been directed to the links between the financial system and the real estate industry. This paper ties to this discussion insofar as it focuses on the relationship between the methodology of property valuation and the recent non-agency or subprime crisis. After a brief discussion of the crisis various questions are raised concerning both the theoretical background and the application of property valuation, property management and automated valuation modelling. Despite the magnitude of the crisis in terms of the financial loss suffered, our observation is that the mainstream real estate academia is still essentially preoccupied with the task of integrating financial and property markets. Apparently, after the crisis financial models based on the concept of perpetual increasing income are still used, and deterministic relations between value and property characteristics still constitute the dominant paradigm. In the hope of avoiding repeating the errors that led to the crisis we identify the need to analyse this crisis from a property valuation point of view. We contend that in-depth analysis of the tools used in property valuation is necessary to understand why and how valuation methods should be improved given recent experiences.

Keywords Property valuation · Non-agency crisis · Automated valuation methodologies

---

Although the work was carried out in strict cooperation between the two authors, we may approximate the following share of responsibilities between the authors: Sects. 1 and 4 were written by Tom Kauko whereas Sects. 2 and 3 were written by Maurizio d'Amato.

M. d'Amato (✉)
DICATECh Technical University Politecnico di Bari, Bari, Italy
e-mail: madamato@fastwebnet.it

T. Kauko
The School of the Built Environment, Oxford Brookes University, Oxford, UK

M. d'Amato and T. Kauko (eds.), *Advances in Automated Valuation Modeling*,
Studies in Systems, Decision and Control 86, DOI 10.1007/978-3-319-49746-4_2

1 Introduction

When the global financial crisis and economic downturn begun 2007–08 attention was directed to the ways in which the financial system was attached to real estate—a hitherto forgotten sector within the world economy. Suddenly property price indices had stopped growing and in many countries a downward turn was a new, frightening reality (see Martin 2011; Kušar 2012). That the property prices will always grow in the long run, was a conventional wisdom; that in the long term property is the safest investment, was another one. Greed however is always about short term, and, as most of the industrialised world would find out soon, greed combined with market ignorance and liberal regulations has devastating consequences—perhaps even in the long term.[1] While the situation caught many real estate investors 'with their pants down', also the actions of intermediating real estate analysts—in particular, the relationship between brokers and the appraisers—became embarrassing to follow. Brokers often made the choice of a specific appraiser conditional upon a specific valuation result. The erroneous idea of an ever growing and stable property price trend had become a fundamental perception and basis for the most part of the methodological income approach to property valuation. At the time of writing the crisis is still going on, has worsened and is getting worse (for reasons that go beyond the scope of this contribution). This has implications for the real estate industry in general and valuation procedures in particular. The new question we want to air is as to whether the crisis has stimulated the real estate industry to be more humble towards 'getting the valuations right'. At the same time valuation automata has become increasingly widespread and is already an established tool in mortgage appraisal as well as tax assessment. The legitimate worry is that such new valuation and mass appraisal tools launched will only replicate the old mistakes made. Given this tendency, two research questions can be formulated: why and how should valuation methods be improved to cope with extreme magnitudes of cyclical market fluctuations? What then exactly went wrong? What would be such a 'better' valuation tool? Are the prospects of developing something entirely different yet feasible realistic at all? This paper raises some issues concerning the relation between real estate appraisal methods and the real estate market crisis. Our contribution focuses on a specific aspect of the so called non-agency mortgage crisis (or subprime crisis). Non-agency mortgages are mortgages borrowed by persons who do not normally have the requisites to obtain financial funds from a bank (through agency mortgages). In several articles, books and documentary TV-programs the origin of the crisis is explained in such a way that it appears to be related only to the financial world. We consider this proposition incorrect; namely, we strongly believe that this crisis may stimulate a discussion on the role of real estate to the extent the (global) financial system is being supported by valuation and automated valuation modelling. The text is organized in four

[1]Munasinghe (2010) estimates the inflated financial values—*toxic* assets—at twice the annual global GDP.

sections as follows. After this introduction Sect. 2 deals with the origin and the characteristics of the crisis. Section 3 takes issue with the inherent limitations of appraisal methods and automated valuation modelling methodologies. Final remarks will be offered at the end (in Sect. 4).

2 A Glance at the 2008 Crisis

According to Garton (2009), the root cause of the crisis is that in 2007 panic occurred due to the lack of knowledge of the details about the institutional settings and design of the 'shadow banking system' which had emerged by the early 2000s. He concludes that this lack of detailed understanding has four elements: (1) the sensitivity of the chain of interlinked securities to house price trends; (2) the creation of symmetric information via complexity; (3) the opaque way of spreading the risk; and (4) the trade in asset backed security indices linked to subprime bonds. It could be said that the financial crisis begun when the real estate prices and price indices in the US—and later more globally—ceased to rise. As it were, irrational exuberance (Greenspan 1996; Shiller 2000) had exposed the limit of the false myth of forever increasing real estate market prices.[2] As is now well-known also for the lay person, the crisis escalated because investment banks, during the growing real estate price trend, discovered the real estate products, and decided to buy more and more real estate mortgages. All these mortgages were securitized in financial products called Collateralized Debt Obligations. Other similar financial vehicles were Collateralized Mortgage Obligations and Collateralized Loan Obligation. The name of the vehicle changed according to the nature of the underlying asset. For example in Collateralized Debt Obligations several kinds of debt from different sources (study loans, mortgages and so forth) were collected together and sold. Three rating agencies gave the highest reliability rating (triple A) for these collateral debt obligations; as a consequence, several financial institutions in the globalized financial market bought an enormous amount of these toxic assets. The high rating guaranteed the reliability of this kind of assets. Subsequently investment banks begun to collect money from investors for investing in these kinds of financial vehicles; this occurred through a procedure where the collateral was treated as a real estate asset that is always regularly paid by the borrower. However, in many cases there was no reasonable expectation of the borrower actually being able to complete the back-payment deal. When analysing the situation it is crucial to note that the rating of the importance of the abovementioned financial tools was tied to the real estate world. As a consequence, estate agents (realtors) begun to increase the

[2]The phrase irrational exuberance was (to our knowledge) used for the first time by the then Federal Reserve Board Chairman, Alan Greenspan, in a speech given at the Annual Dinner of The *American Enterprise Institute for Public Policy Research*, in Washington, D.C., on 5 December, 1996. The phrase was interpreted as a warning that the market prices of stocks might be overvalued. (This is furthermore the title of one of the most important books of Robert Shiller.).

volume of their mortgages. Under the pressure of benefiting from an improved profit opportunity they started buying more and more non-agency mortgages from the realtors. Under the same pressure the realtors manipulated non-agency mortgage relying on the implicit assumption that real estate is an ever growing sector. As a consequence the non-agency mortgage which are composed by three different categories: prime jumbo, alt-A and the well known subprime increased dramatically, especially the last of these categories. The dollar amount in subprime mortgage (one of the three non-agency mortgage types) passed from 332 Billion US Dollars in 2003 to 1.3 Trillion USD in 2007—thus the rate of increase was as high as 292 %.[3] In the meantime the Collateralize Debt Obligations were insured against a possible default by insurance companies. In particular, the insurance company AIG was one of the most important players in this field. In this company decisions were made based on an ever increasing property market price trend. This happened under the surveillance of the President of Federal Reserve. For this reason Abelson (2007) wrote the following statement: "Financial mischief on such a grand scale is not a one man job, and Mr. Greenspan, needless to say, had a lot of help from Wall Street, Washington…and just as the contempt for risk that made possible the gross extravagances in housing and the financial market was sustained by confidence that Mr G would always bail out the participants". The fact that all the financial institutions had this kind of asset in their portfolio assisted to another important change: the introduction of a new accounting standard in 2007 and exactly the Financial Accounting Standard nr. 157 introducing the concept of mark to market value. This new criteria compelled the owner of this asset to adjust its value at the current—thus inflated—price of the asset. All the financial institutions with important quantities of these kinds of assets suffered heavy losses as the property market prices begun to fall. Lehman Brothers was one of them.

3 The Role of Property Valuation Methods Herein

When we speculate about the real estate appraiser's role in the type of market and behavioural context discussed above, a number of methodological issues stand out. Dealing with these issues also adds to the discussion after one of the most remarkable economic crises in the modern history of mankind. It is well known that property valuation relies on three different basic approaches: market-sales, income and cost. In this contribution three different problems are raised concerning

[3]Testimony of Emory W. Rushton, Senior Deputy Comptroller and Chief National Bank Examiner, Office of the Comptroller of the Currency (OCC), before the Senate Banking Committee (March 22, 2007). OCC's primary mission is to charter, regulate, and supervise all national banks and federal savings associations, federal branches and agencies of foreign banks. OCC supervises banks and federal savings associations to control that "they operate in a safe and sound manner and in compliance with laws requiring fair treatment of their customers and fair access to credit and financial products".

property valuation both in the market-sales and the income approach. The first issue is related to the market approach. In this method the opinion of value is reached by observing the price and the characteristics of comparable properties (Appraisal Institute 2011); in particular, property valuers (appraisers) search for comparable data to document in the valuation report (appraisal report). Unfortunately "... [w] hen a broker orders a valuation (an appraisal), he provides an estimate or target value for the property to the appraiser. If the appraiser has problem consistently reaching this number, the broker will hire someone else..." (Bitner 2008, p. 92). In this way appraisers may be tempted to include only those comparables which would allow them to reach the broker's target value in the appraisal. As a consequence, another interesting part of a valuation (appraisal) using the market approach concerns the excluded comparables. Normally, one is obliged to include sufficiently similar observations that are also situated near each other. However, the issue about which ones to leave out because of insufficient similarity is not formally specified.[4] Furthermore, the reason why an appraiser should exclude some of the collected market data from the appraisal report is not obvious. Showing both the comparable selected and the data excluded in the appraisal report may improve the transparency of the valuation, thereby giving a specific justification of the selection and exclusion process. A second issue can be raised about the income approach.[5] The application of income approach is based on different theoretical appraisal models whose background is the financial mathematics. Two models are based on the income approach (IVSC 2011, IVS230, C6-C21): one, direct capitalization which is a process that transforms an infinitive group of rents into a value, and two, yield capitalization which is a process that transforms a finite group of rents into a value (Appraisal Institute 2011). Furthermore, a discount cash flow analysis is an appraisal method which discounts a series of cash flows and then sums these to a direct capitalization (going out value) at the end of a holding period. Normally these methods are applied on income producing properties, when available comparable transactions are unavailable. The applications of the direct capitalization model are based on a constant or variable rent which is transformed into value using a capitalization rate normally extracted by the market. In this process there is a separation by time series analysis of the cycle which requires time and data and the day by day problem of appraising a property. This is particularly true in the commercial market. In some cases the direct capitalization technique is referred to a current rent. In other cases like in the calculation of scrap value,[6] the direct capitalization is often based on a capitalization rate calculated as the difference between the discount rate and a growth factor (g-factor). The value is assumed to increase into the infinity! Neither the direct capitalization nor the yield capitalization model takes the real

[4]To overcome a related methodological difficulty of comparable versus target cases was demonstrated by Kauko (2009), albeit not for valuation context but local housing market analysis.

[5]The general equation $P = R/i$, where P is price, R is rent and i is the reasonable rate of yield expected from the investment.

[6]Scrap value or terminal value is the final capitalization at the end of a holding period in a Discount Cash Flow Analysis.

estate market cycle into account, because of an 'information gap' between appraisal. In the scientific debate an immediate comparison among different approaches may be not the most important issue. We need to explore the limitation and the powerfulness of actual and perspective tools we have and we may decide to adopt or test in the future. Academia has different time horizon than AVM industry. The most important things in this book are not the research results but the attempt to increase our level of knowledge in this sector of research practice and real estate market cycle analysis.[7] The former usually operates in a static setting, despite being influenced by market dynamics too. This procedure is chosen because of convenience rather than theoretical or empirical insight. While such an attitude does not necessary produce incorrect point estimates, it perhaps justified to argue here that financial modelling in real estate appraisal should consider the role of real estate market cycles (d'Amato 2004). On the other hand, this would increase the complexity of the valuation task at hand. A third issue can be raised about the valuation that supported the REI Global, the real estate vehicle of Lehman Brothers. The failure of Lehman brothers in fact relates to the diversification techniques of REI Global, whose 80 % portfolio was composed by commercial property. In an article of the Wall Street Journal this is explained as follows: "When it failed the estates of the collapsed investment bank listed its real estate holding as valued at 23 billion of dollars…The 23 billion of dollars has been written down substantially. In all, Lehman expects to receive some of 13.2 billion dollars between 2011 and 2014…". After reading this it would be interesting to understand how the large real estate position of one of the most important investment banks has been valued. At the moment we know that Lehman and brothers was one of the most important players in commercial property before it collapsed. At the third anniversary of its bankruptcy it still remains the most important owner and seller in this property market segment. Let us consider the problem from a different point of view. It is a simple overturn of the market—the omnipresent force; therefore we can go back to our everyday task without asking ourselves if the methodological tools we use have a 'worm'? Can these big commercial buildings/portfolios be considered 'too big to be valued' with our present methodological valuation tools? These are interesting questions that can be raised in one of the most transparent property markets in the world. Probably we need a stronger integration between valuation methodologies and real estate market cycles (d'Amato 2015) in order to deal with market phases in a clear way, especially when dealing with large commercial assets. In this vein, a related question can be raised about the diversification of a property portfolio? Was the Lehman portfolio increased using a deal-by-deal approach? This means analysing the single risk return profile of each single property investment without analysing the general risk profile of the property portfolio. This approach unfortunately is also included in important institutional documents such as the European Property and Market Rating of TeGOVA (TeGOVA 2003; Renigier Biłozor et al. 2014a, b; Renigier-Biłozor and Biłozor 2016a, b; Biłozor and Renigier Biłozor

[7]This can be seen as part of a broader 'knowledge gap' in market value analysis (see Mooya 2011).

2016; Salvo et al. 2015; Kauko and d'Amato 2011). In this document each property in portfolio receives a rating; unfortunately the most important problem of the portfolio policy is the correlation among several assets belonging to the same portfolio which has been completed omitted. Lehman Brothers may teach us something here. Moreover, yet another question related to the application of multiple regression models as automated valuation methods comes up in this vein. It is to observe that the relationship between price and the property characteristics (including environmental and area specific attributes) is deterministic. Such models have been applied extensively in order to control the property valuation or in combination with the appraisal report, particularly in the valuation of portfolios and secondary mortgage markets (Downie and Robson 2007; Borst et al. 2008; Ciuna et al. 2014; d'Amato 2008, 2010) Probably there is a strong necessity to improve hedonic modelling performance in the downturn of a real estate market cycle, or perhaps a question may be raised as to the deterministic relation that links the property price to the building characteristics. Doubts are being raised as to whether research on the relationship between 'heretic' and 'orthodox' automatic valuation mass appraisal modelling, a term coined previously (d'Amato and Kauko 2008; d'Amato and Siniak 2008; Kaklauskas et al. 2012), leads to any helpful guidance for selecting the appropriate AVM methodologies in a downwards sloping phase of the real estate market. Given the relationship between market dynamics and valuation practice discussed above, it comes hardly as a surprise that the subprime crisis has raised efficiency problems for automated valuation modelling as a method to appraise property. In our view various plausible methodologies—both deterministic and non deterministic ones—ought to be explored, and try to adapt a valid method for a given institutional property context. This relationship furthermore varies significantly across different parts of the world. This stretches economic geographer Martin's (2011) point about price bubbles and busts being highly unevenly distributed in space a step further. Indeed, the downturn makes it much more difficult to track value using any kind of existing automated valuation method—especially the method based on deterministic relation between price and characteristics. In the downturn phase of the market it may be more helpful to deal with relationships between price and the property characteristics in a flexible way than to chase a deterministic relationship within the confines of the current valuation/appraisal modelling methodology. The relationship between AVM and valuation (appraisal) indeed is an interesting one. Fleshing out this connection an article in the Washington Post noted opportunity of replacing all real estate appraisals with automatic valuation (Woodward 2008). However, the President of the Appraisal Institute, Wayne Pugh, countered this idea by noting that "no automated valuation system has successfully replaced human inspection and analysis" (Pugh 2008). Here the legislation has been coherent with Pugh's point of view insofar as the Dodd Frank Act is concerned. Dodd Frank Act implements financial regulatory reform after the financial crisis of 2008. In particular the title XIV subtitle F distinguishes appraisal process from automated valuation modelling, reorganizing both. In particular it was stressed how the role of valuation (appraisal) cannot be replaced by AVM. Our point of view is coherent with the Dodd Frank act (and thereby also Pugh's view but not Woodward's): automated valuation modelling is increasingly adaptable in describing real estate market behaviour without succeeding in replacing local information and human inspection in the valuation (appraisal) procedure. In sum the market approach, the income approach and the AVM based portfolio appraisal approach all have their proponents and opponents. We have showed how each of these three problem areas involves room for error but also optimistic improvement. The last section offers some more philosophical thoughts along these lines.

4 Final Remarks

As the discussion so far has shown, a number of issues concerning appraisal methodology need to be dealt with in order to improve the body of knowledge that govern this important and systemic crisis. This role of the subprime crisis is yet unexplored. However, this paper does not want to offer instant solutions to the problem of sorting out how appraisal methods and the non-agency mortgage crisis are tangled into each other. After all, factoring in any new criteria such as corporate social responsibility will be subject to validation and calibration of the valuation/appraisal model and then we are back in square one. Instead, and as a first step towards an innovative and responsible agenda, we have raised some research questions for the appraisal world.

First we can ask why we have a responsibility to take into account the complexity of the sensitivity to macro cycles when applying a valuation method. After that we can ask how this complexity ought to be dealt with. Both questions should be helpful in understanding the plausible relationship between the way the appraisal world functions and the systematic properties of the crisis. In particular, in the market approach the appraiser could complete the appraisal report by also indicating which comparables were eventually selected and which comparables excluded, thereby offering a clear and transparent process of valuation. The income approach may require an effort to understand how our knowledge on real estate market cycles can be incorporated into financial modelling (cf. d'Amato and Kauko 2012). Yet another issue is the improvement of automated valuation modelling when the phases of the real estate market cycle change (see Downie and Robson 2007; see also AVM-News 2008; Allen 2011). The challenge here is to be able to perform self-correcting behaviour in an environment that may be conservative and demented. Garton's (2009) identification of the difficulties to understand the structure of 'the shadow banking system' led to the panic of 2007. Elsewhere, complexity economics—an evolving subdiscipline that integrates complexity theory onto economics—is promoted as an improvement on the analysis of real estate and capital markets (e.g. Smith 2004; Miller and Page 2007). Other enlightened views surely exist too, but to what extent are they noted by the mainstream? Above all there is a risk we see and subsequently try to emphasize: namely, this great crisis may pass without seeing any change in the methodological background of appraisal process and automated valuation modelling. How this exactly is done is however another matter—one which only recently is beginning to emerge in discussions among the valuation community. Mooya (2011), for instance, argues for a new ontology to take over the current one, as this would be a response to the opportunities opened up in amidst the recent "alterations in the social reality" and that the real issue goes far beyond a comparison of AVM based and human valuation methodology; in fact, to involve alternative conceptualisations of market value. We strongly think that the real estate industry together with the realm of academic analysis needs these kinds of changes in thinking and perhaps many other changes too.

References

Abelson, A. (2007). After the Greenspan Put…, Barron's, August 13.

Allen, S. (2011). The Triage Approach To Estimating Value. Service Management, September, www.sm-online.com. Retrieved January 23, 2012.

Appraisal Institute. (2011). *The Appraisal of Real Estate* (13th ed.).

AVM News. (2008). *e-newsletter.* issue July, August.

Biłozor, A., & Renigier-Biłozor, M. (2016). The procedure of assessing usefulness of the land in the process of optimal investment location for multi-family housing function. In *World Multidisciplinary Civil Engineering-Architecture-Urban Planning Symposium*. WMCAUS 2016 Praga.

Bitner, R. (2008). *Confessions of a subprime lenders. An insider's tale of greed, fraud and ignorance*. London: Wiley.

Borst, R. A., Des Rosiers, F., Renigier, M., Kauko, T., & d'Amato, M. (2008). Technical comparison of the methods including formal testing accuracy and other modelling performance using own data sets and multiple regression analysis. In T. Kauko & M. d'Amato (Eds.), *Mass appraisal an international perspective for property valuers*. Wiley Blackwell.

Brown, E. (2011). Lehman still loom large in commercial real estate. Wall Street Journal, September 12.

Ciuna, M., Salvo, F., & De Ruggiero, M. (2014). Property prices index numbers and derived indices. *Property Management, 32*(2), 139–153. doi(Permanent URL):10.1108/PM-03-2013-0021.

d'Amato. (2010). A location value response surface model for mass appraising: An "Iterative" location adjustment factor in Bari, Italy. *International Journal of Strategic Property Management, 14,* 231–244.

d'Amato, M. (2015). Income approach and property market cycle. *International Journal of Strategic Property Management, 29*(3), 207–219.

d'Amato, M., & Kauko, T. (2012). Sustainability and risk premium estimation in property valuation and assessment of worth. *Building Research and Information, 40*(2), 174–185.

d'Amato, M. (2004). A comparison between RST and MRA for mass appraisal purposes. A case in bari. *International Journal of Strategic Property Management, 8*, 205–217.

d'Amato, M. (2008), Rough set theory as property valuation methodology: The whole story. In Kauko & M. d'Amato (Eds.), Mass appraisal an international perspective for property valuers (Chap. 11, pp. 220–258). Wiley Blackwell.

d'Amato, M., & Kauko, T. (2008). Property market classification and mass appraisal methodology. In T. Kauko & M. d'Amato (Eds.), *Mass appraisal an international perspective for property valuers* (Chap. 13, pp. 280–303). Wiley Blackwell.

d'Amato, M., & Siniak, N. (2008). Using fuzzy numbers in mass appraisal: The case of belorussian property market. In T. Kauko & M. d'Amato (Eds.), *Mass appraisal an international perspective for property valuers* (Chap. 5, pp. 91–107). Wiley Blackwell.
Downie, M. L., & Robson, G. (2007). Automated valuation models: An international perspective. The Council of Mortgage Lenders (CML). London, October.
Garton, G. (2009). The subprime panic. *European Financial Management, 15*(1), 10–46.
Greenspan, A. (1996). The challenge of central banking in a democratic society, speech at The American Enterprise Institute for Public Policy Research, in Washington, D.C., 5 December, http://www.federalreserve.gov/boarddocs/speeches/1996/19961205.htm. Retrieved January 27, 2014.
IVSC. (2011). International Valuation Standards. International Valuation Standard Committee.
Kaklauskas, A., Daniūnas, A., Dilanthi, A., Vilius, U., Lill Irene Gudauskas R., D'Amato M. et al. (2012). Life cycle process model of a market-oriented and student centered higher education. *International Journal of Strategic Property Management, 16*(4), 414–430.
Kauko, T., d'Amato, M. (2011). Neighbourhood effect. In *International encyclopedia of housing and home*. Elsevier Publisher.
Kauko, T., & d'Amato, M. (2008). Introduction: Suitability issues in mass appraisal methodology. In T. Kauko, & M. d'Amato (Eds.), *Mass appraisal an international perspective for property valuers* (pp. 1–24). Wiley Blackwell.
Kauko, T. (2009). Policy impact and house price development at the neighbourhood-level—a comparison of four urban regeneration areas using the concept of 'artificial' value creation. *European Planning Studies, 17*(1), 85–107.
Kušar, S. (2012). Selected spatial effects of the global financial and economic crisis in Ljubljana, Slovenia. *Urbani izziv, 23*(2), 112–120.
Martin, R. (2011). The local geographies of the financial crisis: From the housing bubble to economic recession and beyond. *Journal of Economic Geography, 11*, 587–618.
Miller, J. H., & Page, S. E. (2007). *Complex adaptive systems: An introduction to computational models of social life*. Princeton, New Jersey: Princeton University Press.
Mooya, M. (2011). Of mice and men: Automated valuation models and the valuation profession. *Urban Studies, 48*(11), 2265–2281.
Munasinghe, M. (2010). Can sustainable consumers and producers save the planet. *Journal of Industrial Ecology, 14*(1), 4–6.
Pugh, W. (2008). Appraisal Institute Counters Washington Post's Promotion of AVMs (p. 27). AVM News, September–October.
Renigier-Biłozor, M., Wiśniewski, R., Biłozor, A., & Kaklauskas, A. (2014a). Rating methodology for real estate markets—Poland case study. *Pub International Journal of Strategic Property Management, 18*(2), 198–212. ISNN. 1648-715X.
Renigier-Biłozor, M., Dawidowicz, A., & Radzewicz, A. (2014b). An algorithm for the purposes of determining the real estate markets efficiency in land administration system. *Pub. Survey Review, 46*(336), 189–204.
Renigier-Biłozor, M., & Biłozor, A. (2016a). Proximity and propinquity of residential market area —Polish and Italian case study. In *16th International Multidisciplinary Scientific GeoConferences SGEM*. Bułgaria. (web of science).
Renigier-Biłozor, M., & Biłozor, A. (2016b). The use of geoinformation in the process of shaping a safe space. In *16th International Multidisciplinary Scientific GeoConferences SGEM* Bułgaria.
Salvo, F., Simonotti, M., & Ciuna, M. (2015). Multilevel methodology approach for the construction of real estate monthly index numbers. *Journal of Real Estate Literature, 22*, 281–302. ISSN: 0927-7544.
Shiller, R. J. (2000). *Irrational Exuberance*. Princeton, Princeton University Press.
Smith, L. L. (2004). Complexity economics and alan greenspan. *Post-Autistic Economics Review, 26*, 2 August, article 2.
TeGOVA. (2003). *European property and market rating: A valuer's guide*.
Woodward, S. E. (2008). *Rescued by Fannie Mae*. Washington Post, October, 14, p. 17.

Automated Valuation Models and Economic Theory

Manya M. Mooya

Abstract Though not always apparent, especially to practitioners, all valuation methods represent attempts at application, or operationalisation, of underlying economic theory. Automated Valuation Methods (AVMs) are no exception. AVMs represent the most advanced application of conventional, or neoclassical economic theory, to the valuation of real estate. The purpose of this chapter is to explicate the links between neoclassical economic theory and AVMs. The neoclassical economic foundations of AVMs are used to shed light on two subjects. Firstly, to clarify the relationship between traditional valuation methods and AVMs. Secondly, to show the potential connection between AVMs and the non-agency mortgage crisis of 2008. The chapter hopes to provide a clear basis for the justification, potential and limitations of AVMs as a valuation tool. Implied in the discussion is the idea that the validity of AVMs is highly dependent on the market context. The chapter concludes by cautioning against universalistic notions of the applicability of AVMs.

Keywords Neoclassical economic theory · Valuation methods · Real estate markets · Hedonic model · AVMs · Covering-law model · Regression functions · Non-agency mortgage crisis

1 Introduction

Automated valuation models (AVMs) are mathematical models, which, with the aid of appropriate computer software and a database of property information, are used to provide real estate valuations. Consistent with the majority of valuation assignments, AVMs, for the most part, are designed to generate estimates of market value, albeit for large numbers of properties. This brings AVMs within the purview of core economic theory, as the concept of market value is, and has remained central, to much thinking in economics. Implicit in the use of AVMs is an idea of

M.M. Mooya (✉)
University of Cape Town, Cape Town, South Africa
e-mail: manya.mooya@uct.ac.za

M. d'Amato and T. Kauko (eds.), *Advances in Automated Valuation Modeling*,
Studies in Systems, Decision and Control 86, DOI 10.1007/978-3-319-49746-4_3

market value as an attribute of the (real estate) market and validity of AVMs as a method for its determination or estimation. Market value is arguably the most important concept in real estate markets. Its definition has historically been the subject of contestations and contradictions. As defined by the International Valuation Standards Committee (IVSC) market value is “the estimated amount for which an asset or liability should exchange on the valuation date between a willing buyer and a willing seller in an arm’s length transaction, after proper marketing and where the parties had each acted knowledgeably, prudently and without compulsion” (IVSC 2003). This is essentially a price definition (French 2006). The basic principle is that market value is the exchange price in the market place at the date of valuation (Crosby 2000). The key assumption is that the price is set under competitive market conditions. Even though this definition does not fully resolve the underlying tensions and contradictions regarding the fundamental nature of market value, it provides a common framework to guide the different valuation methods. There is thus a degree of international consensus regarding what market value ultimately is, with the different methods seen as alternative, albeit not mutually exclusive, techniques or tools for its measurement. Though not always apparent, especially not to practitioners, all valuation methods are underpinned by, or applications of, abstract economic theory. Automated Valuation Methods (AVMs) are no exception. AVMs arguably represent the most advanced phase of the application of economic theory to the valuation of real estate. This evolution has, in turn, been a function of historical changes in theory regarding the market, the nature of economic value and methodology in the economics discipline. The primary aim of this chapter is to clarify the links between AVMs and economic theory, so as to provide deeper insights into AVM theory. The intention is neither to bolster nor undermine the case for AVMs as a valuation method. Rather, it is to close a critical gap in the AVM literature. Hitherto, much of the intellectual effort has been directed at improving AVM modelling and, consequently, the accuracy of model outputs. The theoretical basis of AVMs as a valuation method has received scant attention in the literature. This chapter aims to remedy this problem, and addresses three principle objectives. Firstly, by contextualising AVMs in economic theory, the chapter provides a clear basis for the justification of AVMs as a valuation tool. Secondly, by clarifying their common foundations in neoclassical economic theory, the chapter explicates the relationship between AVMs and traditional valuation methods. Thirdly, by articulating the theoretical and methodological influence of the neoclassical framework on conventional valuation theory, the chapter demonstrates how AVMs may have unwittingly contributed to the non-agency mortgage crisis of 2008. The rest of the chapter is arranged as follows. Section 2 provides a conceptual point of entry, by summarising conventional theory regarding the concept of market value and how AVMs relate to it. The section proceeds to discuss the key theoretical pillars underpinning AVMs, namely, marginal utility theory, rational choice and equilibrium. AVMs represent an instance of the use of mathematics for the explanation and prediction of social phenomena. Section 3 presents the theoretical case for the use of mathematical models in economics generally and for property valuation in particular. The relative merit of AVMs and traditional methods is frequently the subject of scholarly debate. Section 4 compares and contrasts AVMs and traditional valuation methods, focussing on their relative validity and accuracy. The aftermath of the non-agency mortgage crisis of 2008 raised questions about the potential role that AVMs might have played in precipitating it. The penultimate section considers the extent to which the foundations in neoclassical economics of AVMs might have created a predisposition for AVMS to contribute to the crisis. The concluding section, noting that spreading use of AVMs around the world, cautions against universalist tendencies regarding the applicability of AVMs.

2 Automated Valuation Models and Economic Theory

2.1 *The Neoclassical Synthesis*

As pointed in the introductory section, all valuation methods are united in their principle object, which is the determination of market value. The debates comparing the methods are never about differing conceptions of market value associated with each, but possibly about their relative validity and accuracy in its determination. It is therefore appropriate to begin the discussion by clarifying the common foundations of conventional market value theory. Only after that could AVMs be placed in their proper analytical context. Conventional theory regarding market value is based on the tenets of neoclassical economic theory. Neoclassical economic theory, which remains the dominant paradigm in economics, came about as a result of the fusion, in the nineteenth and early 20th century, of the 'marginal utility' school of economic thought with that of the 'classical' school. This in turn was the culmination of the long-term evolution of value theory whose roots can be traced to Ancient Greece. This body of principles is called 'neoclassical' because it incorporated the insights of the marginal utility theorists while also retaining the classical relevance of the cost of production in the determination of value. The neoclassical economists were marginalists in the sense that they emphasized decision-making and price determination at the margin (Brue and Grant 2007). The key difference between them and the marginal utility school was that they stressed both demand and supply in the determination of market prices. Alfred Marshall (1842–1924) is widely regarded as the greatest figure in the neoclassical school and is credited with this synthesis, which forms the basis for conventional value theory. Marshall analysed demand and supply separately before combining them to arrive at the idea of equilibrium competitive market price and quantity. In line with the marginal utility school, he conceived demand as based on the law of diminishing marginal utility (Brue and Grant 2007). In addition, he postulated the idea of rational consumer choice which, together with the law of diminishing marginal utility, allowed him to develop his 'law of demand'—to wit, that 'the amount demanded increases with a fall in price, and diminishes with a rise in the price' (ibid.). This is the now familiar downward sloping demand function. Regarding supply, Marshall stated that it was governed by the cost of production. Marshal made a seminal contribution to the development of contemporary value theory. He addressed the question of what determined market price. The classical economists attributed it to the cost of production while the marginalists attributed it to demand. Marshall's key contribution was to recognize that both supply and demand were responsible for establishing market prices, with cost lying behind the former and utility lying behind the latter (Brue and Grant 2007). He maintained that market forces tended towards equilibrium, where prices and production costs meet.

Marshall was the first major economist to consider the techniques of valuation, specifically the valuation of real estate. His theories form the basis of the three traditional methods of valuation, namely the sales comparison, income and the cost methods. The neoclassical theoretic approach is in turn the dominant paradigm in economics, and provides the theoretical foundation for much of the conventional approach towards real estate valuation, including for AVMs.

2.2 Neoclassical Economic Theory of the Market

As stated in the foregoing, conventional market value theory is based on fundamental assumptions of the neoclassical economic model of the market. These assumptions frame traditional conceptions about the nature of real estate market value and the practice of real estate valuation (Mooya 2009). Though there are variations within the different strands neoclassical economics, the following are widely regarded as the key assumptions underpinning the school's theory of the market, namely, rational choice, full information, equilibrium and perfect competition.

2.3 Rational Choice

Neoclassical theory proceeds on the basis of methodological individualism, the idea that social phenomena is built from, and can only be understood as arising from, the motivations and actions of individual agents. The basic building block of the neoclassical economic of the market is rational choice theory, at the heart of which lies the concept of homo economicus, 'the rational economic man' (Dyke 1981). Rational choice theory is the framework for understanding and formally modelling social and economic behaviour based on individual action. To satisfy the requirements of economic rationality, the individual human agent must act, and is hypothesised to act, according to the following principles (Dyke 1981, p. 51): He or she always prefers more to less. This means that, for instance, a house buyer will prefer a bigger house to a smaller one, or a 3-bedrooms house to a 2-bedrooms one. A house seller on the other will prefer to sell at a higher, than lower, price. The marginal utility of any good of interest diminishes as the quantity available increases. This means that agents are faced with a normal downward sloping demand function. When face with a choice between alternatives, the agent will be able to say either one is preferred, or that they are indifferent between them. That is to say, agents have the ability to make choices, and have the requisite information and capacity, including cognitive skills, to do so. The preferences of the agent for alternative commodities, or their indifference between alternatives, must be transitive. That is to say if agents prefer house A to house B, and house B to house C then they must prefer house A to house C. The same would hold for cases of indifference between alternatives. The most preferred alternative available is actually chosen by the agent. That is to say, observed behaviour is an outcome of, and consistent with, the internal rationalisation process. These principles are argued to accord with how individuals, more or less actually behave in reality, i.e. it is a realistic approximation of human behaviour. This is the basis of rational choice theory, which is based on the assumption that agents will always behave in a manner that maximises benefits to them, and minimise costs. That is, the behaviour of agents in the market is motivated by utility-maximising considerations. This implies that agents respond solely to price signals, and are not influenced by mere sentiment or altruism, for example. In concrete terms, rational choice means that for a given transaction possibility, sellers will want to obtain the highest price possible, while buyers will want to pay the lowest price possible. The implication of this for valuation theory in general and AVM theory in particular will be discussed in a following section. Implicit in the doctrine of methodological individualism and the concept of the 'rational economic man' is the notion of a typical or representative agent. This is an atomistic view of society, one that sees human beings as more or less homogenous. An atomistic view allows not only prediction of human behaviour, but of generalisation as well. As Lawson (2009) puts it, this view presupposes that individuals exercises their own separate independent and invariable effect, whatever the context, thus guaranteeing that under some repeated conditions x, the same predictable outcome y will always follow. This provides a basis for prediction of individual behaviour and therefore generalisation of human activity and social phenomena. AVM modelling requires that human behaviour is consistent and predictable.

2.4 *Full Information*

As stated above, the ability of agents to make rational choices presupposes both necessary cognitive ability and the availability of relevant information. The availability of information is of such critical importance that it merits separate treatment. Neoclassical economic theory holds that agents have the full information required to make rational choices. Full information, about the behaviour and dispositions of other agents, and of the attributes of the traded product, allows for 'frictionless' exchange in markets. This assumption ensures that trade will not be constrained by transaction costs, arising mostly from imperfect or missing information, thereby enabling markets to adjust efficiently and solely in response to price signals. Prices observed in such markets can therefore be argued as fully reflecting all the attributes of the commodity in question, thereby limiting the potential for mispricing, or divergence between prices and intrinsic worth or value. Relatedly, neoclassical economic theory requires homogeneity in the nature of the commodity being traded. Homogeneity of the traded good, by easing information problems about attributes of the good in question, reduces the cognitive demands for making rational choices. Further, product homogeneity allows the modelling of marginal changes in quantities of the good demanded or supplied in response to changes in the price incentives i.e. the construction of standard demand and supply schedules.

2.5 Equilibrium

Equilibrium is the notion that economic forces (i.e. supply and demand) are in, or tend towards, balance. It is the logical consequence of numerous, rational and fully informed individuals acting out their choices in the market. The concept can be understood either in a narrow sense ('static equilibrium') or more broadly to include general and dynamic equilibrium. In property, static equilibrium is theorised to account for the level of prices (or 'values') in a given market at a given point in time i.e. supply and demand in equilibrium explain property prices. More significantly, from a valuation point of view, prices established by markets in equilibrium are held to be valid measures of market value. This is the basis for the use of (market) prices as proxies for market value in valuation theory. General equilibrium theory applied to property markets on the other hand implies that the level of prices can be understood as arising from a balance in, and between, markets for the determinants of these prices, such as labour markets, financial markets, rental markets etc. It is important to point out that the concept of equilibrium is not merely descriptive, but that it also carries enormous normative weight. The concept implies a state of rest, a normal, if not right, state. This normative force has significant consequences for conventional economics as a scientific enterprise. Prices set under conditions of market equilibrium provide an external independent reality against which judgement about 'correctness', 'truth' or efficiency and so on about economic phenomena can be made. The notion of valuation accuracy, for example, measures the extent to which a valuation differs from a contemporaneous sale price for the same property. The reference sale price is not just any sale price, but only an equilibrium price, set in competitive markets i.e. one that corresponds with the 'true' market value.

2.6 The Neoclassical Perfect Market

Overall, the assumption of rational, perfectly informed, representative agents, trading in homogenous commodities result in equilibrium tendencies in markets. That is to say, the traditional assumptions about the nature of agents, and of the good being traded, serve to bring about, and reinforce, the notion of equilibrium in markets. Trade in such markets will establish prices that will be meaningful representations of value. This means that prices in such markets are not merely historical facts, but signal something much more significant. These prices provide a reference point in matters of economic analysis and judgment, in line with the normative attributes of equilibrium discussed above. The above assumptions give rise to the well-known neoclassical construct of the perfect market. As is well known, this market presupposes numerous, perfectly informed and mobile participants, trading in a homogenous and perfectly mobile product. Indeed, it is not unreasonable to say that the internationally agreed definition of market value is predicated on the traditional perfectly competitive market of neoclassical economics (Mooya 2009). The equilibrium price (i.e. the market value) set in such a market is both determinate and autonomous. Valuation is the process of attempting to discover this figure, with 'valuation accuracy' a measure of the degree of success in this endeavour.

2.7 Neoclassical Economics, Valuation Theory and AVMs

We are now in a position to examine the influence of neoclassical economic theory on valuation theory in general and AVMs in particular. To start, it is quite obvious that real estate markets are nothing like the perfectly competitive markets of neoclassical economics. Due to the heterogeneous nature of real estate, the lack of centralised trading and relatively few transactions, among other reasons, real estate markets are regarded as highly inefficient. At the centre of this inefficiency lie information problems that market participants face, for example, about the distribution of prices given a property of given specifications, or the reservation prices of potential transaction partners. Indeed, it is precisely because real estate markets do not work like the neoclassical ideal that creates the raison d'etre for the existence of the valuation profession. Information problems in these markets create incentives for institutions specialising in information intermediation to evolve (Miceli 1988), thereby lowering transaction and information costs for market participants (Baryla and Zumpano 1995). This is the role that valuation profession plays. Valuers are producers of pricing information without which the function of real estate markets would be severely constrained. The above points notwithstanding, it still remains the case that conventional valuation theory is strongly anchored on the principles of neoclassical economic theory. At a basic, if not trivial, level, market value is determined by the interaction of supply and demand in real estate markets. Crucially, for market value to emerge, there must be sufficient competition in the market so as to create conditions for equilibrium. That is to say, the market must be sufficiently competitive to establish a pattern of prices consistent with the concept of market value. Without such competition and resulting equilibrium, market value, as a knowable and determinate figure, could not possibly exist. The posited existence of market value, as the equilibrium price, is perhaps the most significant application of neoclassical economics to valuation theory. Market value so conceived is relatively stable, determinate and exists independently of players in the market. This is critical for the valuation profession, because valuation methods, including AVMs, are predicated on the existence of market value with these attributes. An idiosyncratic, fleeting or highly contingent market value would render the valuation enterprise untenable. Under such conditions, there will be no target to aim at and, therefore, no way of determining valuation accuracy. The idea of competitive markets, and of equilibrium price, however, is of much more significance than explaining the concept of market value at an aggregate level. It is also relevant for explaining marginal differences in prices between what might otherwise be similar properties. Individual properties are essentially bundles of utility-forming characteristics, such as size, neighbourhood, physical condition etc. It is conventional theory that the total price paid for each property represents the sum total of the valuations of its individual characteristics. Given that what is traded in the market is the total property bundle of characteristics, which cannot be split and sold separately, conditions must exist so to allow the market to solve the pricing problem, such that the price of each of the characteristics could be accurately determined (Evans 1995). In respect of residential property, for example, the market must be able to price the contributory value of house characteristics, such as neighbourhood, swimming pool, car garage etc. Without these conditions it would be impossible to apply valuation methods like the sales comparison method and AVMs. To bring about conditions whereby individual characteristics of property could be accurately priced, there needs to be sufficient competition for those characteristics. The best way to understand this is to conceive each characteristic as a commodity existing in its own market. A swimming pool might be part of a house, for example, but its contributory value is determined by the demand and supply of swimming pools in the specific market area. The contributory value of the swimming pool is thus an equilibrium price, albeit an implicit one. The same goes for all the other house characteristics. In the end, there must be sufficient competition for all relevant characteristics, such that a reliable pattern of marginal prices could be established. It is because real estate markets are considered sufficiently competitive as to be able to establish patterns of prices for marginal differences which provides the rationale for the price adjustment 'grids' of the sales comparison method, wherein the price implication of differences in characteristics between the 'subject' and 'comparables' are quantified. It also makes possible the use of regression analysis to estimate property values, which in turn is the basis for AVMs. We shall examine this latter point in more detail in the following section. For now, it is sufficient to point out that regression analysis works on the assumption that the price of, for example, a house is a function of various observable and measurable characteristics (Evans 1995). In their linear and additive form, the function could be expressed as in the Formula (1):

$$P = a + bx + cy + dz \quad (1)$$

where P is the price of the house, a the constant, x, y and z independent variables representing various characteristics and b, c and d the regression coefficients. To make what might perhaps be a trivial point, regression coefficients are constants that represent the rates of change of the house price as a function of changes in respective house characteristics. More significantly for our purposes, regression coefficients could legitimately be regarded as representing the implicit equilibrium marginal price, in the defined market area, for each unit of property characteristic. Just like there is a market value for property, representing its equilibrium price in the market, there is equally an implicit market value for each property characteristic, as represented by the regression coefficient. Analogous to property market value, regression coefficients are determinate and knowable. Thus, in the same way that different valuers should, in theory, arrive at similar figures for the market value of a specific property, even when they use different comparable sales, different statisticians should arrive at similar coefficients in a defined market area, even when they use different sample sales data. In addition, like market value, regression coefficients cannot, or should not, be idiosyncratic or ephemeral. The foregoing allows the use of regression analysis not only to estimate market values over a defined are but also it ensures that the regression function is able to remain valid for reasonable periods of time. We have commented that real estate markets do not function like the neoclassical ideal, and that a general lack of information is a big part of the explanation. Be that as it may, information remains a critical pillar of market activity. There must be sufficient information in the market, both about the attributes of property under contemplation and comparative prices being paid, to enable the pricing function to work reasonably well. This speaks to the efficiency of the market, and ensures that prices being paid correlate to the attributes of the property in question. Were this not be the case prices, paid may be a misleading indicator of true value or worth, thereby undermining the use of the former as a proxy for the latter. With respect to AVMs, there must be sufficient information so as to determine the price implications of marginal differences in property characteristics.

Finally, we can now look at the influence of rational choice theory on valuation theory and on professional practice. Though controversial in some respects, most people would accept the thesis that, in the main, participants in real estate markets behave in accordance with the principles of economic rationality, as outlined above. That is to say, participants behave so as to maximise utility (or, conversely, to minimise disutility). Thus sellers will hold out for the highest price possible, while buyers will attempt to pay the minimum possible. Rational choice also means that, holding other things constant, a house with a swimming pool will be preferred (and more valued) than one without, a bigger one to a smaller one, 3-bedrooms to 2-bedrooms, and so on. This interpretation of rationality is especially significant for AVMs and the sales comparison method of valuation. Without such an interpretation, the linear and additive regression functions underpinning AVM models, introduced above, would be confounded. By the same token, under the sales comparison method, this interpretation provides theoretical support for the adjustment of comparable property prices, at the margin, relative to the subject property. It must be emphasised that the ability to act in accordance with the principles of economic rationality requires both appropriate cognitive capacity and necessary information. Real estate markets present participants with a particularly challenging decision-making environment on both counts. In essence, in order to make buy or sell decisions, participants in these markets are required to solve a pricing function, of the type introduced earlier. They must have the capacity not only to price the real estate asset, taking into account all its characteristics, but also to process the price implication for what might be marginal differences in characteristics between comparable properties. The complexity of real estate, the relative infrequency of transactions and/or the generalised lack of transparency create significant cognitive and information demands in this respect. Behaviour that is in accordance with principles of economic rationality in real estate markets is most evident in the work of professional valuers. In fact, the demand for professional valuation services is a result of the inability of individuals in the market to, with confidence, solve the pricing function by themselves. By determining market value, among other types of value, the role of valuers is to solve the pricing problem for market participants. Indeed, it is axiomatic that, in the valuation process, valuers adopt the particular perspective of potential buyers. Valuers thus attempt to replicate the decision-making process by which buyers evaluate the characteristics of subject and comparable property to arrive at a bid price. This is reflected in the adjustment process of the sales comparison method of valuation, in terms of which valuers relate the prices of ‘comparable’ properties to the ‘subject’ property, taking into account the differences in characteristics between the former and the latter. This process could fairly be regarded as the epitome of rationality in the economic analysis of real estate.

3 The Theory and Methodology of AVMs

The previous section has outlined the common foundation in neoclassical economics of conventional valuation theory and AVMs. Notwithstanding the fact that AVMs share these foundations with traditional (i.e. manual) valuation methods, they are distinguished by their own particular theory and methodology. AVMs are an artefact of specific and relatively recent developments in economic theory, supported by advances in statistical analysis and computer technology. This section examines the theoretical and methodological foundations of AVMs, as a precursor to a comparative assessment with traditional methods in section four. To reiterate an earlier point, the essence of AVMs is the use of mathematics, or mathematical techniques such as regression analysis, to determine or estimate the market value of real estate. The software and databases that support AVM models are mere enablers in this enterprise. The interesting issue, from the point of view of theoretical engagement, therefore, relates to the nature of the relationship between mathematics, an 'abstract science of number, quantity, and space', on one hand, and market value, an economic and social phenomenon on the other. AVMs are predicated on the view that one could use the former to determine the latter. This puts AVMs squarely within the domain of the theoretical discourse about the role and place of mathematical sciences in economic analysis. The discussion in this section is in two parts and will proceed as follows. The historical antecedents to the use of mathematics in economics are to be found in the marginalist school of economic thought. The first part therefore briefly reviews the salient features of marginalist thinking, in so far as these relate to of the use of mathematics in the analysis of economic phenomena. The second part examines the specific methodological foundation of AVMs.

3.1 The Marginalist Revolution

The term 'marginalist revolution' refers to the nearly simultaneous, but completely independent, discovery in the early 1870s, by Stanley Jevons (1835–1882), Carl Menger (1840–1925) and Leon Walras (1834–1910), of the principle of diminishing marginal utility as the fundamental building block of a new microeconomics (Blaug 1986). Significantly, for our purposes, is that the first economic theory to be revolutionised by this discovery was the theory of value. The prevailing orthodoxy, from classical economics, was that exchange value was determined by the cost of production, especially labour costs. Exchange value was held to be entirely objective, since production costs could be determined with some accuracy and were independent of individual preferences. Marginalists substituted the 'cost-of-production' theory of valuer with a theory of value based on marginal utility, in terms of which value was linked value to the utility of, and demand for, the marginal unit of an item. Classical economics was devoid of a theory of demand and, therefore, an asymmetrical theory of price determination (Blaug 1986). The marginalists repudiated the cost approach to value by arguing that if one more unit than is needed or demanded appears in a given market, the market became diluted and the cost of production became irrelevant. Th marginalists therefore, regarded value as a function of demand, with utility as its fundamental precept. Marginalists made a contribution to neoclassical economics, and via that, to the theory and methodology of AVMs, in at least three critical respects. Firstly, by developing a demand oriented theory of value, they directed attention to the processes by which humans make economic decisions in a context of finite resources, leading to rational choice theory and all that this entails (the discussion in section two drew the link between rational choice theory and AVMs). Secondly, marginalists laid the foundation of economics as a scientific discipline, with a specific 'value-neutral' methodology that was to support the later emergence of mathematical modelling in the discipline (more of this later). Finaly, marginalist thinkers like Jevons and Walras used mathematics in economic analysis and, in that sense, could be regarded as pioneers of the practice. Jevons, for example, used marginal calculus to analyse marginal utility (Screpanti and Zamagni 2005). Indeed, the term 'marginalist revolution' is closely linked to the mathematical result of the marginal conditions for market equilibrium, as derived by calculus (Stuparu and Daniasa, n.d.). Walras for his part constructed a system of simultaneous equations to describe the interaction between consumers and sellers to aid the elucidation of his theory of the general economic equilibrium (Screpanti and Zamagni 2005). In summary, the following main characteristics of marginalism, listed by Screpanti and Zamagni (Screpanti and Zamagni 2005, p. 166) could be regarded as foundational to the theory and methodology of AVMs: Focus on static equilibrium (rather than on long-run economic growth) and on the idea that the allocation problem, given fixed and scarce resources, was the central problem in economics. Interest in the pricing and allocation of fixed factors of production led to the proposition that at the heart of all economic problems lay a mathematical function to maximise under constraints (see also Blaug 1986; Backhouse 2002). This opened the way for a mathematical treatment of the discipline. Acceptance of the hypothesis of decreasing marginal utility and a utilitarian approach to value, leading to a theory of human behaviour understood to be as aimed at maximising utility. This theory of human motivation lends itself to mathematical treatment (Stuparu and Daniasa, n.d). In the words of Walras, "It is only with the aid of mathematics that we can understand what is meant by the condition of maximum utility" (ibid, p. 385). A method based on 'the substitution principle', in terms of which the substitutability of one basket goods for another is assumed. This is the basis upon which, for example, comparable properties could be regarded as substitutes for each other. Buyers of property should, in theory be indifferent as between properties with comparable characteristics, making it possible to infer the price (i.e. value) of one from that of its comparables. This is what makes sales comparison and AVMs valid valuation methods. In respect of AVMs in particular, the inference has to be made from sample sales to all property in the relevant market area for which the specific model is considered valid. An approach based on methodological individualism. Walras argued that economics could only scientifically deal with the behaviour of individual agents and that it was not possible to speak in a scientific way of economic aggregates. Methodological individualism is, as we have seen before, the foundation stone upon which rational choice theory is built. Immutability of 'economic laws', where economics was likened to the natural sciences, especially physics. In terms of this conception, 'economic laws' assumed the absolute and objective characteristic of natural laws. The concern was to develop economics as a discipline on par with the physical sciences (Stuparu and Daniasa, n.d). Just as the physical sciences were being built up in axiomatic fashion on the basis of units of energy, etc., economics was being built up axiomatically on the basis of units of utility, with mathematics being seen as the vehicle for achieving this goal (ibid.). We take up this matter in the following section, to elaborate how the conception of economics as a scientific discipline has influenced the methodology of AVMs.

3.2 *Methodology of Automated Valuation Models*

3.2.1 A Positivist Methodology and the Covering-Law Model

Having laid out their foundations in neoclassical economics and marginalism, we can low look in a more direct fashion at the methodology of AVMs. As has been pointed out, AVMs are mathematical models used to provide real estate valuations. It is therefore appropriate to view them through the prism of the role of mathematics in the analysis of economic phenomena. This not only helps to frame the subject conceptually, but also locates AVMs within territory that has seen a fair amount of debate in the literature. Mathematics is used in economics in two general ways (Brue and Grant 2007; Backhouse 2002). It is used deductively to derive and state economic theories, enabling the drawing of conclusions that might otherwise not be seen, and with greater rigour than is possible with only verbal reasoning (Backhouse, op cit). Algebra, calculus and topology are the major tools employed in this way (Brue and Grant 2007). The second use of mathematics is as a tool in empirical research, to generalise from observations and to test economic theories using evidence (usually statistical data) about the real world (Backhouse, op. cit). Mathematical techniques such as multiple regression analysis are used for this role (Brue and grant, op. cit). Econometrics, which is the standard methodological approach used within the economics discipline, combines these two types of mathematical economics (ibid). To appreciate the role of mathematics in economics, it is perhaps best to start with the basis of the claim that the discipline is a science. The view that economics is a science in approximately the way that physics is, for example, is now well established. That view is sustained by positivism as the dominant methodological orientation of the discipline. Positivism holds that society, like the physical world, operates according to general laws. These laws provide the basis for explanation and prediction. In their quest to emulate the work of the physical scientists, economists see that the former's (predictive and explanatory) success seems to depend upon the ability to find mathematical relationships between phenomena, relationships reflected in famous equations such as Newton's laws and the laws of thermodynamics (Dyke 198, p. 59). The positivist conception of general laws is, in practice, formulated in terms of 'empirical regularities', defined as correlations between two variables taking the form 'if x, then y'. Thus y, the 'dependent' variable could be explained by, or predicted from, the facts or movements of x, the 'independent' variable. This theory of explanation and prediction is called the covering-law model. According to the covering-law model, to explain an event by reference to another event necessarily presupposes an appeal to laws or general propositions correlating events of the type to be explained ('explanandum') with events of the type cited as its causes or conditions ('explanans') (www.brittanica.com). It is rooted in the philosopher David Hume's (1711–1776) doctrine that, when two events are said to be causally related, all that is meant is that they instantiate certain regularities of succession that have been repeatedly observed to hold between such events in the past (ibid). This doctrine was given more rigorous expression by the logical positivist Carl Hempel (1905–1997) (ibid). The covering law model comes in two species, namely, the deductive–nomological model (or D-N model) and the inductive-statistical model (or I-S model) respectively. Both appeal to general laws. The difference is that under the D-N model the laws are deterministic, that is to say, whenever x, then y will eventuate with a probability of 1. The D-N model is appropriate for explanation of phenomena in natural sciences because these tend to exhibit exact relationships. Laws under the I-S model on the other hand are probabilistic, meaning that whenever x, then y will eventuate with a very high probability, but not necessarily 1. This implies that the I-S model has a lower predictive power than the D-N model. The I-S model is therefore more suited to the explanation of social phenomena, where relationships are highly contingent. We shall see in the following section how AVMs are in fact instances of the I-S model. For now let us illustrate how the covering-law model is used for scientific explanation (and prediction), using the house price as an example. In this example we shall use the I-S model, but either model will do for this purpose, as both have the same structure. The I-S explanatory account may be regarded as an argument to the effect that an event or state to be explained (the explanandum) was to be expected by reason of certain explanatory facts (the explanans), divided into two groups, namely (i) particular facts and (ii) empirical regularities expressed by general laws. Conventionally the explanation is laid out as follows:

Particular facts or initial conditions (F)
General laws (L)
Phenomena explained and/or predicted (P).

Suppose the task is to 'explain' (in the sense of rationalise) the (known) price P at which a particular house has been sold using the I-S model. In this case one needs to specify particular facts relating to the house and at least one general 'covering-law'. Let us assume that the relevant particular facts for this house are: (1) size (2) age and (3) neighbourhood. Finally, let us assume that a 'house price law' of general application in the housing market could be specified (more about this shortly). This explanation can then be structured as follows:

F1 House size
F2 House age
F3 House neighbourhood
L1 House price law
P1 House price P explained and/or predicted.

This model renders a more or less complete and logical explanation why a particular house price is what it is, by showing that it resulted from the particular circumstances specified in facts F1, F2 and F3, acting in accordance with general law L1. It is also quite easy to see how the model could be used to predict (or estimate) the unknown price (or value) of another house. All that would be required is knowledge of the particular facts of that house. On the basis of these facts, and the pre-existing 'universal house price law' the prediction (or estimation) is easily done. This in essence is the methodology of AVMs. We develop this line of thought more fully in the section below.

3.3 The Regression Function as General Law

AVMs in practice come in about five types, namely repeat sales index model, tax assessed models, 'intelligent' systems, econometric forecasts and hedonic models (Downie and Robson 2007). The discussion that follows is based on hedonic models, for two main reasons. Firstly, hedonic models are the most widely used in practice. As a matter of fact, econometric forecasting and 'intelligent' systems, such as 'artificial neural networks', are themselves ultimately based on hedonic models. Secondly, and perhaps more significantly, hedonic models mirror the process used by human valuers, with the sales comparison method the perfect example. Hedonic models, therefore, allows us to place both AVMs and human valuers in a common theoretical framework provided by neoclassical economics, which, as we have seen, has as its foundation the exercise of human choice or judgment. This in turn will facilitate a critical comparison of the two approaches to real estate valuation. The basic premise of hedonic models is that the price (or value) of a marketed good is a function of its constituent characteristics. The implication of this is that the good being valued could be decomposed into its constituent parts, and that the market values those constituent parts. The market price (or value) of a specific residential property, for example, is regarded as a summation of the values of its constituent characteristics, such as size, location, age, and so on. Each of these characteristics is valued in the market. Typically, hedonic models use regression techniques to estimate the contribution of each feature of the property to the overall value. At the heart of hedonic AVMs, therefore, lies the use of regression analysis for the prediction (or estimation) of real estates prices. By regression analysis is meant the various mathematical methods (including econometrics) that aggregate observations into a form in which a dependent variable is a mathematical function of independent variables ($y = f(x1, x2, \ldots, xn)$), often in a way that allows a statistical inference regarding the parameters of the function outside the specific sample (Ron 1999). Crucially, the regression equation has the form of a general law, albeit restricted in application to the relevant population. Thus in much the same way that (natural) scientific laws are expressed in terms of relations between observations, in the form $y = f(xn)$, regression analysis rests on a mathematical theorem that assumes the existence of a true model, in the form $y = f(xn)$ (ibid). It is telling that the discovery of regression analysis, by Francis Galton (1822–1911), was in the context of natural scientific observations involving pea seed sizes (Wiki). The procedure of regression analysis could be therefore considered as an exemplar of the positivist methodological approach in the social sciences (ibid.). To illustrate, using the I-S model and the regression equation P = a + bx + cy + dz (introduced earlier, where P is the house price, and x, y and z stand for the size, age and neighbourhood of the house respectively), the problem of predicting (or estimating) the unknown price (or value) of any house in the relevant market area can be structured as follows:

F1 House size x
F2 House age y
F3 House neighbourhood z
L1 P = a + bx + cy + dz
P1 House price P_1 predicted (or estimated).

The model shows that, given the equation, it is possible to know, for any set of independent values (i.e. property characteristics), the range of values that the dependent variable (i.e. property price) can take. This, of course, is the way AVMs function. We have pointed out that AVMs are based on the I-S, rather than the D-N, model of scientific explanation. The reason is that regression functions are derived inductively from a limited training sample of properties. This has implications for their predictive power. We have noted that I-S models are probabilistic, rather than deterministic, and that they consequently have lower predictive power than D-N models. Unlike D-N explanations, the inductive character of I-S explanations means that the relation between premises and conclusion can always be undermined by the discovery of new information (Mayes, n.d). Consequently, it is always possible that a proposed I-S explanation (or prediction), even if the premises are true, would fail to predict the fact in question (ibid.). Herein lies part of the explanation for the observed inaccuracies in AVMs. This also explains why prediction in the economics discipline (and the social sciences generally) has not, and possibly could not, match the successes seen in the natural sciences. That being said, the practical value of AVMs obviously requires that their predictive power be as high as possible. When theorizing about the predictive accuracy of AVMs, it is useful to conceptualise D-N models as the limit case (as indicated above, D-N models have a predictive probability of 1). The question then becomes how close the predictive power of AVMs are to this limit. The answer lies in the tenets of neoclassical economic theory discussed earlier. AVMs will approach the predictive power of D-N models if the assumptions of neoclassical economic theory are an accurate representation of reality. That is to say, the predictive power of AVMs (i.e. their accuracy) is maximized when buyers and sellers in real estate markets behave in accordance with the principles of economic rationality, have information with which to make meaningful choices, and trade in real estate markets that are competitive. Under these conditions, regression functions with a very degree of accuracy could be specified. Indeed, under the ideal-type of the perfectly competitive market of neoclassical economics, the predictive power of AVMs would practically be indistinguishable from that of D-N models. This suggests that proper knowledge of fundamental economic theory is necessary for a deeper understanding of AVMs.

4 Contrasting AVMs and Traditional Valuation Methods

Given the obvious efficiency advantages of AVMs over traditional or manual methods, it is inevitable that comparisons between these two needs to be made. The standard script is that AVMs have the advantage of speed, cost-effectiveness, consistency and objectivity over traditional methods. Ranged against these advantages are a number weaknesses. For one, because they lack insights provided by onsite inspections, AVMs assume average conditions, which could produce highly misleading estimates in atypical situations. Crucially, AVMs also lack the 'street-level' judgment and intuition provided by human valuers, attributes that may be indispensable for the accurate interpretation of market conditions at the individual property level. Overall, it is matter of debate as to which of the two, between AVMs and traditional methods, is more accurate. Much of the effort to try and settle this case has been by way of empirical studies, and has involved the comparison, relative to a 'true value' benchmark, AVM outputs to that of valuations done by humans. The more interesting (and appropriate) question from our point of view, however, is theoretical, and involves asking if there exists anything in economic theory that predisposes one or the other method to greater accuracy. This is the task that we have set for ourselves in this section. It is perhaps appropriate to preface this discussion with an explanation of the meaning of valuation accuracy. Valuation accuracy is a measure of valuation error and refers to the extent to which a valuation deviates from the 'true' or 'correct' value. The 'true value' is defined as the contemporaneous sales price of the same property under conditions consistent with the definition of market value i.e. an 'arms length' transaction in open, competitive markets. The convention is that, to be deemed accurate, a valuation should lie within 10 % either side of this 'true value'. A closely related concept to valuation accuracy is Valuation variation. This is a measure of the precision or deviation of a number of independent valuations of the same property from each other. Given the conventional assumption that there is, for each property, one correct or true value, too wide a spread of valuations is, prima facie, indicative of inaccurate valuations. Having dispensed with definitions, we can now return to our principle concern. We have, in a previous section, outlined the basis of conventional valuation theory in neoclassical economic theory. To answer the question regarding whether economic theory predisposes one or the other method to greater accuracy, therefore, one would need to show which of the two the underlying theory is more supportive of. The best way to proceed in this endeavour is to split the relevant theory into two parts. The first part is that which deals with the posited attributes of the decision maker, or economic agent in real estate markets, while the second is the part that deals with the hypothesised nature of the (real estate) market and of market value (or price). Regarding the first part, the task is to compare the attributes of AVMs and human valuers as 'decision makers'. Conventional theory holds that the decision maker is rational. It is clear from the explanation of rational choice theory that AVMs are superior to human valuers as far as economic rationality is concerned. The purely instrumental and logical approach to decision-making underlying rational choice theory is one that is perfectly suited for the computer/mathematical algorithms at the heart of AVMs. Because they are more logical than human valuers (who may be prone to sentimentalism), AVMs tend to produce more objective valuations. It is well established that the pricing decisions of buyers and sellers of real estate, whose behaviour human valuers are apt to mimic, may be influenced more by sentimentalism or economic pressure, rather than rationality. Further, we noted that the ability to make rational choices presupposes requisite cognitive ability, i.e. the ability to process fairly large amounts of information and/or to decipher complex situations prior to taking a course of action. Because of the ability of computers to handle large quantities of data and to perform elaborate calculations quickly, AVMS are infinitely superior in this respect to the best human valuers. This particular feature is the source of the former's time and cost advantages. The law-like character of regression functions, on the other hand, explains why AVMs are more consistent than human valuers. As explained above, regression functions perform the role of the 'covering law' in the I-S model of explanation and prediction. This, as we have seen, is the standard positivist methodology of neoclassical economics, which takes the form if x, then y. This implies that under similar repeated conditions, the same outcome will always follow. Thus AVMs will, within one market area, always have the same output in all cases where the property characteristics are identical. This is not usually the case with human valuers. The foregoing shows how the time, cost, objectivity and consistency superiority of AVMs over human valuers can be explained in terms of economic theory and methodology. There is much that can be said in favour of these virtues. They nonetheless do not necessarily imply that AVMs are, on average, more accurate than human valuers. To deal with the question of the relative accuracy of the two approaches requires looking at the second part of neoclassical economic theory, the part that deals with the nature of the (real estate) market and of (market) value. The relevant question to ask then is, what does economic theory say about the nature of market value, and given this posited nature, which, between AVMs and human valuers, is likely to be more accurate in its estimation? The answer to that question is complex. To start with, neoclassical theory espouses a psychological notion of market value. This reflects the legacy of marginalism on mainstream economic theory. We noted that marginalism advanced a human-centred utility-based theory of value, in terms of which market value could be explained as arising from the subjective valuations of all those who constitute the market. A subjective theory of value suggests that the phenomenon is highly contingent and variable, relying as it does on variable psychological dispositions associated with human opinion and judgement. Crucially, a subjective conception of value does not, prima facie, lend itself to a logical or mathematical treatment of the phenomenon. On this understanding, the attributes of objectivity and consistence appear to be inappropriate criteria for the evaluation of valuation methods. Rather, the more appropriate criteria should be the extent to which the valuation method mirrors the decision-making processes of actual market participants, of which intuition and sentiment, among others, play a decisive part. This in turn suggests that human valuers are likely to be more accurate at deciphering market value than AVMs, as the latter are ill-equipped to deal with the more subjective elements of the valuation process. The foregoing discussion suggests that AVMs are better at making (rational) decisions than human valuers, but that they are likely to be less accurate than the latter. There is a paradox here that requires resolution. And it is easily explained in terms of economic theory. It turns out that under conditions of the perfectly competitive neoclassical market, market value (i.e. the equilibrium market price) has an objective, determinate character. This is notwithstanding the fact that it is based on numerous individual subjective valuations. The nature of competition under these conditions (i.e. where there are numerous participants, full information, no barriers to entry, product homogeneity, etc.) is such that a single market price is established for each product and that this price is largely independent of individual preferences and dispositions (see Mooya 2011, p. 2273). This, in turn, means that the fact that market value ultimately rests upon subjective perceptions of individuals does not preclude the use of AVMs, or make AVM valuations less accurate than those done manually. The requirement, of course, is that the market must be sufficiently competitive. The perfectly competitive market is an ideal-type, of course, and is not be found in reality. But it is a useful analytical device with which to understand the workings of actual markets. In particular, it aids the comparison of AVMs and human valuations. The comparison starts with the observation that actual markets found in practice have varying degrees of competitiveness. Alternatively, it can be said that actual markets differ in the degree to which they resemble the neoclassical perfect market. Thus one can envisage, for each class of property in a defined market area, a continuum of market-types, with the perfectly competitive market at one end, and a 'market' consisting of a single property and two parties only (buyer and seller) at the other end. Any concrete market-type can be placed at some point between these two extremes. If we take the ends of the continuum as limiting cases, it can be seen that as market-types approach the perfectly competitive end of the continuum, the number of parties (i.e. transactions) will increase and the properties being traded become more and more homogenous. As individual actors became less and less able to influence prices, and as evidence of transaction prices becomes more and more widespread, market value will become more and more determinate and objective. At the limit, market value will be completely determinate and objective. This means that as one approaches the perfectly competitive market, conditions become more and more favourable for the use of AVMs. In this space, AVMs are superior to manual valuations, because they are more efficient and, in theory, at least as accurate, if not more so. It has to be pointed out, however, that whereas the validity and superiority of AVMs increases as you approach the limit of the perfectly competitive market, its use at the limit itself becomes superfluous. By definition, the properties are homogenous at this point, meaning that any training sample sales will be have identical characteristics to the rest of the properties in the market area, rendering AVM modelling pointless. So in practice, AVMs will be employed in market-types that are well before the limit. These are, of course, the types that are found in reality. Movement along the continuum in the opposite direction, away from the perfect market and towards the single property 'market' implies that the number of parties (i.e. transactions) will decrease and the properties being traded will become less and less homogenous. As individual actors became more and more able to influence prices, and as evidence of transaction prices becomes less and less widespread, market value will become more and more indeterminate and subjective. At the limit, 'market value' will be completely indeterminate and therefore entirely subjective. This means that as one approaches the single property 'market', conditions increasingly require the use of human valuers. In this zone, human valuers are superior to AVMs because they are more accurately. This is because actual transaction prices become to depend more on the internal, psychological, dispositions of market participants, rather than on external objective market evidence. As is repeatedly pointed out, valuers are able to deploy the human qualities of intuition and judgement much more effectively than AVMs, qualities that are indispensable in circumstances where there is great uncertainty. Two points are worth emphasising at this juncture. Firstly, it must be noted that at the limit of the single property 'market' the use of both human valuers or AVMs would be invalid. By definition, there will be no market evidence at this point, making the use of the sales comparison method impossible. The use of AVMs in circumstances when there is only the subject property in the market, on the other hand, is clearly inappropriate (and unnecessary). In practice, therefore, the furthest point at which AVMs and manual valuations could validly or usefully be employed lies well before the limit of the single property 'market'. This, of course, is consistent with the range of market-types that are to be found in reality. The second point to highlight, one that immediately follows from the foregoing, is that, in theory, there is a definite point on the continuum at which the use of AVMs becomes not only unfeasible but also inappropriate. This is the (theoretical) point at which market value shifts from being objective to being subjective, where, due to the lack of market evidence, individual psychological factors preponderates, making them decisive in determining transaction prices. This means that there exist a real limit to the extent to which AVMs may be applied in practice, and that beyond that limit no amount of refinement of models could improve model accuracy. By way of conclusion, let us reprise the principal question we are trying to address in this section. We set out to investigate whether there existed anything in economic theory that predisposed either AVMs or traditional manual valuations to greater accuracy. The answer is in the affirmative. In summary, the more the actual market-type under consideration is to the ideal of the neoclassical perfect market, the more valid and superior AVMs become. AVMs become better here, not because they are necessarily more accurate, but because they are more efficient. In contrast, the less the actual market-type under consideration is to the ideal of the neoclassical perfect market, the more valid and superior manual valuations become. Manual valuations are better in this space because human valuers are able make more accurate valuations in thin markets, where judgment, intuition and sentiment play a critical role in price determination. This analysis explains the common-sense observation regarding where AVMs are most frequently used in practice. As Downie and Robson (2007) explains, the accuracy of AVM results are highly correlated with the level of homogeneity of property types and the number of comparable transactions in a given subject area. That is why AVMs are widely used for residential property and less so for commercial property. The former, especially, standard suburban homes are fairly homogenous and tend to be transact more frequently, on average. Commercial property, on the hand, has more heterogeneity, and may be less actively traded.

4.1 AVMs, Neoclassical Theory and the Non-agency Mortgage Crisis

In the aftermath of the 2008 non-agency mortgage crisis, questions have been raised about the role and potential culpability of the valuation (or appraisal) profession in precipitating it, or at best in the collective failure to avert it. The role of AVMs, in particular, has been the subject of critical commentary, with some going as far as to attribute the crisis to their widespread usage in the home-loan origination process. It is a well-known fact that the period leading up to 2008 saw AVMs taking an increasingly larger share of valuations for mortgage purposes, at the expense of human valuers. In the popular imagination, appraisers in the US have been placed squarely centre in the cast of villains responsible for the crisis of 2008, together with mortgage originators, 'greedy' bankers and estate agents. The main charge levelled the valuation profession has been that of the making of inflated appraisals. According to Abernethy and Hollans (2010) inflated appraisals were common during the housing boom and helped contribute to the crisis. It is said that deliberate overvaluations were motivated either by the attraction of higher fees from higher appraisals or by lender and/or loan broker pressure to hit the 'right number'. In their final report, the Financial Crisis Inquiry Commission (FCIC) found that property values were being inflated to maximize profit for appraisers and loan originators (FCIC 2011). The Commission heard evidence about lenders opening subsidiaries to perform appraisals, "allowing them to extract extra fees from "unknowing" consumers and making it easier to inflate home values. Further, the report cites evidence of pressure on appraisers to place artificially high prices on properties, and the blacklisting of those who resisted these unethical practices. Its our contention that explaining the culpability of valuers for the financial crisis solely, or mainly, in terms of unethical and unprofessional conduct is, however, neither an adequate account nor particularly helpful. In does not, for instance, account for the behaviour of numerous honest valuers doing valuations in the heady days of the property boom. In an environment of continually rising prices did these honest valuers interpret the market in a manner that fuelled further rises? Could they have done otherwise, considering the theoretical apparatus at their disposal? It is argued here that foregrounding unethical conduct may have served to detract attention from the insidious role played by conventional valuation theory and methodology on which AVMs are based. The causes that led to the non-agency crisis are many and complex. It is neither possible nor desirable to provide a comprehensive review in this chapter. It is however appropriate, given the nature of our subject matter, to consider if any links could be made between the neoclassical economic framework, AVMs and the crisis. It turns out that such a link could indeed be established. We shall, in this section, argue that any contribution that AVMs may have played towards the crisis stems from an underlying theoretical framework that is ill-equipped to handle market crises and instability, and to a methodological orientation that precludes, or eschews, the possibility of valuer influence in the pricing of property. The key elements of conventional theory that are relevant to our argument are the positivist methodology and the notion of equilibrium. The latter is, as we have seen, a necessary consequence of the neoclassical assumptions of rational choice, full information and perfect competition. The positivist methodological orientation of conventional valuation theory requires, firstly, that observers (i.e. valuers) be independent from the object of observation (market values). It is thus conventional, in terms of this framework, that valuers merely report, or reflect the market, but do not influence or predetermine the magnitude of market prices or market values. In terms of this view, therefore, valuers are held to be dispassionate observer of property markets, objectively weighing evidence to arrive at appropriate value conclusions, leaving no ripples or trails in their wake (Mooya 2009). Their role is seen as merely to interpret the market to provide an opinion of market value, and not to make judgements as to whether the property is worth the price (Peto 2009). Secondly, positivism requires a reliance on empirical facts. Since market value cannot be observed directly, valuers must rely on transaction prices as evidence of market value. The problem, off course, is that observed prices may not correspond to market value, as conventionally defined. Price bubbles are a classic example of market circumstances where transaction prices may diverge from what may be regarded as the true worth of the real estate. To ensure a correspondence between the former and the latter requires the further assumption that observed prices represent 'arms–length' transactions, and that the relevant market is in equilibrium. That is to say, transaction prices are indicative of market value only under conditions of competitive markets and equilibrium. As noted above, the assumption that markets are, or tend towards equilibrium is a central tenet of neoclassical economic theory. This assumption is responsible for the deference with which the market is held in conventional theory. In terms of this view, a lot of faith is placed on the efficiency of markets. In practice, valuers can never be certain that transaction prices actually represent 'arms-length transactions' or that the relevant market is in fact in equilibrium. For pragmatic reasons, therefore, valuers must simplify, by accepting, at face value, observed prices as evidence of market value. This give rise to the erroneous (if explicable) tendency of conflating prices and market value, and the imbuing of the former with normative attributes of the latter. In this way, valuers and conventional valuation methodology may attach far greater significance to prices than is probably warranted. Given the foregoing, it is easy to see how the neoclassical economic framework may have unwittingly resulted in AVMs contributing to the non-agency mortgage crisis. The process could be explained as follows. First, the sample sale prices used for the training of AVM models are obtained from a market that, for reasons explained above, is, or must be, assumed to be in equilibrium. Working within the neoclassical economic framework, valuers would lack the theoretical tools with which to question whether these sample prices are higher than they should be. Since the market is 'always right' the prices must therefore be accepted at face value. As pointed above, neoclassical theory places a lot of faith in the efficiency of markets. Secondly, because AVM valuations influenced lending, and in view of their widespread, these elevated valuations could influence subsequent transaction prices in a material way. Finally, such transaction prices might eventually be used as sample sales in subsequent AVM modelling. By this process, a positive feedback mechanism could be set in motion, leading to ever higher and higher transaction prices, and to a price bubble, with AVMs fuelling this process. Critically, it is important to note that the neoclassical framework allows little scope for valuer agency to bring about a moderating or alternative outcome, even if this were to be desirable. The conflation of price and market value leaves denies valuers the theoretical apparatus with which to perceive price bubbles. Further, even if bubble conditions were to be perceived, any predisposition by valuers to 'correct' the valuations would be blunted by a positivist methodological orientation, which eschews interference with the phenomena under observation. Indeed the common refrain from the valuation profession, in response to criticism about the collective failure to call the price bubble, was that valuers could not be blamed because they were merely interpreting the market at the time. To put it simply, it is conventional theory that the market could not be wrong. To sum up our argument in this section, the widespread usage of AVMs per se could not be blamed for the non-agency mortgage crisis of 2008. Rather, the problem could be placed with the underlying neoclassical theoretical framework. By emphasising equilibrium and stability, neoclassical economic models of the market are, by definition, ill equipped to handle crises and crashes. That the 2008 crisis was largely unanticipated by many economists may have something to do with this. The FCIC, in their final report, conclude, appropriately enough that "the sentries were not at their posts, in no small part due to the widely accepted faith in the self-correcting nature of the markets" (FCIC 2011, p. xviii). More directly for our purposes, the conflation of price and value, together with assumptions that valuers merely report on the market, and that markets tend towards equilibrium, have had the practical effect of denying valuers the theoretical apparatus with which to perceive bubbles, while simultaneously constraining any predilections that they might have to do anything about it To conclude this section, it is important to stress that at the core of the relationships between AVMs, neoclassical economic theory and the non-agency mortgage crisis is an insidious form of circular reasoning. It takes the form: "property X has been sold for price Y, therefore Y is its market value", followed by "the market value of property X is Y, because it has been sold for Y". This, of course is an epistemic fallacy. More significantly, it has the potential for setting in motion a positive feedback mechanism that could lead to, or exacerbate, price bubbles.

5 AVMs for All Seasons? Concluding Remarks

The preceding discussion, especially in Sect. 5, implies that the nature of the market environment may present limits to the applicability AVMs. This is a point worth remembering in many of the developing countries, where there is much pressure to adopt AVMs. Real estate markets in developing countries tend to be less mature than those of the developed countries. This lack of maturity is reflected in factors such as small market size, severe information problems, weak property rights, poor regulatory environment and poor supporting infrastructure (such as financial and professional services). In short these markets tend to be less efficient their developed country peers, and therefore not conducive for the introduction of AVMs. Another way of putting it is to say real estate markets in developing countries are likely to be further way from the neoclassical ideal (of full information, perfect competition and equilibrium) than those of developed countries. Traditional manual valuations, therefore, are going to be the reality for many of these countries for many years to come. By way of conclusion, it is hoped that we have presented a more or less coherent account of the foundation in economic theory of AVMs. Some of the material covered and linkages are perhaps not well known, while some of it is common sense. We would be satisfied with our effort if this chapter has provided a clear explanation of the latter and provided theoretical grounding to commonsense knowledge. The intention has been to encourage a deeper and more reflective understanding of AVMs as a valuation method. Finally, it is important to state that in this chapter we have advanced a particular viewpoint, one that is based on neoclassical economic theory. It is the dominant paradigm in economics, but it is by no means the only one. We have deliberately skirted around the numerous controversies associated with the fundamental assumptions of neoclassical economics, as advanced in alternative theoretical perspectives. We have, however, hopefully revealed, or suggested, useful points of entry for a more a critical engagement with AVMs.

References

Abernethy, A. M., & Hollans, H. (2010). The home valuation code of conduct and its potential impacts. *The Appraisal Journal, Winter*, 81–93.

Backhouse, R. E. (2002). *The ordinary business of life*. Princeton NJ: Princeton University Press.

Baryla, E. A., & Zumpano, L. V. (1995). Buyer search duration in the residential real estate market: The role of the real estate agent. *Journal of Real Estate Research, 10*(1), 1–13.

Blaug, M. (1986). *Economic history and the history of economics*. Brighton: Wheatsheaf Books.

Brue, S. L., & Grant, R. R. (2007). *The history of economic thought*. Mason, OH: Thomsun South-Western.

Crosby, N. (2000). Valuation accuracy, variation and bias in the context of standards and expectations. *Journal of Property Investment and Finance, 18*(2), 130–161.

Downie, M. L., & Robson, G. (2007). *Automated valuation models: An international perspective*. London: Council of Mortgage Lenders.

Dyke, C. E. (1981). *Philosophy of economics*. Englewood Cliffs, NJ: Prentice Hall.

Evans, A. W. (1995). The property market: ninety per cent efficient? *Urban Studies, 32*(1), 5–29.

FCIC. (2011). *The Financial Crisis Inquiry Report*. Washington, DC: US Government Printing.

French, N. (2006). Freehold valuations: The relationship between implicit and explicit DCF methods. *Journal of Property Investment & Finance, 24*(1), 87–91.

IVSC. (2003). *International valuation standards*. London: IVSC.

Lawson, T. (2009). The current economic crisis: Its nature and the course of academic economics. *Cambridge Journal of Economics, 33*, 759–777.

Mayes, R.G. (n.d.). *Theories of explanation*. Retrieved June 7, 2014 from http://www.iep.utm.edu/explanat/, accessed 7 June 2014.

Miceli, T. J. (1988). Information costs and the organisation of the real estate brokerage industry in the US and Great Britain. *AREUEA Journal, 16*(2), 173–188.

Mooya, M. (2011). Of mice and men: Automated valuation models and the valuation profession. *Urban Studies, 48*(11), 2265–2281.

Mooya, M. M. (2009). Market value without a market: Perspectives from transaction cost theory. *Urban Studies, 46*(3), 687–701.

Peto, R. (2009). The value of worth. *RICS Commercial Property Journal*, February–March, 5.

Ron, A. (1999). *Regression analysis and the philosophy of social sciences—A critical realist view*. Retrieved September 15, 2014. http://polmeth.wustl.edu/media/Paper/ron99.pdf

Screpanti, E., & Zamagni, S. (2005). *An outline of the history of economic thought*. New York: Oxford University Press.

Stuparu, D., & Daniasa, C. I. (n.d.). Significance of mathematics for economics. Retrieved September 15, 2014 from http://fse.tibiscus.ro/anale/Lucrari2009/064.%20Stuparu,%20Danaiasa.pdf, www.Brittanica.com

Part II
AVM, Valuation and Non-Agency Mortgage Crises: Experiences in AVM

Automated Valuation Models for the Granting of Mortgage Loans in Germany

Franz Eilers and Andreas Kunert

Abstract The general framework for the use of automated valuation models, based on the specific legal and regulatory standards for the valuation of property used as collateral in Germany, is described below. With these requirements in mind, the calculation methods for Market Value and Mortgage Lending Values are examined first. There are three methods available in Germany, which are used as either the principal or a secondary method depending on the type of property being valued. There then follows a summary of the principal databases. Without high quality, reliable and up-to date property data, it would be impossible to produce reliable Market Values and Mortgage Lending Values. How is this information used in automated valuation models and what role do valuers assume in the process? The answers form a further focal point of this discussion.

Keywords Automated valuation · Mortgage · Mortgage lending value

1 Introduction

An important component in the granting of mortgage loans is the valuation of real estate as loan collateral. It is necessary in order to assess credit risk and to make the conditions of the loan appropriate for the level of risk involved. Very few automated valuation models (AV models) were used in Germany until just a few years ago. This has changed significantly, at least in terms of financing owner-occupied houses and condominium apartments. The majority of these types of residential properties are now valued using such models. Automated valuation models can be seen as software programs comprising three principal components; the central calculation module, a comprehensive database and a matching process. The

F. Eilers (✉) · A. Kunert
vdpResearch, Berlin, Germany
e-mail: eilers@vdpreseacrh.de

A. Kunert
e-mail: kunert@vdpresearch.de

M. d'Amato and T. Kauko (eds.), *Advances in Automated Valuation Modeling*,
Studies in Systems, Decision and Control 86, DOI 10.1007/978-3-319-49746-4_4

calculation module is a system of deterministic equations, which reflect the typical valuation methods in the various countries. In Germany, the typical valuation methods are the Depreciated Replacement Cost, Income Capitalisation and Sales Comparison approaches. The second component comprises market and property databases. After entering unique property information, the user obtains specific valuation data from the database, from which the calculation module automatically generates suggested valuations. The importation of valuation data from the market and property databases is made using the property address and selected property characteristics. The important components of the matching and search functions are checking the address and the correlation of comparable prices, rents, capitalisation rates and cost estimates. The principal driving forces for using automated valuation models were the push for rationalisation within financial institutions and the progress made in terms of internet-based services. It is not just the speed and cost savings which have to be addressed, but more importantly, the quality of the valuation results generated. These depend on the proper implementation of valuation methods and the quality of the input data from the databases used for the valuation calculations. Both are critical elements of the explanations below, which are structured as follows. After the introduction, we will address a number of legal and regulatory peculiarities relating to real estate valuation, which need to be considered when granting mortgages in Germany. There then follow a few statistics relating to the size of the market. Section 4 deals with valuation methods used in this country, which in some cases differ significantly from those used in other countries. The Sect. 5 explains the specific databases, which are essential for the generation of valuation data. The principal information sources are the purchase price collections carried out by the land valuation boards (Gutachterausschüsse für Grundstückswerte), which provide land values across the whole of Germany, and the vdpResearch transaction database, which comprises sales information collected through the real estate financing process. Section 6 describes the matching process, i.e. the use of property-related data in automated valuation models. Section 7 comprises the concluding remarks.

2 Valuation for the Granting of Mortgage Loans

Properties are valued for a variety of reasons. In Germany this could be, for example, for determining rateable value for land tax purposes, a Market Value for inheritance tax and capital transfer tax purposes or an Insurance Value for building insurance. This study will refer to the computer-aided valuation of standard residential property (owner-occupied houses[1] and condominium apartments) for the

[1]The terms ‘owner-occupied house’ and ‘single-family and two-family dwellings’ are used to mean the same thing in the following text. The property types ‘single-family and two-family dwellings’ include ‘detached single-family and two-family dwellings’, ‘terraced houses’ and ‘semi-detached houses’.

provision of finance secured by a land charge. It is assumed here that financial institutions will refinance the loans via secured interest-bearing bonds (mortgage bonds—Hypothekenpfandbriefe). This type of refinancing, commonly used in Germany, inevitably places emphasis on the German Pfandbrief Act (Pfandbriefgesetz) and the Regulations for the Determination of Mortgage Lending Value (BelWertV) and the valuation regulations and valuation definitions contained within them, which serve to protect the creditors of financial institutions. For the purposes of mortgage bond cover, financial institutions which refinance via the issue of secured bonds have to ensure that mortgage land charges are limited to a maximum of 60 % of the valuation of the property. The valuation amount therefore determines the Mortgage Lending Value (Beleihungswert). This value is defined in §16 (2) of the German Pfandbrief Act. This provides that 'The Mortgage Lending Value must not exceed the value resulting from a prudent assessment of the future marketability of a property by taking into account the long-term, sustainable aspects of the property, the normal regional market conditions as well as the current and possible alternative uses. Speculative elements must not be taken into consideration. The Mortgage Lending Value must not exceed the Market Value calculated in a transparent manner and in accordance with a recognised valuation method.' The definition raises two issues. Firstly, the Mortgage Lending Value is an underlying value which is adjusted to exclude speculative market and property appraisal factors to make it relevant well beyond the valuation date. Secondly, the definition of Mortgage Lending Value refers expressly to the Market Value as its reference parameter and, as such, this must not be lower than the Mortgage Lending Value. The second issue has resulted in the customary banking practice of always obtaining a Market Valuation in addition, as part of the process of determining the Mortgage Lending Value. According to the German Pfandbrief Act, the Market Value is 'the estimated amount for which a property serving as collateral could be sold on the valuation date in an arm's length transaction between a willing seller and a willing buyer after a proper marketing period, where the parties had each acted knowledgably, prudently and without compulsion.'[2] Similar to the Mortgage Lending Value, this is an anticipated future value calculated on the basis of relevant property and market information using the 'correct' valuation model. But, in contrast to the Mortgage Lending Value, the Market Value relates to a specific valuation date; it varies over time, whilst the Mortgage Lending Value does not vary over time or is at least fixed temporarily. The way in which this difference between the two values is typically reflected in valuation practice in Germany will be explained below in Sect. 4. Now a few statistics relating to the size of the market: how many condominium apartments and owner-occupied houses are now being purchased or newly constructed in Germany and must, in principle, be valued?

[2]German Pfandbrief Act (Pfandbriefgesetz) §16 (2). There is a similar definition in German Building Code (Baugesetzbuch) §194.

3 A Few Statistics on Market Size

When private households build new owner-occupied houses or condominium apartments, or make a purchase in the existing market, they typically require debt financing because of the high capital sums required for this. The great majority of such funds are made available by financial institutions, which carry out a valuation of the mortgaged property as part of the financing process. Therefore, the number of transactions in the existing market plus the number of owner-occupied houses built by owners are approximate indicators for the number of valuations of standard residential properties instigated by financial institutions. For a number of reasons, the actual number could be a lot higher: firstly, there are regular revaluations for the purposes of debt restructuring and loan extensions. Secondly, financing in many cases involves various financial institutions who often carry out their own valuations. Thirdly, there are also other special reasons such as the sale of loan portfolios, which means a requirement to revalue the collateral properties. Fourthly, some financing enquiries require an initial estimate of value, but do not result in a loan because the client opts for another financial institution, which then carries out a further estimate of value and a full property valuation. Between 2007 and 2013, around 70,000 new owner-occupied houses p.a. were built by private households in Germany, the majority of which would have required a valuation. The number of single-family and two-family dwellings transacted in the same period was more than three times as high at 240,000 p.a. In addition, around 250,000 condominium apartments p.a. are traded in the existing market and these are also generally purchased by private households. The aggregate of the three sums is around 560,000 new-build properties and transactions in the existing market p.a. There is some double counting here, as not only developers but also private households sometimes build single-family and two-family dwellings, and then sell them in the existing market after completion. The number does not include newly constructed condominium apartments. This comprises part-ownership in multiple occupancy apartment blocks, which the developer either keeps in its own portfolio and rents out after completion, or sells in the existing market and therefore features in the relevant statistics (Table 1).

4 Selection of Valuation Method

4.1 Preliminary Remarks

According to the Regulations for the Determination of Mortgage Lending Value (Beleihungs-wertermittlungsverordnung—BelWertV), there are basically three methods available for the determination of Mortgage Lending Value, namely the Income Capitalisation, Sales Comparison and Depreciated Replacement Cost approaches. The Income Capitalisation Approach comprises a feasibility analysis

Table 1 A few statistics on market size of the German residential property market

Year	Single-family and two-family dwellings		Condominium apartments	Total
	New-build[(1)]	Transactions in existing market		
	Number of properties ('000)			
2007	90	221	213	524
2008	69	225	213	507
2009	61	226	234	521
2010	63	240	258	561
2012	72	253	287	612
2012	63	255	290	608
2013	78	250	278	606

[(1)]The terms 'owner-occupied house' and 'single-family and two-family dwellings' are used to mean the same thing in the following text. The property types 'single-family and two-family dwellings' include 'detached single-family and two-family dwellings', 'terraced houses' and 'semi-detached houses'.
Number of single-family and two-family dwellings newly constructed by private developers
Source Federal Statistical Office, vdpResearch

which examines the present values of the rent, yield and remaining useful lifespan of a property. The Depreciated Replacement Cost Approach examines the net asset value of the property comprising the land value, the value of external appurtenances and the value of the improvements. Finally, the Sales Comparison Approach looks at the value of a property in terms of the price now being paid in the market for comparable properties. Which of the approaches must be used in Germany, whether just one method or several simultaneously, which one is relevant to the valuations when using several methods, and which method is used only for the cross-check, depends on the property type and usability of the subject property. In terms of

Table 2 Methods used for the valuation of residential property in Germany

		Property type		
		Owner-occupied houses	Condominium apartments	
			Suitable for owner-occupation	Not suitable for owner-occupation
Method	Income capitalisation	B	B	A1
	Depr. replacement cost	A1	A2	B
	Sales comparison	A2	A1	B

Notes
A1 Principal method; *A2* Alternative principal method; *B* Secondary method
Explanations
The principal valuation method A1 and A2 may be used alternatively; the secondary valuation method B is used only to cross-check the results from the alternative valuation methods A1 and A2

property type, owner-occupied houses and condominium apartments therefore differ in terms of usability depending whether or not they are suitable for owner-occupation. A property is suitable for owner-occupation if the layout, fitout and location could suit a purchaser for his own use over the long term. Both owner-occupied houses and condominium apartments may or may not, in principle, be capable of owner-occupation. One normally assumes that owner-occupied houses are always suitable. The following are the basic rules for the choice of valuation method (see also Table 2):

5 Owner-Occupied Houses

The relevant valuation method can be either the Depreciated Replacement Cost Approach or the Sales Comparison Approach, although the Depreciated Replacement Cost Approach clearly takes precedence. If a cross-check is carried out by way of support, the Income Capitalisation Approach must be used.

6 Condominium Apartments

In the case of condominium apartments, it must be decided whether these are suitable for owner-occupation. In the case of condominium apartments which are not suitable for owner-occupation, the Income Capitalisation Approach will always be the principal valuation method; as a cross-check, either the Depreciated Replacement Cost or Sales Comparison Approaches may be used. In the case of condominium apartments which are suitable for owner-occupation, it must be decided which of the Depreciated Replacement Cost and the Sales Comparison Approaches is relevant to the valuation; it is generally the Sales Comparison Approach which is used. If a cross-check is carried out by way of support, then the Income Capitalisation Approach is used. In the following example, the three valuation methods for the Market Valuation and Mortgage Lending Valuation are shown separately. The illustration is limited to the basic principles inherent in the three methods; much of the detail has been dispensed with. However, it clearly shows the differences which are relevant to the determination of Market Value and Mortgage Lending Value.

7 Income Capitalisation Approach

7.1 Basic Principles

The Capitalised Income Value of a property comprises both the land value and the Capitalised Income Value of the improvements. Both values are determined separately. The land value element comprises mainly the product of the site area and the guideline land value (Bodenrichtwert). Guideline land values are average values for a location stated as €/m^2 for the land, and these are regularly updated and published. There is a relevant guideline land value for almost every property in Germany, and these values may be adjusted to match the characteristics of a particular property according to the specific guidelines. Statistically, they are based on the purchase prices collected by the land valuation boards (Gutachterausschüsse für Grundstückswerte), which, together with the calculation of the guideline land values, are discussed in more detail in Sect. 5.2. The Capitalised Income Value of the improvements comprises the Capitalised Income Value of the improvements and any other building components (external appurtenances). This value is calculated as the present value (Barwert) of the anticipated future net rental income generated by the building.[3] This net rental income comprises the potential gross rental income less the assumed operating costs; the latter of which includes a number of different components such as the costs of maintenance, operation and management and a provision for loss of rent. The Income Capitalisation Approach as a whole may be represented by the following Formula (1):

$$Capitalised\ Income\ Value = (Net\ income - land\ value * r) * V + land\ value \quad (1)$$

$$\text{where : r} = \text{yield for annual return on land value}$$
$$\text{V} = \text{multiplier}$$

The multiplier is calculated as in the Formula (2) below:

$$V = \frac{q^n - 1}{q^n \times (q - 1)} \quad (2)$$

$$\text{where : q} = (1 + \text{r}) \text{ and}$$
$$\text{n} = \text{remaining useful lifespan in years}$$

8 Differences Between the Market Value and Mortgage Lending Value Calculations

The principal differences in calculation method between the Market Value and the Mortgage Lending Value are as follows: calculation of net rental income assumed yield r, which is the main determinant of the multiplier V. The net income per m^2 residential space is mainly dependent on the level of rent per m^2. For the Market Value calculation, the typical market rent for comparable properties is used for unlet apartments, and for apartments which are let this would generally be the rent in the

[3]The net rental income of the building is the total annual net rental income of the property less the annual return on land value (Verzinsung des Bodenwertes).

lease contract. In contrast, the sustainable rent is used for the Mortgage Lending Value calculation. The sustainable rent is the achievable existing rent in the local market for comparable properties; however, if this rent is above the current agreed rent according to the lease contract, then the contractually agreed rent should normally be used. The yield used to determine the present value of the net rental incomes for the Market Value is called the property yield (Liegenschaftszinssatz) and for the Mortgage Lending Value it is the capitalisation rate (Kapitalisierungszinssatz). The property yield can be determined by reference to the remaining useful lifespan of a residential property based on current sales. For the Mortgage Lending Value calculation, the capitalisation rate used is that 'with which the future achievable prudently estimated net rental income from a property is typically discounted over the period during which it is anticipated that it will be received. It must be determined by reference to long-term trends in the applicable regional market.'[4] For residential properties, the capitalisation rate must not be under 5 %. The Appendix 3 to Regulations for the Determination of Mortgage Lending Value §12 (4) contains a suggested range of yields for residential property between 5 and 8 %.

9 Depreciated Replacement Cost Approach

9.1 Basic Principles

In Germany, the Depreciated Replacement Cost Approach is generally used as the principal valuation method for the financing of single-family and two-family dwellings. The Depreciated Replacement Cost comprises the land value and the value of the improvements, which also includes the external appurtenances. Similar to the Income Capitalisation Approach, the land value is the product of the site area and the guideline land value. The value of the improvements is calculated incrementally. First, the replacement cost of the improvements is calculated as at the valuation date. The calculation is based either on typical building costs per m^2 gross external area (Bruttogrundfläche—BGF) or on relevant building costs for 1 m^2 of residential space. To the value calculated in one of these ways is then added the value of the external appurtenances, which are typically included in the calculation as a flat rate addition. The amount resulting from this calculation is then reduced by depreciation due to age and by any deduction for building defects or damage. The result is the calculated present value of the improvements. This value is then added to the land value to produce the provisional Depreciated Replacement Cost. As the costs and prices may differ significantly, the provisional Depreciated Replacement Cost is then adjusted to the price conditions in the local market via a relevant multiplier. This multiplier is based on an empirical analysis of comparable

[4]See Crimmann, Mortgage Lending Value, p. 157.

properties. If the achievable prices on the local market are above the replacement cost, the multiplier is greater than 1; if the opposite is true, then the multiplier is less than 1.[5] Ultimately, the Depreciated Replacement Cost is therefore similar to the Sales Comparison Approach but, in addition to the analysis of comparable prices, the replacement cost and the land value of the subject property (net asset value) must also be calculated.

10 Differences Between the Market Value and Mortgage Lending Value Calculations

Figure 1 shows a summary of the basic structure of the Depreciated Replacement Cost Approach and demonstrates the differences between the Market Value and Mortgage Lending Value calculations. The first difference is that the Mortgage Lending Value—value of improvements (5b) differs from the Market Value—value of improvements (5a) by a flat rate deduction. This deduction takes into account the risk of possible decreases in construction costs, and must be at least 10 %. The second difference is the transition from the Depreciated Replacement Cost to the Market Value and Mortgage Lending Values: The Market Value is derived from the product of the Depreciated Replacement Cost (Market Value) and a Depreciated Replacement Cost factor, whilst the Mortgage Lending Value is the product of the Depreciated Replacement Cost (Mortgage Lending Value) and a sustainability factor. The two multipliers differ in that the sustainability factor is generally limited in two ways. Firstly, it must not exceed the Depreciated Replacement Cost factor and secondly, it must not be greater than 1.[6]

11 Sales Comparison Approach

11.1 Basic Principles

In the Sales Comparison Approach, the property is valued on the basis of purchase prices for appropriate comparable properties, which have transacted in the recent past. These properties are characterised by the fact that they are sufficiently comparable with the subject property in terms of their principal characteristics affecting value.[7] These characteristics principally include the macro and micro locations and

---

[5]Therefore, like 'Tobin's q', this multiplier also indicates whether it is worthwhile to purchase an existing building or construct a new-build. If the multiplier is greater than 1, then a new-build will be more beneficial than purchasing an existing property. If it is less than 1, the opposite is true.

[6]In exceptional cases, there may be a limited departure from the second rule.

[7]What is 'sufficient' in terms of comparability unfortunately remains unclear.

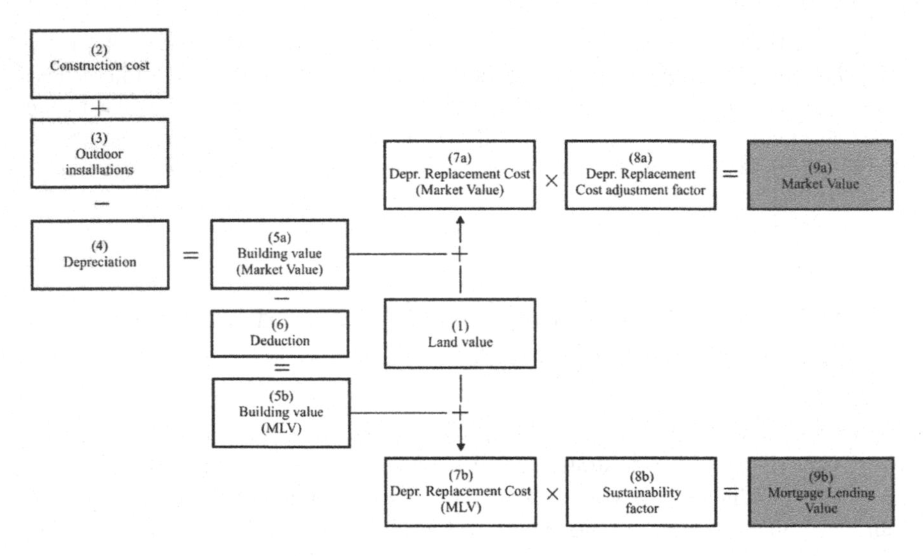

Fig. 1 Depreciated replacement cost approach

the age, state of repair and fitout of residential properties, which can typically form the basis of the stated price per m^2 of residential space. Remaining differences between the comparable property and the property now being valued may be reflected, if possible, by compensatory additions and deductions. The Market Value is then calculated as the product of the residential area and a price per m^2 of residential area. Differences between the Market Value and Mortgage Lending Value calculations

Market Value and Mortgage Lending Value calculations based on the Sales Comparison Approach differ in that the Market Value is adjusted downwards by a safety margin (see Fig. 2). The safety margin is dependent on cyclical market

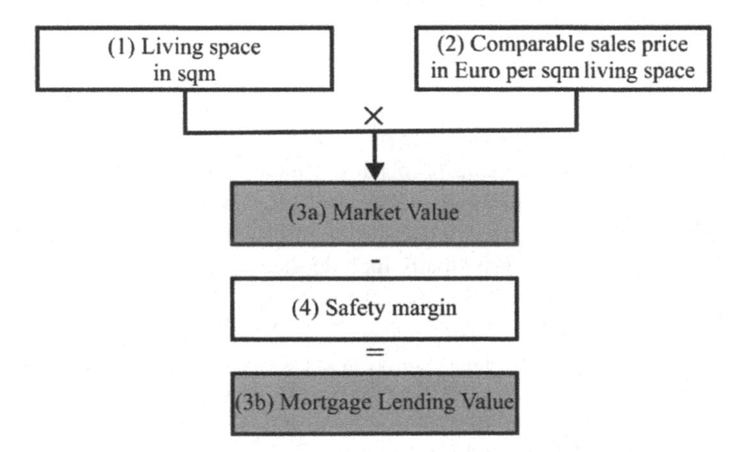

Fig. 2 Sales comparison approach. Data sources as evidence for model variables

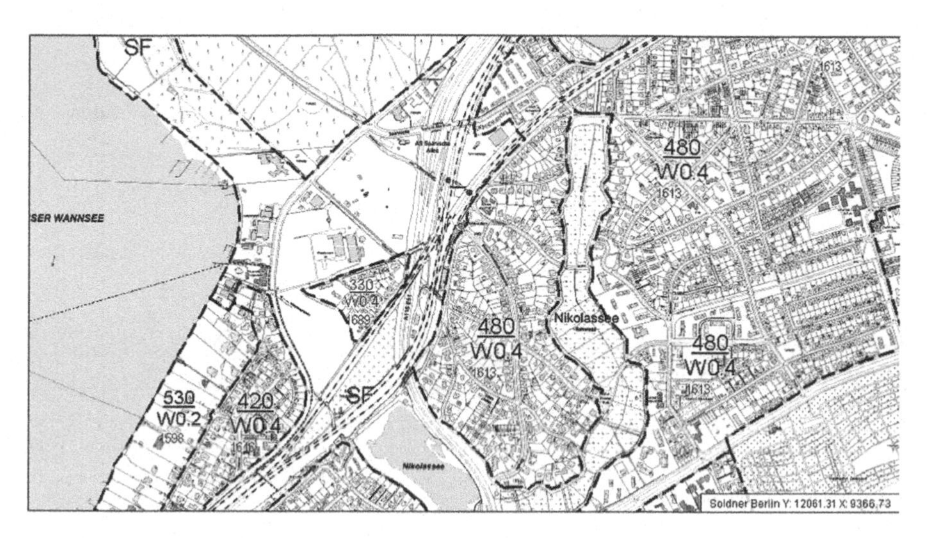

Fig. 3 Extract from the Guideline Land Value Map for Berlin Dated 1 January 2014. *Notes* $\frac{480}{W0A}$—Guideline land value €480/m^2 for residential building land with a typical plot ratio (Geschossflächenzahl–GFZ) of 4.0. The plot ratio defines the ratio between the total gross external area of all complete floors in the building on a site and the total site area. 1613—Identification number of the individual guideline land value zone (e.g. by using the automated purchase price collection (AKS) for Berlin). SF—Special Use Zones, which are used as, for example, allotments, cemeteries, sports facilities and airports. [ ]—Guideline land value zone boundary

conditions; in boom times, the deduction may be higher than in weaker markets. Part 3 of Regulations for the Determination of Mortgage Lending Value (BelWertV) §19 (1) states that the safety margin must be at least 10 %.

12 Preliminary Remarks

Automated valuation models started out as software programs comprising three principal components, namely the central calculation module, a comprehensive database and a matching process. Having described the calculation module in terms of the valuation methods, we now come to the topic of data. Each of the three valuation methods contains a series of variables, which require relevant information to be collected in order to be called an 'automated' valuation model at all. It would be too much at this stage to present a comprehensive and co-ordinated data catalogue. The empirical bases of a small number of factors, which are particularly important to the valuation, will be discussed below. These factors include land value, rents, prices for apartments and houses as well as property yields and capitalisation rates. Table 3 shows these together with other key variables for the valuation methods for the calculation of Market Value and Mortgage Lending Value. With this in mind, three

Table 3 Key variables for the three valuation methods

Variable	Unit	Value	
		Market value	Mortgage lending value
Income capitalisation approach			
Land prices	€/m^2	X	X
Comparable rents	€/m^2 residential space	X	
Sustainable rents			X
Property yields	%	X	
Capitalisation rates			X
Depreciated replacement cost approach			
Land values	€/m^2	X	X
Building costs	€/m^2 residential space	X	X
Depreciated replacement cost factors	Multiplier	X	
Sustainability factors			X
Sales comparison approach			
Comparable prices	€/m^2	X	X

Note An 'X' denotes whether each variable is necessary for the Market Value and/or Mortgage Lending Value calculation

sources which relate directly to individual properties are discussed below. These are the purchase price collections carried out by the land valuation boards, the vdpResearch transaction database and property-specific databases showing asking prices. The purchase prices are used primarily for generating land values. The vdpResearch transaction database is relevant to the generation of comparable rents and comparable prices, and to the calculation of property yields and Depreciated Replacement Cost variables. The property-specific asking price databases serve to complement the information from the transaction database.

13 Purchase Prices Collected by the Land Valuation Boards

Every property sale and purchase agreement in Germany must be notarised. Copies of these documents are delivered to the land valuation boards.[8] These then enter the key data of the sale and purchase agreements into the so-called purchase price

[8]Land valuation boards are autonomous and independent panels, which according to German Building Code (Baugesetzbuch) §195 Sect. 1 receive copies of all notarised real estate sale & purchase agreements.

collections, which are then used to create data for property valuations in line with the market. Land values are key components here.

14 Land Values (Guideline Land Values)

The prices made available by the land valuation boards based on their analysis of the purchase price collections are called 'guideline land values'. By its very name, the term emphasises that these are suggested values which, by their nature, are not legally binding. All the same, they are accepted almost unreservedly by property market participants and in valuations as the relevant value. Guideline land values are published in the form of so-called guideline land value maps. Some of the values relate to individual properties and some comprise geographically contiguous guideline land value zones, which are estimated to be comparatively homogenous in terms of their land value.[9] The guideline land values are calculated on the basis of the Sales Comparison Approach. The inclusion of the guideline land values into the automated valuation models is fraught with difficulties for a number of reasons. The land valuation boards generally only update the guideline land values every two years. When prices are changing fast, this may result in more or less significant over-estimates and under-estimates of the actual price level in the property markets. The guideline land values are calculated by the land valuation boards on the basis of the Sales Comparison Approach. In illiquid markets when transaction activity is limited, there are often not enough sales from which to make reliable use of the Sales Comparison Approach. The guideline land values derived from the sales serve both as a comparable and as an orientation benchmark to establish current prices in the property market. In particular, because the values come from an authoritative source, they have significant influence on price negotiations between buyer and seller. Just like the limited frequency of updates, the self-referential nature of guideline land values smoothes cyclical changes. The organisation and working methods of the land valuation boards and their branch offices are the responsibility of the federal states. There are in fact 16 different implementation and land valuation board ordinances in Germany. In the federal state of Baden-Württemberg, the land valuation boards are organised at municipal level, whilst in the other states they mainly operate at administrative district and autonomous town level. The organisational differences and the differences between the ordinances affect both the quality of the data and also access to the information. Despite these difficulties, guideline land values are widely accepted and are an integral component of property valuation. The inclusion of guideline land values in automated valuation models for properties located in Germany is now obligatory.

[9]See also Fig. 3. This is an extract from the guideline land value map for Berlin dated 1 January 2014.

15 Further Valuation Variables

In addition to guideline land values, the land valuation boards also provide other data sources for use in real estate valuation. These include comparable prices, particularly Depreciated Replacement Cost multipliers and property yields. However, this data does not have anything like the same relevance as the guideline land values in terms of the automated valuation models. There are two reasons for this. The first relates to the differences in the organisation of the land valuation boards described above, which make it almost impossible to obtain comprehensive valuation variables. Secondly, the calculation of the valuation variables varies from state to state and often within the states themselves. This is particularly relevant when calculating the Depreciated Replacement Cost multipliers. These are defined as the ratio of the purchase price to replacement cost. Various assumptions are required for the calculation of the replacement costs. Despite all efforts to standardise the production of the data, these assumptions vary significantly between the individual land valuation boards, so that the Depreciated Replacement Cost multipliers produced by them are not comparable. The result is that, at present, the factors cannot be used in automated valuation models.

16 vdpResearch Transaction Database

As described in Sect. 3, around 550,000 owner-occupied houses and condominium apartments are purchased every year in the existing market in Germany. These transactions feed not only into the purchase prices collected by the land valuation boards, but also in principle into the vdpResearch transaction database. The database comprises property data collected in the real estate financing process and the associated property valuations. If, in order to purchase a property, a loan is taken from a financial institution which participates in the transaction database, the purchase will be picked up by the database. More than 330 financial institutions belonging to various groups currently participate in the transaction database (as at 2014). These include the major banks, direct banks, savings banks, co-operative banks and the Raiffeisen banks. The database is highly diversified in terms of the types of institutions feeding in data and its regional coverage is representative of the market. The database grows by around 60,000 new records every quarter. The individual variables of these records are clearly defined and recalibrated as necessary to standardise them. At the end of 2014, the transaction database comprised around two million records, which are filtered for the purposes of analysis. The most important contents of the database are firstly, data on rents and prices and the associated contract dates. Secondly, the database contains information allowing statistical analysis relating to important characteristics of the individual property, which substantiate the rent or the price. Important information of this type principally includes the residential location, the property fitout, its age, size and state of repair. Further important factors, depending on the property type, include the plot ratio and site area, and also suitability for third party use. As the transaction database contains comprehensive documented purchase prices and rents, as well as the accompanying principal characteristics affecting the price of the properties, it is possible to analyse the data using hedonic regression models. In particular, the following information can be generated by reference to the database: comparable prices and comparable rents, Depreciated Replacement Cost factors and property yields The generation of data by a co-ordinated and consistent model is governed by the following considerations: the individual valuation variables are produced on the basis of the general valuation rules (such as the Regulations for the Determination of Mortgage Lending Value) according to a unified method nationwide. This ensures firstly, that the contents of the variables are not affected by the differing understanding of the individual users and secondly, that the variables are comparable across all locations and can be used by regional banks all over the country. From an international perspective, the German residential market appears sluggish. Nonetheless, in order to describe the market dynamic adequately, the valuation variables are updated quarterly. This is viewed as sufficient to observe the ups and downs of the markets whilst not asking too much of the existing records. The resulting valuation variables must be appropriate for in-house automated valuation models, as well as for models operated by others. Others include financial institutions which have developed and use their own automated valuation models. These must take sufficient account of differences in understanding the contents of the individual variables. The data is generated such that the detail of the results is consistent with information from other sources. This means above all that information from the transaction database can be analysed in conjunction with existing analyses of information from the purchase prices collected by the land valuation boards. The features shown here have made it possible to develop a standard analysis on the basis of the Regulations for the Determination of Mortgage Lending Value, which has a comparatively short update cycle (quarterly), can be adjusted by way of small changes to suit existing automated valuation models and allows for the revision of valuation parameters using other statistics.

17 Asking Price Databases

One further source for providing evidence of valuation variables for automated valuation models are databases of advertised prices, of which there are a number operating in Germany. In addition to asking prices and asking rents, as a rule these extensive data collections often contain a description of the property. This means that the statistics can be analysed using traditional hedonic regression models. The use of asking prices and asking rents as evidence for valuation variables is controversial. Particularly because asking prices may differ significantly from transaction prices, an empirically reliable property valuation based on advertised prices is fraught with significant problems. With this in mind, asking prices are used solely for the purpose of cross-check analyses.

18 Summary

Ultimately, automated valuation models require databases which comprise a large number of transaction prices. The purchase prices collected by the land valuation boards and the vdpResearch transaction database provide two data sources in Germany, which together satisfy this requirement comprehensively. The guideline land values derived from the purchase price collections provide comprehensive details relating to land values for individual sites. The vdpResearch transaction database is useful for correlating prices based on the characteristics of the individual properties. The regression models used for these produce implicit prices in the form of estimated coefficients, which can then be used in the valuation of real estate.

19 Matching Process and Presentation of Results

19.1 Preliminary Remarks

In order to use guideline land values, implicit price and other information in automated valuation models, the information must be directly linked to the subject property. The link usually relates to the address and characteristics of the property, which form the principal components of the property's value. The address is used to derive the relevant land value and the characteristics are used to derive implicit price information. The quality and extent of this link depend on how well the characteristics can be described and on the accuracy of the information available. The quality of the valuations therefore depends equally on the description of the property and the databases (or the analyses based on these). The databases were described in the preceding section. We now come to the question of how the characteristics of subject properties must be input in order to achieve high-quality valuations. A regulatory provision relating to the responsibility of the user (valuer) must also be addressed here. The effect of this provision is that the models used in Germany would be better described using the adjective 'computer-assisted' than the expression 'automated'. As users of automated valuation models, financial institutions are often faced with the problem that their initial information relating to a property is incomplete or incorrect. The reasons for this can lie in a combination of the lending process and the competitive situation. In the following example, client C makes a personal or online approach to Bank B with the minimum package of documents required with a request for mortgage finance for an owner-occupied apartment he wishes to purchase. The client usually approaches not just one bank but rather a number to secure the most reasonable financing terms. Before making an offer, the banks willing to provide finance will require an indication of the creditworthiness of the client and the value of the property to be used as collateral. The offer has to be provided quickly in order to remain competitive. Unless these issues are addressed, the provision of finance is risky.[10] In order to be competitive, the 'valuation calculation' must be carried out quickly and inexpensively. This means that the bank will have to make its decision based on the documents provided by the client, which may be only partly verifiable. Against this background, the CIB[11] automated valuation model used by vdpResearch and discussed as an example below may be divided into two sub-models (see Fig. 4). Both sub-models may be used separately from each other or incrementally in combination. The first sub-model comprises an Indicative Value. This is an initial quick and inexpensive estimate of the value of the collateral property for the purposes of the sales team, and for which only outline information on the property is required; there is no Market Value or Mortgage Lending Value calculation. This is a matter for the second sub-model, which is specifically intended for use at the back office stage. The second sub-model is significantly more detailed and, correspondingly, makes greater demands on information sources. All valuation factors can be adjusted manually, which is necessary in order to conform to regulations. At his own discretion, the user can choose to take responsibility by bringing to the valuation his own insight and information based on his professional experience. Both sub-models are the same in that initially they make use of the same information. This means that the only variations between the Indicative Value and the detailed Market Value and Mortgage Lending Value calculations will be those arising from the property information and the use of the model. The Indicative Value and the Market Value and Mortgage Lending Value calculations will be described below.

Step 1: Indicative Value
In principle, the Indicative Value and the full Market Value and Mortgage Lending Value calculations are made by the same valuation module, but they vary significantly in the range of control and selection options available. In the case of the Indicative Value, just the address and a few property characteristics are required to generate an automatic value indication. All other characteristics are determined automatically from the address itself or, for the time being, made the subject of a general assumption for all purposes. When the user has entered the address, the first step is to select the property type. If this is a single-family or two-family dwelling, the house type must next be entered; if the property is a condominium apartment, then it must be decided whether the apartment is suitable for owner-occupation or not. There are seven further factors required in the case of single-family or two-family dwellings and six for a condominium apartment (see Table 4). With the exception of fitout and state of repair, this information may be taken directly and quickly from the building description or the sale and purchase agreement, which

[10]With the emergence of finance provision via the internet and the increased presence of direct banks in the market, the proportion of enquiries resulting in a loan may have decreased significantly. Unfortunately, the only available information available is from confidential discussions; there is no information based sufficiently on fact.

[11]The acronym CIB stands for ***C***omputergestützte ***I***mmobilien-***B***ewertung (computer-assisted real estate valuation).

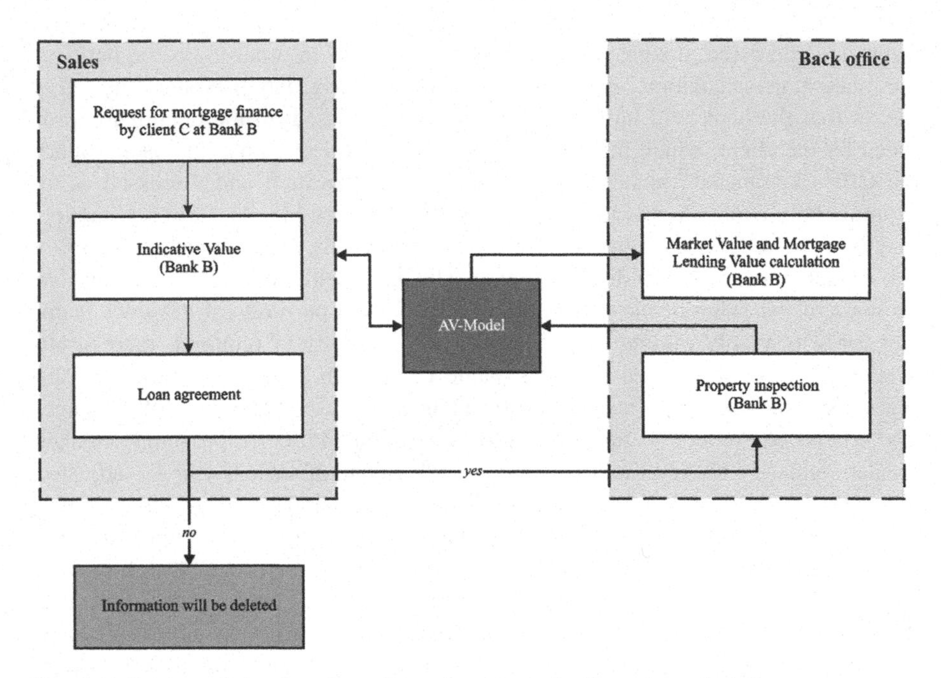

Fig. 4 Indicative value and market value and mortgage lending value calculations

greatly reduces the risk of error. Fitout and state of repair are estimated based on information provided by the client, but information provided by the client is supported if possible by photographs or other documentation. The estimates are made on the basis of a scale with five categories, from very good to poor. All other information is entered quantitatively, with the exception of the building types which are categorised as follows: solid, pre-fabricated (solid), pre-fabricated (wood), low-energy house, passive house, zero energy house and pre-fabricated (concrete–Plattenbau).

The Indicative Value serves as a proposed value for the Market Value and Mortgage Lending Value calculations. In addition, the quality of the residential location is assessed on the basis of the address. In the case of single-family or two-family dwellings, Indicative Valuations are carried out on the basis of the Depreciated Replacement Cost Approach. Indicative valuations for condominium apartments suitable for owner-occupation are carried out on the basis of the Sales Comparison Approach, with comparative values produced using hedonic regression analyses.[12] For condominium apartments which are not suitable for owner-occupation, Indicative valuations are made on the basis of the Income

[12]An alternative method would be to classify comparable prices from the databases using appropriate selection filters to ensure comparability. In Germany, this alternative is fraught with significant problems because of data protection laws, which means that this is not common practice at present.

Table 4 Indicative value input data

	Property type							
	Single-family and two-family dwellings						Condominium apartment	
							Suitable for owner-occupation?	
	Type 1	Type 2	Type 3	Type 4	Type 5	Type 6	Yes	No
Site area	Quantitative						X	
Year of construction	Quantitative						Quantitative	
Type of construction	Qualitative						Qualitative	
Residential space	Quantitative						Quantitative	
Garage	Quantitative						Quantitative	
Fitout	Qualitative						Qualitative	
State of repair	Qualitative						Qualitative	

Notes
Type 1 Free standing single-family dwelling; *Type 2* Single-family dwelling with granny flat; *Type 3* two-Family dwelling; *Type 4* Semi-detached house; *Type 5* End terrace; *Type 6* Mid-terrace

Capitalisation Approach. The residential location is assessed using a geographical statistical approach comprising guideline land values and comparable prices.

Step 2: Market Value and Mortgage Lending Value Calculations
As explained above, the automated Market Value and Mortgage Lending Value calculations conceived for the back office process are more detailed than the Indicative Value calculation, and the level of detail can be controlled by the user. The process begins with the input of the address and property type, which are the primary drivers of the valuation process.[13] The next inputs are the selected property characteristics, for which more detail is required than for the Indicative Value. The most important of these are the energy performance rating and assessment of the residential location, which are generated by the model in the case of the Indicative Value.

Table 5 shows these additional items. In addition, there are further inputs which are not shown in the table. Firstly, it is possible to enter a more detailed description of the building fitout. This optional additional function requires a qualitative estimate of individual sections on a 5-point scale from very high quality to basic. Secondly, the Market Value and Mortgage Lending Value model allows for the input of refurbishment status for all component sections. This function is also optional and, if required, can be shown in place of the general categorisation of the state of repair.

Thirdly, the rights from section II of the land register can be integrated into the valuation calculation. The land register is a public register which publishes the rights attaching to a site. It shows details relating to the ownership of the site and any encumbrances and restrictions affecting it. These encumbrances and restrictions include easements, pre-emption rights, heritable building rights and restrictions on disposal. Typical encumbrances include easements such as rights of way and personal easements such as usufruct.

In contrast to the items 1 and 2 of the additional data, No. 3 is not optional. If there are any encumbrances and restrictions, they must be entered for the Market Value and Mortgage Lending Value calculation. Here, there are variances between the Market Value and Mortgage Lending Value. In contrast to the Market Value, the ranking of the encumbrances and restrictions is relevant to the Mortgage Lending Value.[14] If the land charge is ranked after the mortgage, then the bank can avoid this lower ranking right by way of a compulsory auction.[15] In the Market Value and Mortgage Lending Value calculations, the encumbrances and restrictions are generally investigated for the Market Value and Mortgage Lending Values, but

[13]The inputs from the Indicative Value can of course be used here if this is preferable and appropriate.

[14]If there are a number of rights entered in the land register, these are organised in order of ranking. The order of ranking reflects the order in which they are entered in the land register, as they indicate the order in which any proceeds from a compulsory auction of the property would be distributed.

[15]See Wolfgang Crimmann, Mortgage Lending Value, p. 115.

Table 5 Principal input data for market value and mortgage lending value calculations

	Property type							
	Single-family and two-family dwellings						Condominium apartment	
							Suitable for owner-occupation?	
	Type 1	Type 2	Type 3	Type 4	Type 5	Type 6	Yes	No
Site area	Quantitative							
Year of construction	Quantitative						Quantitative	
Type of construction	Qualitative						Qualitative	
Residential space	Quantitative						Quantitative	
Cellar	**Quantitative**							
Garage	Quantitative						Quantitative	
Fitout	Qualitative						Qualitative	
State of repair	Qualitative						Qualitative	
Residential location	**Qualitative**						**Qualitative**	
Energy efficiency	**Qualitative**						**Qualitative**	
Balcony, (roof) terrace	**Qualitative**						**Qualitative**	
Monument listing	**Qualitative**						**Qualitative**	

Notes

Type 1 Free standing single-family dwelling; *Type 2* Single-family dwelling with granny flat; *Type 3* Two-family dwelling; *Type 4* Semi-detached house; *Type 5* End terrace; *Type 6* Mid-terrace

Remarks

The cells marked bold are the additional input data compared with the Indicative Value. It is not possible to enter these in the case of the Indicative Value in the case of the Mortgage Lending Value they are shown for information purposes only. If the inputs necessary for the Mortgage Lending Value calculations are entered, then the model will automatically calculate the valuation variables. Depending on the valuation method being used, these may take the form of inherent prices for particular characteristics, derived yields and/or costs. The values created are suggestions, which can be accepted or modified by the user. This means that all valuation cells may be edited by the user. As explained above, editing needs to be possible for regulatory reasons. Whether and to what extent this option is of purely formal or actual practical relevance depends on a number of factors including the quality of the suggested valuation data. The result of the valuation calculation is a Market Value or Mortgage Lending Value based on the relevant (primary) valuation method. In addition, Market Values and Mortgage Lending Values are calculated using the secondary valuation method. These are used only to cross-check the results. The Market Values and Mortgage Lending Values calculated on the basis of the relevant valuation method are generally close to the results of the Indicative Value. The Indicative Value and the Market Value and Mortgage Lending Value calculations refer to one database and use the same valuation module, but, in the case of the Market Value and Mortgage Lending Value calculations, the user just has far more room for manoeuvre. The high level of concurrence is good in that typically the loan is granted on the right terms from the start.

20 Concluding Remarks

Since the Millennium, automated valuation models have become an integral part of the valuation of owner-occupied houses and condominium apartments in Germany. This development was accompanied by the establishment and publication of transaction databases across the whole of Germany. The purchase prices collected by the land valuation boards and the vdpResearch transaction database are of great importance in this regard. The latter was set up for our own use in deriving empirically accurate valuation data and to create transaction-based price indices. Despite initial reservations on the part of the banking sector, automated valuation models have now gained a high level of acceptance. Against this background and the high level of competitiveness in the banking industry, these models will no become increasingly widespread in the next few years. This appears not least from the increasing number of associations in the banking industry which are actively encouraging the integration of automated valuation models into their own IT processes. The use of automated valuation models in the mortgage lending process in Germany is significantly affected by regulatory rules. These rules relate to both the selection of valuation method and the interaction between the valuer and the automated valuation model. Whether a residential property is valued using the Income Capitalization, Depreciated Replacement Cost or Sales Comparison Approach is entirely determined by the type of property to be valued and its suitability for owner-occupation; the availability and quality of data play no role at this point. The continued high importance of valuers in the valuation of real estate is recognised in the fact that automated valuation models, by definition, generate merely suggestions in terms of Market Value and Mortgage Lending Value. The appraisal carried out by the valuer then turns these suggestions into substantive Market Values and Mortgage Lending Values. This is achieved technically by allowing all automatically generated cells to be adjusted by the user. It is for this reason that the term computer-assisted valuation model is generally used in Germany rather than automated valuation model.

Reference

Crimmann, W. (2011). *Mortgage lending value* (Vol. 49). Berlin: Association of German Pfandbrief Banks.

An Estimative Model of Automated Valuation Method in Italy

Marina Ciuna, Francesca Salvo and Marco Simonotti

Abstract The Automated Valuation Method (AVM) is a computer software program that analyzes data using an automated process. It is related to the process of appraising an universe of real estate properties, using common data and standard appraisal methodologies. Generally, the AVM is based on quantitative models (statistical, mathematical, econometric, etc.), related to the valuation of the properties gathered in homogeneous groups (by use and location) for which are collected samples of market data. The real estate data are collected regularly and systematically. Within the AVM, the proposed valuation scheme is an uniequational model to value properties in terms of widespread availability of sample data, allowing the use of statistical models, and in the opposite conditions of the absence of data of comparable properties. Under these conditions the 'appraisal model' has a unique shape, when its coefficients are calculated with a mathematical-statistical model and when they are determined by an estimative process. The main part of the appraisal model is unique and the universal in the valuation, for which the mathematical-statistical and estimative procedures are the underlying part. Of course, the accuracy of the valuation increases with the number of available data, other conditions being equal, and the valuations, carried out in the absence of data (but in the presence of other market information), require extra-statistical appraisal procedures involving a complete knowledge of the real estate market (Ciuna and Simonotti 2011). However such knowledge is also required in the AVM performed by quantitative models with regard to the data sampling and the verify of the results (Kauko and d'Amato 2008a, b). The appraisal model is based on uniequational 'appraisal functions', on indices measured in the market and on tests. In first approximation, the linear form is preferred for simplicity, for the modularity (the majority of models are linear or linearized or attributable to additive forms), for

M. Ciuna (✉) · M. Simonotti
University of Palermo, Palermo, Italy
e-mail: ing_ciuna@hotmail.com; marina.ciuna@unipa.it

M. Simonotti
e-mail: marco.simonotti@unipa.it

F. Salvo
University of Calabria, Rende, Italy

M. d'Amato and T. Kauko (eds.), *Advances in Automated Valuation Modeling*,
Studies in Systems, Decision and Control 86, DOI 10.1007/978-3-319-49746-4_5

the understanding of the calculations, the intermediate elements (e.g. marginal prices of the real estate characteristics) and the results of the valuation.

Keywords AVM · Valuation · Market segment · Appraisal function

1 Introduction

The AVM (Automated Valuation Method) regards the process of valuation of a set of properties, and takes place with models and methods based on the direct detection of punctual data of individual contracts and other market indications, mainly concerning the parameters of the market segments, and the statistical and economical indicators. The appraisal procedures may be statistical, statistical-estimative and estimative, in accordance with the purposes of the valuation, the availability of data, the type of property and the means and the time put forward by the valuer (d'Amato and Kauko 2008). The purpose of this work is to present an appraisal model of AVM characterized by power to operate even in cases in which there is in the conditions with few data, only one property datum or no data, according to a unique pattern represented by the appraisal function. The valuation model proposed is based on the appraisal function that in the basic form establishes a relationship of cause and effect between the market price, the characteristics of the property and the parameters of a given segment in a specific context of the real estate market. The peculiarity of the model is the possibility to construct functions with a minimum amount of information available, starting from a sample of market prices to a single datum or no data, in the latter case doing the valuation with the other market information, in a structured way. The appraisal model is in fact able to use any kind of real estate information, perfecting with the knowledge of the market and considering the result according to the purpose of the valuation.

2 Basis of the Valuation

The bases of the valuation are the economic variables taken as the foundation of real estate valuation. The main foundations of the valuations are the market prices and the rents as true references to the real estate market. The respective bases of the valuation are then the market value and the market rent. The appraisal model poses as the basis of valuation the market value, the market rent and the net income referred to the market rent. These bases of valuation are defined according to the current valuation standard the market value is "the estimated amount for which an asset or liability should exchange on the valuation date between a willing buyer and a willing seller in an arm's-length transaction after proper marketing and where the parties had each acted knowledgeably, prudently, and without compulsion." (IVS 2011, IVS Framework); the market rent is "the estimated amount for which a property would be leased on the valuation date between a willing lessor and a willing lessee on appropriate lease terms in an arm's length transaction, after proper marketing and where the parties had each acted knowledgeably, prudently and without compulsion" (IVS 2011, IVS Framework); the net market income is the amount obtained by subtracting to the market rent, as defined, the operating expenses incurred by real estate. Generally the AVMs are applied to appraise the market value, the market rent and the net income of the residential properties, for which there are a number of sample data. Although AVMs development requires skilled analysis and attention to quality assurance, AVMs are characterized by the use and application of statistical and mathematical techniques. This distinguishes them from traditional appraisal methods in which an appraiser physically inspects properties and relies more on experience and judgment to analyze real estate data and develop an estimate of market value. Provided that the analysis is sound and consistent with accepted appraisal theory, an advantage to AVMs is the objectivity and efficiency of the resulting value estimates (Salvo et al. 2015). Of course, sound judgment is required in model development and an appraiser should review the values produced by the model (IAAO 2003). However, the valuation on large scale may be related to properties with destinations other than residential. These properties are intended for special uses or special property (agricultural, industrial and commercial real estate, secondary real estate, etc.) that have a limited market and which often show specific structure, shape and size. The market prices and rents of instrumental properties include the effects of the a typicalness of these properties, the system of non-competitive market, the difficulty of building renovation and production conversion, as well as the effects induced by the dynamics of the productive sector. The appraisal model aims to provide a process for uniform valuation of residential and special properties (Salvo et al. 2015).

3 Market Segment and Area

The real estate market is segmented for the diversity, the atypical nature and complexity of the properties. The real estate market is divided into segments, which represent the basic unit, not further divisible, of the economic-estimative analysis. The market segment is defined with respect to a set of parameters for which two or more housing units fall in the same market segment if they have similar economic-estimative parameters, whether if they are similar housing units for valuation. A segment of the real estate market is classified by: the type of contract (rent, purchase, etc.); the destination; the location; the type of property; the building type; the characters of supply and demand; the shape of the market and the price level. The market segments are defined not only by the parameters but by a series of indicators, ratios and market indices. A market segment may include a single property, a group of real estate, a block or a neighborhood. The segments are defined by the boundaries within which the properties are subject in the same way to the economic forces that largely determine the market price and rent. The properties of the same real estate market segment may have a spatial discontinuity even within restricted geographical areas (Ciuna et al. 2014).

In the present study we consider the contract of sale for the price and the lease for the rents, as they are typically the most popular, there is a ratio between them with appraisal meaning (the gross rent multiplier, the capitalization rate, etc.) and they can be easily extended to other contracts and rights (lease, leasehold, etc.) or to other situations (comparable market analysis). The market segment is typically a small unit of analysis (parts of the building, blocks, neighborhoods, etc.) but with the definition of the parameters it is possible to group similar or comparable segments in a larger unit (also spatially), taking into account the different parameters. For example, the market data of real estate from two spatially neighboring segments can be combined into a single sample introducing between the characteristics of collected properties, the different localization (e.g. central and semi-central). Likewise, the market data of real estate segments for the different building types can be grouped in the same sample data, taking into account the type (e.g. for apartments: the multi-storey buildings and chalet). Based on the study of the segmentation process, the model proposes as a basic unit of application of the AVM in the market area. The market area delimits, by means of a continuous line, a set of market segments, for the purposes of the construction and application of the functions of the model (Renigier-Biłozor and Biłozor 2016a, b). The market area is defined according to four main parameters: the type of contract, the destination, the type of building and the type of property (Simonotti 1998). The boundaries of the market area varies according to the parameters taken into consideration. In practice, the market areas, individual by the main coordinates, are layered in a sectional for the purpose of simplifying the valuation procedure, and their mapping follows perimeter lines that correspond physically to the boundary lines between buildings and areas. The definition of the market area can be done by aggregation of similar properties with inductive procedure, taking into account the fact that more widen is the perimeter of the market area, greater is the variability of the properties. This implies a priori knowledge of the market. The market area may include: segments in which there are market prices and rents; segments in which there are only market prices and no rents; segments in which there are only the rents and no market prices; segments in which there are not neither the market prices nor the rents. The model aims to settle the principles of international standards, because it is based on the detection of market data on a uniform appraisal methodology and on the ability to operate quality controls.

4 Appraisal Function

In the appraisal model the general form of the function used to appraise the market value, referred to the market area, the characteristics and parameters of the real estate segment, can be proposed as follows in the following Formula (1):

$$V = L_0 + \sum_{f=1}^{n} p_f \cdot x_f + \sum_{f=1}^{n} q_g \cdot X_g \quad (1)$$

where:

V market value of the property;
L_0 constant term;
p_f marginal price of the generic real estate characteristic f (con f = 1, 2, ..., n);
q_g marginal price of the generic market segment parameter g (con g = 1, 2, ..., m);
x_f generic real estate characteristic;
X_g generic market segment parameter.

The marginal price of the real estate characteristic represents the variation in the market value varying the characteristic. The marginal price of the parameter segment expresses the variation in the market value varying the parameter. The function used to appraise the market value is presented in the deterministic form. The general form of the function used to appraise the market rent refers to the market area, to the segment parameters and the real estate characteristics can be proposed as follows in the following Formula (2):

$$R = l_0 + \sum_{f=1}^{n} r_f \cdot x_f + \sum_{g=1}^{m} v_g \cdot X_g \quad (2)$$

where:

R annual market rent of the property;
l_0 constant term;
r_f marginal income of the generic real estate characteristic;
v_g marginal income of the generic market segment parameter;
x_f generic real estate characteristic;
X_g generic market segment parameter.

The marginal income of the characteristic expresses the variation in the real estate market rent varying the characteristic. The marginal income of the segment parameter expresses the variation in the real estate market rent varying the parameter. The function used to appraise the market rent is presented in the deterministic form. The equivalent statistic appraisal function of the market value and the market rent is the following Formula (3):

$$y_j = b_0 + \sum_{f=1}^{n} b_f \cdot x_{jf} + \sum_{g=1}^{m} B_g \cdot X_{jg} + e_j \quad (3)$$

where:

y_j total market price or annual market rent of the generic real estate j (with j = 1, 2, ..., m);
b_0 constant term (euro);

b_f coefficient of the generic real estate characteristic;
B_g coefficient of the generic market segment parameter;
x_{jf} generic real estate characteristic;
X_{jg} generic market segment parameter;
e_j stochastic error.

The statistical appraisal function of the market value and the market rent is presented as a multiple linear regression equation according to the most commonly used symbols. The regression equation considers the price or the market rent collected as explained variable and the parameters of the real estate segment and the characteristics as explanatory variables. The statistical appraisal function of the value and market rent is submitted in stochastic form.

There is complete identity between the appraisal Formulas (1) and (2) and the statistical Formula (3) in the constant component (L_0, l_0 and b_0) and in marginal prices and income of real estate characteristics (p_f, r_f and b_f) and in marginal prices and incomes of the parameters (q_g, v_g and B_g).

The general formula of net market income can be proposed as follows in the Formula (4):

$$R_N = R - c_E \tag{4}$$

where:

R_N annual net income of the property;
R annual total market rent of the property;
c_E annual operating cost of the property.

The net income is appraised by subtracting to the market rent, obtained by the estimation function, the operating expenses incurred by the property owner.

5 Construction of the Appraisal Function

The peculiarity of the appraisal model consists in the possibility of constructing the prediction function with the statistical models and estimation procedures (the market comparison method, the method of the depreciated reconstruction cost and direct capitalization method) according to the valuation standards (d'Amato 2008; d'Amato and Siniak 2008).

To this purpose, the model considers four specific situations: in the first situation there is a sample of market prices or a sample of market rents sufficiently numerous for the construction of a statistical model; in the second situation there is a market price or a market rent of a real transaction; in the third situation there is a sample of market prices or a sample of market rents of comparable properties, in itself few to be treated statistically, but perfectly suitable for use in the appraisal process; in the fourth situation, finally, there are not any real estate data (market prices and market rents), but we know the functions of market areas similar and close to that for which we want to estimate the function.

6 Function of the Market Value

The construction of the appraisal function of the market value follows four situations regarding the availability of market data. In the first situation, collected a statistical sample of prices related to the market area, known the characteristics of the contracted real estate property and the parameters of the market segment, the appraisal function of the market value can be presented as multiple linear regression equation according to the general Formula (3). The uniequational model can also be calculated in a mathematical-statistical way different by the regression analysis, provided that the locational factor and the marginal prices of the characteristics and of the segment parameters are specified. The model in this form then provides directly the constant term and the marginal prices of the characteristics and parameters of the real estate segment. In the first situation for the individual property being appraised the market value V_0 is equal to Formula (5):

$$V_0 = b_0 + \sum_{f=1}^{n} b_f \cdot x_{0f} + \sum_{g=1}^{n} B_g \cdot X_{0g} \tag{5}$$

where:

x_{0f} generic real estate characteristic of the property being assessed;
X_{0g} generic segment's parameter of the property being assessed.

In principle, the appraisal function is able to estimate individually by the interpolation all properties of the market area. In the second situation, detected in a given area of the market, the market price P_j, note the real estate characteristics and parameters of the market segment of the contracted property, for the individual property being appraised according to the general Formula (1) the market value V_0 is equal to Formula (6):

$$V_0 = L_0 + \sum_{f=1}^{n} p_f \cdot x_{0f} + \sum_{g=1}^{m} q_g \cdot X_{0g} \tag{6}$$

The constant term of the appraisal function of the market value is mainly related to the localization of the property and the effect of other characteristics and segment parameters different from those reported in the function. The constant term of the appraisal function of the market value can be calculated by setting an appraisal comparison equation for which the difference in the price between the two properties is a function of the differences presented by their characteristics and their parameters of segment. The functional relationship between the known market price

P_j of the generic comparable property j and its characteristics and parameters of real estate segment according to the Formula (1) is the following in the Formula (7):

$$P_j = L_0 + \sum_{f=1}^{n} p_f \cdot x_{jf} + \sum_{g=1}^{m} q_g \cdot X_{jg} \tag{7}$$

The appraisal comparison equation of the market value refers to the comparison between the property being appraised and the generic comparable property respectively according to the Formulas (6) and (7), as follows in the Formula (8):

$$V_0 - P_j = \sum_{f=1}^{n} p_f \cdot x_{0f} + \sum_{g=1}^{m} q_g \cdot X_{0g} - \sum_{f=1}^{n} p_f \cdot x_{jf} - \sum_{g=1}^{m} q_g \cdot X_{jg} \tag{8}$$

Consequently the market value V_0 of the individual property being appraised is equal to the Formula (9):

$$V_0 = (P_j - \sum_{f=1}^{n} p_f \cdot x_{jf} - \sum_{g=1}^{m} q_g \cdot X_{jg}) + \sum_{f=1}^{n} p_f \cdot x_{0f} + \sum_{g=1}^{m} q_g \cdot X_{0g} \tag{9}$$

where the constant term is setted equal to Formula (10):

$$L_0 = P_j - \sum_{f=1}^{n} p_f \cdot x_{jf} - \sum_{g=1}^{m} q_g \cdot X_{jg} \tag{10}$$

The appraisal model in this form requires an exogenous valuation of the marginal prices of the characteristics and real estate segment parameters. In principle, the apraisal function is able to estimate individually, by interpolation, all properties of the market area. In the third situation, detected an appraisal sample (not very large) of price P_j ($j = 1, 2, \ldots, k$) referred to the market area, known the parameters of the market segment and the characteristics of the traded properties, the appraisal function can be achieved by setting k equations according to the Formula (9) in the following way (Simonotti 1985), Formula (11):

$$\begin{cases} V_0 = P_1 - \sum_{f=1}^{n} p_f \cdot x_{1f} - \sum_{g=1}^{m} q_g \cdot X_{1g} + \sum_{f=1}^{n} p_f \cdot x_{0f} + \sum_{g=1}^{m} q_g \cdot X_{0g} \\ V_0 = P_2 - \sum_{f=1}^{n} p_f \cdot x_{2f} - \sum_{g=1}^{m} q_g \cdot X_{2g} + \sum_{f=1}^{n} p_f \cdot x_{0f} + \sum_{g=1}^{m} q_g \cdot X_{0g} \\ \cdots = \cdots \\ V_0 = P_k - \sum_{f=1}^{n} p_f \cdot x_{kf} - \sum_{g=1}^{m} q_g \cdot X_{kg} + \sum_{f=1}^{n} p_f \cdot x_{0f} + \sum_{g=1}^{m} q_g \cdot X_{0g} \end{cases} \tag{11}$$

In the third situation, for the property being appraised, the market value V_0 is then equal to Formula (12):

$$V_0 = \frac{1}{k} \cdot \left(\sum_{j=1}^{k} P_j - \sum_{f=1}^{n} p_f \cdot \sum_{j=1}^{k} x_{jf} - \sum_{g=1}^{m} P_g \cdot \sum_{j=1}^{k} X_{jg}\right) + \sum_{f=1}^{n} p_f \cdot x_{0f} + \sum_{g=1}^{m} q_g \cdot X_{0g} \tag{12}$$

where the divisor is the number of sampled data. In the third situation the constant term L_0, according to the Formula (12) is equal to Formula (13):

$$L_0 = \frac{1}{k} \cdot \left(\sum_{j=1}^{k} P_j - \sum_{f=1}^{n} p_f \cdot \sum_{j=1}^{k} x_{jf} - \sum_{g=1}^{m} \Pi_g \cdot \sum_{j=1}^{k} X_{jg}\right) \tag{13}$$

The appraisal model, in this form, requires an exogenous valuation of the marginal prices of the characteristics and of the parameters of the real estate segment. In order to consider the possibility of resolution of the equations system of the Formula (9), the system can be presented in the following Formula (14):

$$\begin{cases} V_0 + \sum_{if=1}^{n} p_f \cdot (x_{1f} - x_{0f}) + \sum_{g=1}^{m} q_g \cdot (X_{1g} - X_{0g}) = P_1 \\ V_0 + \sum_{f=1}^{n} p_f \cdot (x_{2f} - x_{0f}) + \sum_{g=1}^{m} q_g \cdot (X_{2g} - X_{0g}) = P_2 \\ \cdots = \cdots \\ V_0 + \sum_{f=1}^{n} p_f \cdot (x_{kf} - x_{0f}) + \sum_{g=1}^{m} q_g \cdot (X_{kg} - X_{0g}) = P_k \end{cases} \tag{14}$$

The unknowns of the equations system are the market value of the property being appraised and the marginal prices of the characteristics and parameters of the real estate segment. The known terms of the system are represented by the recorded market prices. To the valuation of the market value are valid the mathematical conditions of resolution of the equations system, including, eventually, appropriate solution conditions mathematically approximate. In principle, the appraisal function is able to estimate individually by the interpolation all properties of the market area. In the fourth situation it is assumed that there are not any data of market prices, characteristics and parameters of the real estate market segment, but we know the appraisal functions of market areas next to the one for which we want to build the function. In this circumstance the appraisal model allows the interpolation of the appraisal functions of the next two areas, or of their neighbors segments, by varying marginal prices (if with the same parameters of the segment) or by introducing the parameter or parameters that are different. The appraisal function can be achieved by setting the equations of the market areas A and B selected in accordance with the Formula (1) in the following Formula (15):

$$\begin{cases} V_A = L_{0A} + \sum_{f=1}^{n} p_{fA} \cdot x_f + \sum_{g=1}^{m} q_{gA} \cdot X_g \\ V_B = L_{0B} + \sum_{f=1}^{n} p_{fB} \cdot x_f + \sum_{g=1}^{m} q_{gB} \cdot X_g \end{cases} \tag{15}$$

where the locational factor L_{0A} of the market area A according to the Formula (7) is equal to Formula (16):

$$L_{0A} = P_{jA} - \sum_{f=1}^{n} p_f \cdot x_{jfA} - \sum_{g=1}^{m} q_g \cdot X_{jgA} \tag{16}$$

and the locational factor L_{0B} according to the Formula (7) of the market area B is equal to Formula (17):

$$L_{0B} = P_{jB} - \sum_{f=1}^{n} p_f \cdot x_{jfB} - \sum_{g=1}^{m} \Pi_g \cdot X_{jgB} \tag{17}$$

In the fourth situation, the market value V_0 of the individual property being appraised is equal to Formula (18):

$$V_0 = (\frac{L_{0A} + L_{0B}}{2}) + \frac{p_{fA} + p_{fB}}{2} \cdot \sum_{f=1}^{n} x_{0F} + \frac{q_{gA} + q_{gB}}{2} \cdot \sum_{g=1}^{m} X_{0g} \tag{18}$$

The appraisal model in this form requires an exogenous valuation of the marginal prices of real estate properties. The inclusion in the appraisal function of the parameter of segment solves the case of the segments without data in the valuation of the marginal prices of the parameters. In principle, the interpolated appraisal function is able to estimate individually, by extrapolation, all properties of the market area.

7 Function of the Market Rent

The construction of the appraisal function of the market rent follows four situations regarding the availability of market data. In the first situation, collected a statistical sample of the rents related to the market area, known the real estate characteristics and the parameters of segment, of the contracted properties, the appraisal function of the market rent can be presented as multiple linear regression equation according to the general Formula (3). The uniequational model can also be calculated with mathematical-statistical analysis different from the regression, provided that in the form there is specified the locational factor and the marginal income of the real estate characteristics and of the parameters of the segment. The model in this form then provides directly the constant term and the marginal income of real estate characteristics and parameters of the segment. In principle, the appraisal function is able to estimate individually, by interpolation, all properties of the market area. In the second situation, detected in a given market area, a market rent R_j, known the characteristics and parameters of the real estate segment of the contracted property, for the individual property being appraised, according to the general Formula (3), the market rent R_0 is equal to Formula (19):

$$R_0 = l_0 + \sum_{f=1}^{n} r_f \cdot x_{0f} + \sum_{g=1}^{m} v_g \cdot X_{0g} \tag{19}$$

The constant term of the appraisal function of the market rent is mainly due to the location of the property and the effect of other characteristics and segment parameters different from those specified in the function. The comparison equation of the market rent refers to the comparison between the generic comparable property and the property to be appraised, according to the Formula (2) and the Formulas (7), (8) and (9), is the following Formula (20):

$$R_0 = (R_j - \sum_{f=1}^{n} r_f \cdot x_{jf} - \sum_{g=1}^{m} q_g \cdot X_{jg}) + \sum_{f=1}^{n} r_f \cdot x_{0f} + \sum_{g=1}^{m} q_g \cdot X_{0g}) \tag{20}$$

where the constant term is equal to Formula (21):

$$l_0 = R_j - \sum_{f=1}^{n} r_f \cdot x_{jf} - \sum_{g=1}^{m} q_g \cdot X_{jg} \tag{21}$$

The appraisal model, in this form requires the exogenous valuation of marginal income of the real estate characteristics and of the parameters of the segment (paragraph 6). In principle, the apprasial function is able to estimate individually, by interpolation, all properties of the market area. In the third situation, detected an appraisal sample (not numerous) of market rents R_j ($j = 1, 2, \ldots, k$) refers to the market area, known the characteristics of the traded properties and the parameters of the market segment, the appraisal function can be achieved by setting k equations according to the Formula (11) in the following way, Formula (22):

$$\begin{cases} R_0 = R_1 - \sum_{f=1}^{n} r_f \cdot x_{1f} - \sum_{g=1}^{m} v_g \cdot X_{1g} + \sum_{f=1}^{n} r_f \cdot x_{0f} + \sum_{g=1}^{m} v_g \cdot X_{0g} \\ R_0 = R_2 - \sum_{f=1}^{n} r_f \cdot x_{2f} - \sum_{g=1}^{m} v_g \cdot X_{2g} + \sum_{f=1}^{n} r_f \cdot x_{0f} + \sum_{g=1}^{m} v_g \cdot X_{0g} \\ \cdots = \cdots \\ R_0 = R_k - \sum_{f=1}^{n} r_f \cdot x_{kf} - \sum_{g=1}^{m} v_g \cdot X_{kg} + \sum_{f=1}^{n} r_f \cdot x_{0f} + \sum_{g=1}^{m} v_g \cdot X_{0g} \end{cases} \tag{22}$$

In the third situation, for the individual property being appraised, the market rent R_0 is then equal to Formula (23):

$$R_0 = \frac{1}{k} \cdot \left(\sum_{j=1}^{k} R_j - \sum_{f=1}^{n} r_f \cdot \sum_{j=1}^{k} x_{jf} - \sum_{g=1}^{m} q_g \cdot \sum_{j=1}^{k} X_{jg}\right) + \sum_{f=1}^{n} r_f \cdot x_{0f} + \sum_{g=1}^{m} q_g \cdot X_{0g} \tag{23}$$

In the third situation the constant term I_0, according to the Formula (23), is equal to Formula (24):

$$I_0 = \frac{1}{k} \cdot \left(\sum_{j=1}^{k} R_j - \sum_{f=1}^{n} r_f \cdot \sum_{j=1}^{k} x_{jf} - \sum_{g=1}^{m} q_g \cdot \sum_{j=1}^{k} X_{jg}\right) \tag{24}$$

The appraisal model in this form requires an exogenous valuation of the marginal income of the real estate characteristics and of the parameters of the segment. In order to consider the possibility of resolution of the equations system of the Formula (14), the system can be presented in the following form, Formula (25):

$$\begin{cases} R_0 + \sum_{f=1}^{n} r_f \cdot (x_{1f} - x_{0f}) + \sum_{g=1}^{m} q_g \cdot (X_{1g} - X_{0g}) = R_1 \\ R_0 + \sum_{f=1}^{n} r_f \cdot (x_{2f} - x_{0f}) + \sum_{g=1}^{m} q_g \cdot (X_{2g} - X_{0g}) = R_2 \\ \ldots = \ldots \\ R_0 + \sum_{f=1}^{n} r_f \cdot (x_{kf} - x_{0f}) + \sum_{g=1}^{m} q_g \cdot (X_{kg} - X_{0g}) = R_k \end{cases} \tag{25}$$

The unknowns of the equations system are the market rent of the property being appraised and the marginal income of real estate characteristics and of the parameters of the segment. The known terms of the system are the collected market rents. For the valuation of the market rent are valid the mathematical conditions of resolution of the equations system, including, where appropriate, solution conditions mathematically approximate. In principle, the appraisal function is able to value individually, by interpolation, all properties of the market area. In the fourth situation it is assumed that there are not any information on market rents, property characteristics and parameters of the market segment, but we know the appraisal functions of market areas next to the one for which we want to build function. In this circumstance the appraisal model allows the interpolation of the appraisal functions of the next two areas, or neighbors segments, by varying the marginal income (if with the same parameters of the segment) or by introducing the parameter or parameters of difference. The appraisal function can be achieved by setting the equations of the market areas A and B selected in accordance with the Formula (1) in the following way, Formula (26):

$$\begin{cases} R_A = l_{oA} + \sum\limits_{f=1}^{n} r_{fA} \cdot x_f + \sum\limits_{g=1}^{m} q_{gA} \cdot X_g \\ R_B = l_{oB} + \sum\limits_{f=1}^{n} r_{fB} \cdot x_f + \sum\limits_{g=1}^{m} q_{gB} \cdot X_g \end{cases} \quad (26)$$

where the locational factor l_{0A} of the market area A according to the Formula (21) is equal to Formula (27):

$$l_{oA} = R_{jA} + \sum_{f=1}^{n} r_f \cdot x_{jfA} + \sum_{g=1}^{m} q_g \cdot X_{jgA} \quad (27)$$

and the locational factor l_{0B} of the market area B according to the Formula (19) is equal to Formula (28):

$$l_{oB} = R_{jB} + \sum_{f=1}^{n} r_f \cdot x_{jfB} + \sum_{g=1}^{m} q_g \cdot X_{jgB} \quad (28)$$

In the fourth situation, the to market rent R_0 of the individual property being appraised is equal to Formula (29):

$$R_0 = (\frac{l_{0A} + l_{0B}}{2}) + \frac{r_{fA} + r_{fB}}{2} \cdot \sum_{f=1}^{n} x_{0f} + \frac{q_{gA} + q_{gB}}{2} \cdot \sum_{g=1}^{m} X_{0g} \quad (29)$$

The appraisal model in this form requires an exogenous valuation of marginal income of the real estate characteristics. The inclusion in the appraisal function of parameter of segment solves the case of segments with the absence of data in the valuation of the marginal income of the parameters (Renigier-Biłozor et al. 2014a). In principle, the appraisal function is able to estimate individually, by interpolation all properties of the market area (Ciuna 2014a, b).

8 Appraisal Procedures

The procedures to the valuation of the market value and market rent refer to current valuation standard, which indicate the market comparison method, the depreciated reconstruction cost method and the income capitalization method.

9 Market Comparison Method

The market comparison method is an appraisal procedure of the market value or the market rent of the property, by comparing the property being appraised and a set of similar comparable properties, recently traded and with price or rent known (Salvo et al. in print). The market comparison method applies: in the situation where we have the market price or the market rent of a real transaction (depending on the situation) (Salvo and De Ruggiero 2011, 2013); and in the situation where we have a sample of market prices or a sample of market rents of comparable properties, in itself not very numerous, but suitable for use in the valuation process (the third situation) (Kaklauskas et al. 2012a, b). The framework of the coefficients of the appraisal function concerns: the constant factor, the marginal rents and prices of the characteristics and parameters (Borst et al. 2008).

10 Constant Term

The extra-statistical valuation of the constant term concerns the appraisal procedures. The constant term expresses the characteristics for which two samples of real estate data, taken from two different market areas, while presenting the same characteristics and the same parameters of the segment, show two different price levels. For the market area, the appraisal function assigns to the constant term an abstract meaning, as it is not referred to the market segment, as in traditional valuations, but to the market area. The same is for the marginal price of real estate characteristics. The calculation of the constant term in the construction of the appraisal functions of the market value and the market rent is divided: in the situation where we have the market price or market rent for a single transaction (depending on the situation); and in the situation of the appraisal sample of prices or market rents of comparable properties (third position).

11 Analysis of the Marginal Prices of the Characteristics

The marginal price of a real estate characteristic represents the variation of the total price of the property varying the characteristic. The marginal prices can be expressed in terms of value and in percentage terms. In the appraisal analysis the marginal prices are accounting prices, i.e. prices that perform instrumental tasks and are estimated a priori according to the purposes of the valuation. The analysis of the marginal prices of the main characteristics of the properties can be carried out for all of the real estate characteristics. However, for the purposes of the MA, the interest is reduced to the real estate characteristics for which the valuation of the marginal prices and income applies directly available market information. They are generally recurring ratios between economic-estimative sizes of the real estate market and applied by the operators. These commercial ratios are naturally available and are recorded directly from the market. The marginal price of the main area of the property is calculated by multiplying the average price for the position ratio σ_P, which locates on the cartesian floor the relative position of the curve of marginal price and that known of the average price. If the curve of the average price is above the marginal price, then the position ratio is less than unity (Simonotti 2001). The position ratio σ_P between the marginal price p_i and the average unit price $\bar{p}_i$ of the same surface characteristic s_i, it indicates the following Formula (30):

$$\sigma_p = \frac{p_i}{\bar{p}_i} \tag{30}$$

Therefore, the marginal price of the surface characteristic is calculated as the product of the unit price directly calculable on the market data and the position ratio σ_P, as follows in the Formula (31):

$$p_i = \bar{p}_i \cdot \sigma_P \tag{31}$$

To calculate the average price of the main surface in the presence of secondary and accessory surfaces is necessary to detect the market ratio of these surfaces with the main surface. The surface ratio π_{Pf} expresses the ratio between the marginal price p_f of the generic secondary surface (with f = 2, 3, …, h) and the marginal price of the main surface p_1, as follows in the Formula (32):

$$\pi_{Pf} = \frac{p_f}{p_1} \tag{32}$$

In the ratios of the secondary area is generally supposed that these are worth less than the main surface that have the greater importance in the property ($\pi_{Pf} < 1$). However, sometimes the secondary surfaces are more relevant and more useful of the main surface because they can be the subject of significant real estate valuations ($\pi_{Pf} > 1$). The commercial ratios are explicitly indicated by the market. Knowing the total price of the property P_j, being s_{j1} the main surface, s_{jf} the generic secondary surface, the unit price $\bar{p}_{j1}$ of the main surface is equal to Formula (33):

$$\bar{p}_{j1} = \frac{P_j}{s_{j1} + \sum_{f=2}^{h} \pi_{Pf} \cdot s_{jf}} \tag{33}$$

then the marginal price p_{j1} of the main surface, according to the Formula (31), is equal to Formula (34):

$$p_{j1} = \frac{P_j}{s_{j1} + \sum_{f=2}^{h} \pi_{Pf} \cdot s_{jf}} \cdot \sigma_p \tag{34}$$

In the ratio is calculated the average price, in the denominator appears a fictitious surface called commercial area. The marginal price p_{jf} of the secondary areas according to the Formula (32) is equal to Formula (35):

$$p_f = \pi_{Pf} \cdot p_{j1} \tag{35}$$

i.e. the product of the marginal price of the main surface and the market ratio of the secondary surface in consideration. The marginal price of the outer surface of the property is obtained by multiplying the average price for the appropriate position ratio. The unit price of the built area is based on the survey: (a) the market prices of the land built when these soils have an independent market; (b) the market prices of the building areas, considering the potential transformation of the land in building area and subtracting the cost of demolition; (c) the market prices of the buildings constructed by including the impact of developed land on the market value of the property. The marginal price p_T of the outer surface, based on the survey of market prices of built lands (a), calculated their average price $\bar{p}_T$ and appraised the position ratio σ_T, is equal to Formula (36):

$$p_T = \bar{p}_T \cdot \sigma_T \tag{36}$$

The marginal price p_T of the outer surface, based on the survey of market prices of building lands (b), calculated their average price $\bar{p}_E$, the cost of demolition unit c_E and appraised the position ratio σ_E, is equal to Formula (37):

$$p_T = (\bar{p}_E - c_E) \cdot \sigma_e \tag{37}$$

The marginal price p_T of the outer surface, based on the survey of market prices of the properties (c), is based on the impact of built-up land, which expresses the ratio between the market value of the built land and the market value of the property (including the building and the land). Calculated the average unit price $\bar{p}_I$ of property, measured the impact of built land λ and appraised the position ratio σ_I, the marginal price p_T of the outer surface is equal to Formula (38):

$$p_T = \bar{p}_I \cdot \lambda \cdot \sigma_I \tag{38}$$

For example, considering only the surface's characteristics indicated, other things being equal property characteristics and parameters of the market segment, the appraisal function of the market value V_0 according to the Formula (5) is as follows in the Formula (39):

$$V_0 = \left[P_j - p_{j1} \cdot (s_{j1} + \pi_{P2} \cdot s_{j2} + \lambda \cdot s_{j3})\right] + p_{j1} \cdot (s_{01} + \pi_{P2} \cdot s_{02} + \lambda \cdot s_{j3}) \tag{39}$$

where:

P_j market price of the generic property;
s_{j1} main area of the generic property;
s_{j2} secondary area of the generic property;
s_{j3} outer surface area of the generic property;
s_{01} main area of the property being appraised;
s_{02} secondary area of the property being appraised;
s_{03} outer surface area of the property being appraised;
p_{j1} marginal price of the main surface according to the Formula (34);
π_{P2} market ratio of secondary surface according to the Formula (32);
λ incidence of the built area according to the Formula (38).

The possibility of inserting in the function other real estate characteristics over the surface's characteristics may cover characteristics such as the date, the level of the floor, the technological installations, the maintenance status, the number of toilettes and all those characteristics for which the market expresses a ratio or other indications of the market, and can apply the comparison appraisal procedures, such as the paired data analysis (PDA) (Ciuna and Simonotti 2014).

12 Analysis of the Marginal Incomes of the Characteristics

The marginal income of a real estate characteristic is the change in the total market rent of the property varying the characteristic. The marginal income can be expressed in terms of value and in percentage terms.

In the appraisal analysis the marginal income are prices that perform instrumental tasks and are estimated a priori, according to the purposes of the valuation. The analysis of the marginal income of the main characteristics of the properties can be carried out for all of the real estate. However, for the purposes of the MA interest is reduced to the real estate characteristics for which the valuation of marginal income applies directly available market information. The marginal income of the main area of the property is calculated by multiplying the average income for the position ratio σ_R, which locates on the Cartesian floor the relative position of the curve of marginal income with respect to that known of the average income. To calculate the average income of the surface in presence of primary and secondary accessory surfaces is necessary to detect the market ratio of these surfaces with the main surface. The market ratio π_{Rf} expresses the ratio between the marginal income r_f of the generic secondary surface (with f = 2, 3, …, h) and the marginal income of the main surface r_1, as follows in the Formula (40):

$$\pi_{Rf} = \frac{r_f}{r_1} \tag{40}$$

Knowing the total market rent R_j of the generic property, and the main surface s_{j1}, the generic secondary surface s_{jf}, then the marginal income r_{j1} the main surface is equal to Formula (41):

$$r_{j1} = \frac{R_j}{s_{j1} + \sum_{i=2}^{h} \pi_{Rf} \cdot s_{jf}} \cdot \sigma_R \tag{41}$$

In the fraction is calculated the average rent, in the denominator appears the commercial area. The marginal income r_{jf} of the secondary areas of the property, in accordance with the general Formula (36), is equal to Formula (42):

$$r_{jf} = \pi_{Rf} \cdot r_{j1} \tag{42}$$

i.e. the product of the marginal income of the main surface and the market ratio of the secondary surface considered. The marginal income of the outer surface of the property we can get from the marginal price of outer surface according to the Formulas (36), (37) and (38) multiplying by the capitalization rate of the soil i_T as follows in the Formula (43):

$$r_T = p_T \cdot i_T \tag{43}$$

If we have the market incomes of properties, the marginal income of the outer surface, can be calculated, according to the incidence of the built land. Known the average income $\bar{r}_I$ of the real estate properties, the impact of the built land λ and the position ratio σ_I, the marginal income r_T of the outer surface is equal to Formula (44):

$$r_T = \bar{r}_I \cdot i_T \cdot \lambda \cdot \sigma_I \tag{44}$$

As an example, considering only the surfaces characteristics indicated, being coeteris paribus the other characteristics and parameters of real estate market segment, the appraisal function of the market rent R_0, according to the Formula (39), is the following Formula (45):

$$R_0 = \left[R_j - r_{j1} \cdot (s_{j1} + \pi_{R2} \cdot s_{j2} + \lambda \cdot s_{j3})\right] + r_{j1} \cdot (s_{01} + \pi_{R2} \cdot s_{02} + \lambda \cdot s_{03}) \tag{45}$$

The possibility of inserting in the function other real estate characteristics over the surfaces characteristics may cover characteristics such as the date, the level of the floor, the technological installations, the maintenance status, the number of toilettes and all those characteristics for which the market expresses a market ratio or other indications of the market, and can apply the appraisal comparison procedures, such as the PDA.

13 Analysis of the Marginal Prices and Incomes of the Parameters

In the market comparison method, the appraisal functions are based on the collection of the market prices and rents (type of contract) in a definite market area (location), second the use of the property (destination). Consequently the parameters of the market segment insured into the appraisal function are essentially: the type of property and the type of building. The type of property indicates whether it is a contract relating to land and buildings; if the property is in a market of used property, the renovated or restored, new or almost new; whether it is in a condominium unit or exclusive ownership. The type of building refers to the character of the building or construction, or indicate if it is reinforced concrete structure, masonry, metal structure, or mixed; if it is multi-storey building, house, warehouse, shed, building complexes or other. Also other parameters of the market area can be taken into account (d'Amato 2010). The appraisal comparison is conducted between market segments that make up the market area and is carried out with the set parameters, including the parameters for which market segments differ in the area and excluding the parameters coeteris paribus. The marginal prices and incomes of the segment parameters are generally expressed in percentage and are taken positive or negative in relation to their effect on the market price. The valuation of the adjustment of a parameter answers the question: in what percentage the market price or income level of comparable segment differs from the level of the price or income of the segment for which we want to build a function, being other parameters coeteris paribus? To answer this question one must consider that quantitative market information is available about the parameters, the market ratios and the average levels indicative of prices and incomes. In practice, this information may take various forms and different employment opportunities and are assessed with the information and knowledge of the market operators. They are in fact indicative measures sometimes referred to wider contexts of the market area, which are not considered in the valuation for their absolute amount but in a relative sense compared to their mutual ratios. The marginal prices and incomes of the segment's parameters are generally estimated with the PDA, by comparing two or more segments that have equal amounts for all parameters except for the one that we should estimate the marginal price or income. It is a practical procedure that can be applied in situations of data availability and of substantially equal conditions. Specifically, the marginal price and the income of the segment's parameter could be based on the ratio between the supply prices or the quotations of properties represented by insertions (Active Listings) (Conditionally Sold Listings) and possibly by other atypical lists (Expired/Suspended/Terminated Listings), while not recognizing any interest in the valuation, because they don't meet the definition of market value and rent, but admitting a meaning to their ratio that is the adjustment percentage (Salvo et al. 2013a, b). So for example, the marginal price of the parameter type of property can be based on the ratio between the average supply prices $\bar{p}_N$ for new buildings and the supply prices of used buildings $\bar{p}_U$, being the other parameters coeteris paribus, as follows in the Formula (46):

$$\Pi_{NU} = \frac{\bar{p}_N - \bar{p}_U}{\bar{p}_U} \tag{46}$$

The PDA can be used in an extended form to calculate the marginal price of a parameter of the segment. As an example, in the comparison between two market segments 1 and 2, in which were recorded market prices P_1 and P_2 of the two properties, which differ in the parameter segment g, considering only the surface characteristics, being coeteris paribus other real estate characteristics and parameters of market segment (excluding the parameter), the marginal price of the parameter of segment q_g according to the Formula (39), is equal to Formula (47):

$$q_g = \frac{P_1 - P_2 + p_1 \cdot [s_{21} - s_{11} + \pi_2 \cdot (s_{22} - s_{12}) + \lambda \cdot (s_{23} - s_{13})]}{X_{1g} - X_{2g}} \tag{47}$$

where:

s_{11} main surface of the first property;
s_{12} secondary surface of the first property;
s_{13} outside surface of the first property;
s_{21} main surface of the second property;
s_{22} secondary surface of the second property;
s_{23} outside surface of the second property;
X_{1g} parameter of the segment of the first property;
X_{2g} parameter of the segment of the second property;
p_1 marginal price of the main surface;
π_2 market ratio of secondary surface;
λ impact of the built area.

The PDA then work adjustments of the parameters of the market segments comparable to that reference. The appraisal comparison between the market segments is carried out on the parameters specified in the definition of the market areas. The marginal prices and incomes relate to the specific knowledge of the market area, of the segments and then of the market as a whole.

14 Cost Method

The application of the cost method falls into the fourth situation in which we do not have any data property (market price and market rent).

The method of the depreciated reconstruction costs, or cost method, consider a property built into its component parts: the built land and buildings. The cost method is based on a comparison of the property to be appraised and similar properties, considering the characteristics of the area and differences in age, in the state and utility buildings (Salvo et al. 2015; Ciuna 2010, 2011).

The cost method is composed of three elements:

- the valuation of the market value of the built land;
- the valuation of the reconstruction cost of buildings, which insists on the ground;
- the valuation of the depreciation.

The main procedures for appraising the market value of a built land are based on the survey of: (a) the market price of the land built; (b) the market price of the building areas; (c) the market price of the constructed real estate properties (defined as land and buildings).

The appraisal function of market value, based on observed market prices of land built (a), adds to the general function of the Formula (1) referred to the built land, the cost of rebuilding of the property net of depreciation. The appraisal function of the market value V_0 of the property to be appraised according to the Formula (5) is equal to Formula (48):

$$V_0 = V_{T0} + c_R \cdot (1 - d) \cdot z = L_0 + \sum_{f=1}^{n} p_f \cdot x_{0f} + \sum_{g=1}^{m} q_g \cdot X_{0g} + c_R \cdot (1 - d) \cdot z \tag{48}$$

where:

V_{T0} market value of the built land;
c_R unit cost of rebuilding of the building;
z consistency of the building (area, volume, etc.);
d depreciation percentage presented by the building.

The appraisal function of the market value of the built land based on the collection of market prices of building land (b) considers the potential transformation of the land into building area by deducting the cost of demolition. The appraisal function of the market value of the land built subtracts to the general function of the Formula (5) refers to the building, the cost of demolition and adds the rebuilding cost of the property net of depreciation. The appraisal function of market value V_0 of the property to be appraised according to the Formula (5) is equal to Formula (49):

$$V_0 = V_{A0} - c_D \cdot z + c_R \cdot (1 - d) \cdot z \qquad (49)$$

where:

V_{A0} market value of the built land;
c_D unit cost of demolition of the building.

The appraisal function for the market value of the built land based on the collection of market prices of the properties (c) includes the impact of the built land on the market value of the property (land and building). The appraisal function of the market value of the built land considering the function of the general Formula (1) refers to the property, multiplying it by the ratio of built land and adding the cost of rebuilding of the property net of depreciation. The appraisal function of the market value V_0 of the property to be appraised according to the Formula (5) is equal to Formula (50):

$$V_0 = V_{I0} \cdot \lambda + c_R \cdot (1 - d) \cdot z \qquad (50)$$

where:

V_{I0} market value of the property;
λ percentage of built land on the market value of the property.

In the cost method the appraisal function for the market rent is based on the appraisal function of the market value, considering the capitalization rate of the built land and the capitalization rate of the building. The appraisal function of the market rent in the cost method follows the cases of the market value of the built land, based on the survey of market prices of built land, building areas and constructed buildings (Salvo et al. 2014). In synthetic terms according to the previous cases, in the cost method the appraisal of the market rent R_0, based on the survey of market prices of land built (a), according to the Formula (48) is as follows in the Formula (51):

$$R_0 = V_{T0} \cdot i_T + c_R \cdot (1 - d) \cdot z \cdot i_F \qquad (51)$$

where:

i_T annual capitalization rate of built land;
i_F annual capitalization rate of the building.

In the cost method, the appraisal function of the market rent R_0, based on the survey of market prices of building land (b) according to the Formula (49) is the following Formula (52): on the survey of prices in the real estate mar

$$R_0 = (V_{A0} - c_D \cdot z) \cdot i_T + c_R \cdot (1 - d) \cdot z \cdot i_F \qquad (52)$$

In the cost method, the appraisal function of the market rent R_0, based on the detection of market prices of the properties (c) according to the Formula (50) is the following Formula (53):

$$V_0 = V_{I0} \cdot \lambda \cdot i_T + c_R \cdot (1 - d) \cdot z \cdot i_F \tag{53}$$

In the cost method, the appraisal function of the market value and the market rent is based on the survey of prices in the real estate market (Ciuna et al. 2015a, b).

15 Income Capitalization Method

The income capitalization method is applied in a situation where there are not prices of comparable properties, but there are one or more market rents. The income capitalization method (or approach based on the expected results) considers the ability of a property to generate an income. The method is based on a comparison of the property to be appraised and market rents for similar properties known and on the conversion of income into capital value using a capitalization rate. The direct capitalization method consists of two elements: the valuation of market rent of the property through the appraisal function; the search of the capitalization rate. In the direct capitalization method, the general form of the function to estimate the market value according to the Formula (1) considers the capitalization rate of the property. The function of the estimated market value V_0 of the property to be appraised according to the Formula (19) is equal to Formula (54):

$$V_0 = \frac{R_0}{i} = \frac{l_0 + \sum_{f=1}^{n} r_f \cdot x_{0f} + \sum_{g=1}^{m} q_g \cdot X_{0g}}{i} \tag{54}$$

where:

i annual capitalization rate of the property.

The capitalization rate is the ratio between the income and the market price of a property. His research is based on data collection in the real estate market. According to the valuation standards, the search of the capitalization rate should reflect only the data and information of the real estate market (Biłozor and Renigier-Biłozor 2016). The direct capitalization method can be used in the inverse formulation designed to calculate the market rent once we know the market value and the capitalization rate. In this circumstance the appraisal function of the market rent in accordance with the Formula (1) considers the capitalization rate of the property. The appraisal function of market rent R_0 of the property to be appraised according to the Formula (5) is equal to Formula (55):

$$R_0 = V_0 \cdot i = (L_0 + \sum_{f=1}^{n} p_f \cdot x_{0f} + \sum_{g=1}^{m} q_g \cdot X_{0g}) \cdot i \quad (55)$$

In the direct capitalization method, the appraisal functions of the market value and the market rent are based on the survey of prices and rents of the real estate market (Renigier-Biłozor et al. 2014b; d'Amato 2015).

16 Appraisal Test

In general, the accuracy of the valuation of the market value and rent is related to the uniformity of conditions related to the market segment and the degree of similarity of the characteristics of comparable properties. Under these conditions the segment parameters (location, type, etc.) and for some real estate properties (e.g., access, age, etc.) are coeteris paribus. Those characteristics and parameters therefore are excluded from the estimative analysis; while for the characteristics that have different modality (e.g. the surface, the outer area, etc.) and which are introduced in the analysis, the accuracy of the valuation increases with decreasing variability. In other words, the valuation is much more accurate than most similar are the properties comparable to the property being appraised. The accuracy of the appraised value and the market rent in the market area, which include multiple market segments, it is linked to the characteristics of the comparable properties collected and the parameters for which the segments differ. For the estimative analysis, the parameters of the market segment can be considered macro-characteristics, the choice and the measurements of which influence the precision of the valuation. One the real estate market area should therefore include market segments, different for the fewest number of parameters. The appraisal procedures must be tested to ensure that they have achieved the required standard for their use. This is done through the ratio study and the diagnostic statistics, in which the estimated values are compared to observed prices (prices and market rents). It is a subsequent verification of the discrepancy between the predictions and the data collected (d'Amato and Kauko 2012). The appraisal error can be measured by various indices and percentages. The measures of the performance valuations take into consideration: the appraisal level, referred to the difference between the values (or incomes) and the appraised market prices (or rents) of a defined group of properties; and the valuation consistency, which regards the equity within a group of properties and between groups of real estate properties. The uniformity between the groups can be analyzed in terms of horizontal equity and vertical equity. The measures of performance are tested with the minimum and maximum indices and coefficients of verification (IAAO 1999; Kauko and d'Amato 2011).

17 Conclusions

The appraisal model of indicates the basis for the valuation in the market value and the market rent of residential properties and special destination. The appraisal model proposes the use of the universal "appraisal functions" of the market value and rent related to the market area, consisting of one or more market segments. The shape of the appraisal function is linear and report the real estate characteristics and the parameters of real estate segment. The peculiarity of the appraisal model is the ability to build the prediction function with the statistical models and with appraisal procedures depending on the availability of market data (d'Amato 2004). For a sufficiently large number of data samples for the construction of a statistical model, the appraisal function is an equation of multiple linear regression. The uniequational statistical model can also be calculated in a mathematical-statistical manner different by the analysis of regression, provided it contains the marginal prices and incomes of real estate properties and parameters of the segment. For samples of market data, few in number, which can not be treated statistically, for only one datum or in the absence of data, the appraisal function is determined with the appraisal procedures (market comparison method, cost method and direct capitalization method) (Simonotti et al. 2015). The appraisal model in fact allows to determine the appraisal function for the areas of real estate market without data operating on the marginal prices of the parameters of the segment, on the appraisal functions of the market areas, upstream, on prices and rents of the next market areas. The appraisal model aims to provide a uniform procedure for value through the modular appraisal functions, which form a system of interrelationships between the market areas, between the data and the market information, including the estimative and statistical procedures and between statistical and estimative verification tests. The appraisal model is in accordance with the international valuation standards. The appraisal model, based on the appraisal functions, presents: uniformity of application regarding the estimation of market values and incomes for all properties taking place solely and evenly with the appraisal functions; immediate understanding by the operators for which the linear prediction functions, which show the locational factor and prices and incomes, the property's characteristics and parameters of the segment, allowing comparisons and tests; modularity regarding the possibility of calculating the appraisal function with statistical procedures and appraisal methods. The appraisal model based on the appraisal functions is offered as a simple and economical model for the estimation of the market value and rent of the property.

References

Biłozor, A., & Renigier-Biłozor, M. (2016), The procedure of assessing usefulness of the land in the process of optimal investment location for multi-family housing function. In *World Multidisciplinary Civil Engineering-Architecture-Urban Planning Symposium*. WMCAUS 2016 Praga.

Borst, R. A., Des Rosiers, F., Renigier, M., Kauko, T., & d'Amato, M. (2008). Technical comparison of the methods including formal testing accuracy and other modelling performance using own data sets and multiple regression analysis. In T. Kauko, M. d'Amato (Eds.), *Mass appraisal an international perspective for property valuers*. Wiley Blackwell.

Ciuna, M. (2010). L'Allocation Method per la stima delle aree edificabili. *AESTIMUM, 57*, 171–178.

Ciuna, M. (2011). The valuation error in the compound values. *AESTIMUM* [S.l.], 569–583, Aug. 2013. ISSN:1724-2118.

Ciuna, M., & Simonotti, M. (2011). Linee guida per la rilevazione dei dati del mercato immobiliare. *Geocentro*, nn.15 e 16.

Ciuna, M., & Simonotti, M. (2014). Real estate surfaces appraisal. *AESTIMUM, 64*, Giugno, pp. 1–13.

Ciuna, M., Salvo, F., & Simonotti, M. (2014a). The expertise in the real estate appraisal in Italy. In *Proceedings of the 5th European Conference of Civil Engineering (ECCIE '14). Mathematics and Computers in Science and Engineering Series* pp. 120–129, North Atlantic University Union, ISSN:2227-4588, Firenze, 22/24 November 2014.

Ciuna, M., Salvo, F., & De Ruggiero, M. (2014). Property prices index numbers and derived indices. *Property Management, 32*(2), 139–153. doi:(Permanent URL):10.1108/PM-03-2013-0021.

Ciuna, M., Salvo, F., & Simonotti, M. (2014b). Multilevel methodology approach for the construction of real estate monthly index numbers. *Journal of Real Estate Literature, 22*(2), 281–302.

Ciuna, M., Salvo, F., & Simonotti, M. (2015a). Compensation appraisal processes for the realization of hydraulic works in an agricultural area. In *Proceedings of XLIV INCONTRO DI STUDI Ce.S.E.T. Il danno. Elementi giuridici, urbanistici e economico-estimativi* (pp. 69–82). Bologna, Italy. November 27–28, 2014. ISBN:978-88-99459-21-5.

Ciuna, M., Salvo, F., & Simonotti, M. (2015b). Parametric measurement of partial damage in building. In *Proceedings of XLIV INCONTRO DI STUDI Ce.S.E.T. Il danno. Elementi giuridici, urbanistici e economico-estimativi* (pp. 171–188). Bologna, Italy. November 27–28, 2014. ISBN:978-88-99459-21-5.

d'Amato, M. (2004). A comparison between RST and MRA for mass appraisal purposes. A case in Bari. *International Journal of Strategic Property Management, 8*, 205–217.

d'Amato, M. (2008). Rough set theory as property valuation methodology: The whole story, Chap. 11. In T. Kauko, M. d'Amato (Eds.), *Mass Appraisal an International Perspective for Property Valuers* (pp. 220–258). Wiley Blackwell.

d'Amato, M. (2010). A location value response surface model for mass appraising: An "iterative" location adjustment factor in Bari, Italy. *International Journal of Strategic Property Management, 14*, 231–244.

d'Amato, M. (2015). Income approach and property market cycle. *International Journal of Strategic Property Management, 29*(3), 207–219.

d'Amato, M., & Kauko, T. (2008). Property market classification and mass appraisal methodology, Chap. 13. In T. Kauko, M. d'Amato. (Eds.), *Mass appraisal an international perspective for property valuers* (pp. 280–303). Wiley Blackwell.

d'Amato, M., & Kauko, T. (2012). Sustainability and risk premium estimation in property valuation and assessment of worth. *Building Research and Information, 40*(2), 174–185 (March–April 2012).

d'Amato, M., & Siniak, N. (2008). Using fuzzy numbers in mass appraisal: The case of belorussian property market. In T. Kauko, M. d'Amato (Eds.), *Mass appraisal an international perspective for property valuers* (Chap. 5, pp. 91–107). Wiley Blackwell.

IAAO. (1999). *Standard on ratio studies*.

IAAO. (2003). *Standard on automated valuation models (AVM_S)*.

IVSC. (2011). *International valuation standards*. London: IVSC.

Kaklauskas, A., Zavadskas, E. K., Kazokaitis, P., Bivainis, J., Galiniene, B., d'Amato, M., et al. (2012a). Crisis management model and recommended system for construction and real estate. In N. T. Nguyen, B. Trawi´nski, R. Katarzyniak, & G.-S. Jo (Eds.), *Advanced methods for computational collective intelligence in studies in computational intelligence series edited by Janusz Kacprzyk* (pp. 333–343). Berlin: Springer Verlag.

Kaklauskas, A., Daniūnas, A., Dilanthi, A., Vilius, U., Irene, L., Gudauskas, R. et al. (2012b). Life cycle process model of a market-oriented and student centered higher education. *International Journal of Strategic Property Management, 16*(4), 414–430.

Kauko, T., & d'Amato, M. (2008a). Introduction: Suitability issues in mass appraisal methodology. In T. Kauko, M. d'Amato. (Eds.), *Mass appraisal an international perspective for property valuers* (pp. 1–24). Wiley Blackwell.

Kauko, T., & d'Amato, M. (2008b). Preface. In T. Kauko, M. d'Amato, (Eds.), *Mass appraisal an international perspective for property valuers* (p. 1), Wiley Blackwell.

Kauko, T., & d'Amato, M. (2011). *Neighbourhood effect, international encyclopedia of housing and home*. Elsevier Publisher.

Renigier-Biłozor, M., & Biłozor, A. (2016a). Proximity and propinquity of residential market area—Polish and Italian case study. In *16th International Multidisciplinary Scientific Geoconferences SGEM*. Bułgaria. (web of science).

Renigier-Biłozor, M., & Biłozor, A. (2016b). The use of geoinformation in the process of shaping a safe space. In: *16th International Multidisciplinary Scientific Geoconferences SGEM*. Bułgaria.

Renigier-Biłozor, M., Wiśniewski, R., Biłozor, A., & Kaklauskas, A. (2014a). Rating methodology for real estate markets—Poland case study. *Pub. International Journal of Strategic Property Management. 18*(2), 198–212. ISNN:1648-715X.

Renigier-Biłozor, M., Dawidowicz, A., & Radzewicz, A. (2014b). An algorithm for the purposes of determining the real estate markets efficiency in Land administration system. *Public Survey Review, 46*(336), 189–204.

Salvo, F., De Ruggiero, M. (2011). Misure di similarità negli Adjustment Grid Methods. *AESTIMUM, 58*, 47–58. ISSN:1592-6117.

Salvo, F., De Ruggiero, M. (2013). Market comparison approach between tradition and innovation. A simplifying approach. *AESTIMUM, 62*, 585–594, ISSN:1592-6117.

Salvo, F., Ciuna, M., & d'Amato, M. (2013). Appraising building area's index numbers using repeat values model. A case study in Paternò (CT). In *Dynamics of land values and agricultural policies*. (p. 63–71), Bologna: Medimond International Proceedings, Editografica, ISBN:978-88-7587-690-6, Palermo, 22–23/11/2012.

Salvo, F., Ciuna, M., & d'Amato, M. (2013). The appraisal smoothing in the real estate indices. In: (a cura di): *Maria Crescimanno, Leonardo Casini and Antonino Galati, Dynamics of land values agricultural policies* (pp. 99–111), Bologna: Medimond Monduzzi Editore International Proceeding Division, ISBN:978-88-7587-690-6, Palermo, 22–23/11/2012.

Salvo, F., Ietto, F., & Cantasano, N. (2014). The quality of life conditioning with reference to the local environmental management: A pattern in Bivona country (Calabria, Southern Italy). *Ocean and Coastal Management*, ISSN:0964–5691, doi:10.1016/j.ocecoaman.2014.10.014.

Salvo, F., Francini, M., & Palermo, A. (2015). The "Urban Damage" into the description of the plan's alternative equalization addresses for mitigate the effects. In (a cura di): Alessandra Castellini Lucia Devenuto, *Proceedings of XLIV INCONTRO DI STUDI Ce.S.E.T. Il danno. Elementi giuridici, urbanistici e economico-estimativi* (pp. 189–206). Mantova: Universitas Studiorum S.r.l., Bologna, 27/28 November 2014.

Salvo, F., Ciuna, M., Simonotti, M. (2015). Compensation appraisal processes for completion of hydraulic works in an agricultural area. In (a cura di): *Alessandra Castellini Lucia Devenuto, Proceedings of XLIV INCONTRO DI STUDI Ce.S.E.T. Il danno. Elementi giuridici, urbanistici e economico-estimativi* (pp. 69–82). Mantova: Universitas Studiorum S.r.l., Bologna, 27/28 November 2014.

Salvo, F., De Ruggiero, M., Zupi, M. (2015). The valorization of public real estate. A first outcome of the experiences in progress and a methodological proposal. In *XLIII INCONTRO DI STUDI Ce.S.E.T. Sviluppo economico e nuovi rapporti tra agricoltura, territorio e ambiente. AESTIMUM* (Vol. 67, pp. 135–146). FIRENZE: Firenze University Press, ISSN:1592-6117, Verona, 21/23 november 2013.

Simonotti, M. (1985). La comparazione e il sistema generale di stima. *Rivista di economia agraria, 4*, 543–561.
Simonotti, M. (1998). *La segmentazione del mercato immobiliare per la stima degli immobili urbani*. Roma: Atti del XXVIII Incontro di studio CeSET.
Simonotti, M. (2001). I rapporti estimativi e le funzioni di stima. Estimo e territorio, n. 9.
Simonotti, M., Salvo, F., Ciuna, M. (2015). Appraisal value and assessed value in Italy. *International Journal of Economics and Statistics, 3*, 24–31. ISSN:2309-0685.
Salvo, F., Simonotti, M., Ciuna, M., & De Ruggiero, M. (in print). Measurements of rationality for a scientific approach to the Market Oriented Methods. *Journal of Real Estate Literature*. ISSN:0927-7544.

Emerging Markets Under Basel III: Can Moral Hazard Lead to Systematic Risk and Fragility? Analysis of REIT's in Turkey

Kerem Yavuz Arslanlı and Dilek Pekdemir

Abstract This paper is an attempt to explain Emerging market of Turkey REIT's performance, concerning finance sector reforms and REIT's with respect to Basel III requirements; as government interference in the sector. Findings presents performance comparison of Turkish direct real estate investments and real estate investment companies (REICs) by three property types (residential, retail and office) using risk-adjusted return. Extant literature on performance of direct and indirect real estate has been investigated by different property types and time periods using various methods. The common data used in the studies are appraisal-based (IPD) and transaction-based (NCREIF) indices for direct real estate, while EPRA and FTSE/NAREIT indices for REITs covering the major international markets. In this paper, first REICs are classified regarding their property portfolio by type (residential, retail and office) to compare benchmark direct real estate investment. Since no index (such as NCREIF, INREV) is available for direct real estate investments in Turkey, quarterly return is calculated for direct commercial investments based on transaction indices while for direct residential investments based on valuation indices. Finally, Sharpe Ratio is used to compare the performance of REIC versus direct real estate for each property types. Two different time period is used; first period covers 43 quarters from 2002Q1 to 2012Q3. Second period runs for 11 quarters from 2010Q1 to 2012Q3, which data for all three property types are available. The performance of direct real estate investments in all property types is quite well compared to REICs in Turkey. In other word, direct real estate provides great return for less risk, however REICs provided less return for the same risk. The weak performance of REICs can be attributed to their portfolio allocation. The current asset composition of REICs emphasizes development of

This Paper is a part of Research Project titled "Performance Analysis of Real Estate Investment Companies and Direct Real Estate Investments in Turkey" funded by EPRA.

K.Y. Arslanlı
Istanbul Technical University Institute of Social Sciences, Taksim, İstanbul, Turkey

D. Pekdemir (✉)
Cushman & Wakefield, İstanbul, Turkey
e-mail: dilekpekdemir@eur.cushwake.com

M. d'Amato and T. Kauko (eds.), *Advances in Automated Valuation Modeling*,
Studies in Systems, Decision and Control 86, DOI 10.1007/978-3-319-49746-4_6

their own assets, due to the lack of investment grade property portfolios in Turkey. REICs become a “developer’s vehicle” for construction companies and contractors. They act like “developer” instead of “investor” and also focus on “developer’s profit” instead of “rental income and “capital gains”. Their behaviours indicate unique characteristics of Turkish REICs, and therefore they may call as “Real Estate Development Companies—REDCs”. Among direct real estate investments, residential provide higher return. It can be attributed to the unique residential investment characteristics in Turkey. Residential are considered for sale instead of income producing asset, compared with mature markets. No large companies invest in residential portfolios for leasing purposes, because residential properties provide much more return from capital gains and less income yield. Also considering two different periods in the study the negative effect of the global credit crunch observed especially on commercial markets in Turkey. Both office and retail markets witnessed a slowdown in rentals and property values. Due to low housing loan ratio in Turkish market, the effect was quite limited on residential market and no sharp decline was observed in residential prices.

Keywords Emerging market · Basel III · CAPM · REIC

1 Introduction

The relationship between private (unsecuritized) and public (securitized) real estate markets are well documented in literature by both academics and industry practitioners. Primary reason for this interest has been to assess portfolio allocations between private and public assets and to evaluate the substitutability of these two assets.

Investors and portfolio managers may generally prefer public real estate investment exposure to direct, private real estate investment due to the transparency, liquidity, and, in terms of management, simplicity of listed investment in real estate investment trusts (REITs) (Yunus et al. 2012). Although recent research consistently documented a long run equilibrium relation between public and private returns, the answer is still not conclusive in the literature. The transition from private to public market real estate investment raises the question of whether REITs share the same investment performance characteristics as the underlying direct property. The numerous studies have examined the accurate indicator of the risk and return characteristics of investment vehicles, and have compared performance of direct and indirect real estate by property types. REICs were introduced as a capital market institution in Turkey several years ahead of many developed countries, including Germany, France, UK, Japan, Singapore and Hong Kong. The legal framework for Turkish Real Estate Investment Companies (REICs)—Gayrimenkul Yatırım Ortaklıkları (GYO)—was prepared by the Capital Markets Board (CMBT) in 1995. First REIC was established in 1996 and REICs became publicly listed in Istanbul Stock Exchange (ISE) starting from 1997. Turkish Real

Estate Investment Companies are established in the form of joint-stock corporations and they have a legal personality. They don't have a trust status and are not managed by a board of trustees. REICs may be constituted by establishing new joint stock companies, or existing joint stock companies can convert into REICs by amending their articles of association in accordance with the procedures of the Communiqué and Capital Market Law. For either the establishment or the conversion of a company into an REIC, CMBT approval must be obtained. The company's name must include "real estate investment company". Similarly of REITs around the world, Turkish REICs must deal primarily with portfolio management. In accordance with the Communiqué, the REICs portfolio is required to be diversified based on industry, region and real estate and is to be managed with a long-term investment purpose. It is also required that 75 % of the portfolios of the companies, established with the purpose of operating in certain areas or investing in certain projects, must consist of assets mentioned in their titles and/or articles of association. A REIC must invest at least 50 % of its portfolio value in real estate, rights to real estate and real estate projects. At most, 10 % of its portfolio value may be invested in time deposits or demand deposits. Investments in foreign real estate and capital market instruments regarding may only constitute no more than 49 % of REICs portfolio value. The land and lots in the portfolio of the REIC, on which any project has not been realized for five years as of the acquisition date, may not exceed 10 % of its portfolio value. In order to promote the growth of the Turkish REIC industry, the significant tax incentives have been granted to REICs. Profits generated from the portfolio management activities of REICs are exempt from the general applicable 20 % corporate tax. In addition, although an official exemption has not been granted, the income tax rate has been determined to be "zero" for REICs. Aside from these two incentives, REICs are subject to all other applicable taxes, such as VAT, title deed fee, except stamp duty. An important difference of Turkish REICs from other REITs in the developed economies is that Turkish REICs do not have to pay out dividends to the shareholders on an annual basis. The Turkish real estate market has entered an upward trend, especially from 2004 onwards, following the political stabilization, economic improvements and declining interest rates. As illustrated in Fig. 1 the number of REICs increased in line with these developments, and their portfolios specialized in certain sectors, as well (Pekdemir and Soyuer 2012).

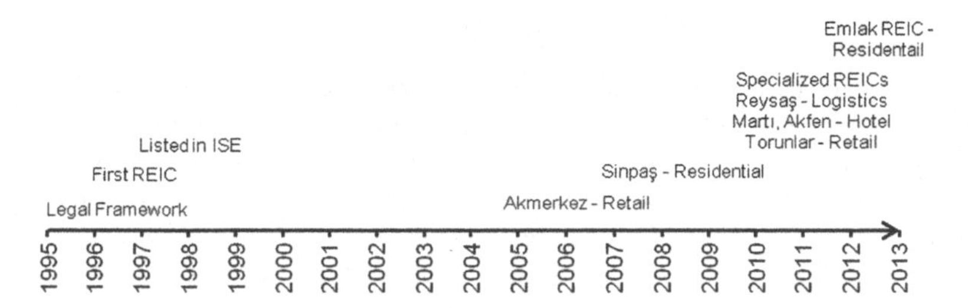

Fig. 1 Historical background of Turkish REICs

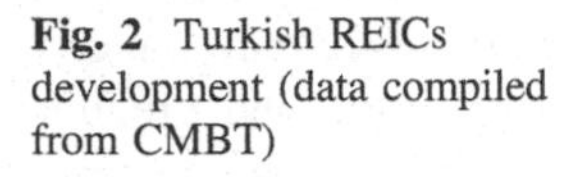

Fig. 2 Turkish REICs development (data compiled from CMBT)

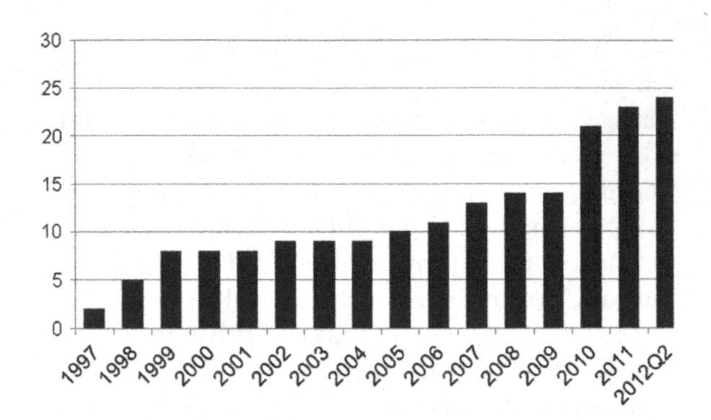

Most Turkish REICs have a particular portfolio of properties, in many cases the REIC management had been operating the properties for a period of time and was, in effect, transferring ownership and refinancing existing assets. As Fig. 2 shows, since 1997 REIC IPOs in Turkey have shown both "hot" and "cold" market waves. The first "hot" market occurred between 1997 and 1999 when 8 REICs came to the market. There then followed a "cold" period until 2007 when very few REICs came to the market, followed by a new "hot" market from 2007 and especially 2010 (Arslanlı et al. 2011).

The REICs portfolio composition, which came to the market in the different stages, is displayed a notable difference (Table 1). The REICs that came to the market in the first "hot" market from 1997–1999, following introduction of the Communiqué, are more evenly spread across on the traditional real estate sectors than the REICs that came later. In contrast, the REICs from the "cold" period are focused on the retail and office sectors. The REICs from the latest "hot" market in comparison are much more divers, with relatively large holdings in hotels, logistics and warehouse properties as well as state owned residential REICs.

Turkish REICs have growth potential, although total REIC market capitalization is relatively small with a share of 3 % of total stock market capitalisation, compared with REITs in developed capital markets. As of the end of 2012Q3, 24 REICs were listed on the ISE with a total net asset value of USD 7.48 billion, while market capitalization was USD 12.59 billion (see Appendix 1, Table 13).

REICs have played an important role in institutionalisation of Turkish real estate market. The legal framework makes REICs more transparent providing reliable and quality information. Furthermore, their structure brought international standards and professionalism to the broader real estate industry and fostered foreign investments in Turkey, especially at the institutional scale (Erol and Tırtıroğlu 2008). Turkish REICs present an alternative investment vehicle for both individual and institutional investors. The aim of this study is to evaluate the performance of direct real estate investments and REICs in Turkey. For this purpose, the market performances of REICs and direct real estate investments by different property types; residential, retail and office, are compared by measuring risk-adjusted

Table 1 Asset allocation of Turkish REICs

Asset types	97–11 (%)	97–99 (%)	00–06 (%)	07–10 (%)	2012Q3 (%)
Land	12	14	0	13	25
Retail	23	13	64	21	16
Office	9	14	14	6	5
Residential	10	8	0	13	9
Hotel and entertainment	5	4	3	6	3
Logistics and warehouse	8	0	0	13	3
Development commercial	4	2	0	6	9
Development residential	9	10	0	12	15
Securities and money market	14	24	20	7	10
Participation	5	11	0	2	5
Total	100	100	100	100	100

performance of return. The rest of the paper is organized as follows. In the following section a literature review is presented to provide evidence on the link between direct and public real estate. In the third section the method and data used to compare risk adjusted return between direct real estate and REICs. Section 4 presents preliminary tests and the empirical results on comparison of risk-adjusted return of REICs and direct real estate. In the final section, concluding remarks are presented.

2 Literature Review

Several papers have investigated the linkage between real estate investments alternatives, by both academics and industry practitioners. Recent studies are concentrated on the linkage between direct real estate and REITs (Tuluca et al. 2000; Pavlov and Wachter 2011; Boudry et al. 2012; Hoesli and Oikarinen 2012; Yunus et al. 2012; Gyamfi-Yeboah et al. 2012). Although no consensus on the relation in the short time, the latest research confirmed that the relation between REITs and direct real estate returns appears to be stronger at longer horizon (Boudry et al. 2012; Hoesli and Oikarinen 2012; Yunus et al. 2012). Most of the studies focus on the U.S market, due to longer time series data availability. In recent years, more papers investigate international markets including European markets (Baum 2006; Daveney et al. 2012; Yunus et al. 2012). Although these studies mainly concentrate on large and mature markets, a few new studies focus on developing markets (Hamzah et al. 2010; Pham 2012). Extant literature on performance of direct and indirect real estate has been investigated by different property types and time periods using various methods. The common data used in the studies are appraisal-based (IPD) and transaction- based (NCREIF) indices for direct real estate, while EPRA and FTSE/NAREIT indices for REITs covering the major international markets. The recent studies indicated that transaction based indices may facilitate a greater understanding of the relationships between direct and listed real estate (Pavlov and Wachter 2010; Boudry et al. 2012; Yunus et al. 2012). Return and risk characteristics of REITs compared to their direct market benchmarks by different property types to allocate capital between these alternatives (Pavlov and Wachter 2010; Boudry et al. 2012; Hoesli and Oikarinen 2012; Gyamfi-Yeboah et al. 2012).

To explore the dynamic relationship between assets, various methods have been utilized in these studies. Vector error correction model (VECM) is one of the common methods (Tuluca et al. 2000; Ling and Naranjo 2003; Boudry et al. 2012; Hoesli and Oikarinen 2012; Gyamfi-Yeboah et al. 2012) and correlation analysis as well (Baum 2006; Boudry et al. 2012). Also recent studies adopted alternative methods, such as spectral and cross-spectral analysis (Daveney et al. 2012) and wavelet method (Zhou 2012) to examine the cyclical attributes of the data or correlations between time series data. The performance of REITs is investigated by analyzing return and volatility dynamics in the literature, based on the capital asset pricing model (CAPM). It was improved to calculate risk-adjusted return in the mid-1960. Three standard performance measurement methods are employed to evaluate the performance of REITs; Treynor and Jensen index consider systematic risk while Sharpe considers total risk of portfolio. As one of the common risk-adjusted index, some researchers used Sharpe ratio not only to measure performance of REITs, but also to compare performance of direct real estate and REITs. Springer and Cheng (2006) use Sharpe ratio to test for the effects on risk and risk-adjusted return of REITs by property level using operational, ownership and financial characteristics. The office and industrial models have the best results, however other three models for apartment, healthcare and hospitality are not overly revealing. In addition, portfolio age and demographic trend measures inconsistently explain risk and risk-adjusted returns.

Baum (2006) examines the performance characteristics of direct and unlisted indirect property markets of the UK, U.S.A, Germany and Netherlands using Sharpe ratio. The result confirms that the performances of direct and unlisted markets are largely similar with the exception of the UK, there is not a significant difference between the ratio of returns to volatility in the direct and unlisted sectors. Chou et al. (2013) analyse the diversification effects of real estate investment trusts in order to give improvement for investment opportunities by additional REIT into the portfolio. Including international REITs for equity portfolios experience significant diversification by improved Sharpe ratio and reduction of overall risk. Hamzah et al. (2010), employ three standard performance measurement methods (Sharpe Treynor and Jensen Indexes) on the performance of REITs on pre-crisis, during-crisis and post-crisis. REITs are found to outperform market portfolio during-crisis but underperform in the pre and post-crisis periods. The average systematic risks of REITs are slightly higher than market portfolio during pre-crisis but lower in the post crisis periods. The studies on Turkish REICs are very limited, although REICs legal framework was prepared in 1995. These studies concentrate on characteristics of REICs, relationship with capital markets and also return performance regarding their unique structure. Despite certain number of studies, research on REICs performance and underlying reasons are incomplete. Erol and Tırtıroğlu (2008) examine the Turkish REICs inflation hedging abilities over a period of 1999–2004. Different from developed capital markets, Turkish REICs tax incentives and flexibility in managing portfolios result to provide better hedge against actual and expected inflation than the ISE common indices. Study provides good results to test for hedging behaviour of real estate stocks on high and moderate inflation rates. Authors report that number of the REICs is not adequate to analyze into sub-sectors such as hotels or apartments in the study. High inflation rate period is performed better hedging abilities than moderate inflation period. Another finding of the study is strong inflation hedging performance of REICs influenced by idiosyncratic risk on the ISE REIT price index. Results suggest that REIC managers form accurate inflation expectations and write lease contracts for office and shopping malls with clauses for rapid rent adjustments. However, these results could be criticized that commercial rentals are quoted in Dollar or Euro as a common practice in Turkey against high inflationary pressure on the market. Aktan and Öztürk (2009) investigate the risk-return relationship of Turkish REICs within the framework of modern portfolio theory (MPT) using the standard version of the capital asset-pricing model (CAPM) and the single index model (SIM) over the period 2002–2008. Results indicate that linearity assumption for both the CAPM and the SIM are rejected. The coefficient of ex-post beta has negative explanatory power on average asset returns that is contradictory with the fundamental relationships between risk and return under the framework of the CAPM and the SIM.

Erol and Tırtıroğlu (2010) analyze capital structure of Turkish REICs, where they don't have to pay dividends and exempt from corporate and income taxes. Findings are revealed that Turkish REICs employ low long term debt in their capital structure. Turkish REICs tend to reduce short-term debt and also not to borrow in the long-term market. Thus, they appear to use inexpensive internal equity resources only for their short-term financing needs. The firm size, REICs engagement in development and stock market development have influence debt ratios positively where tangibility, ownership and country specific determinants appear to have no influence. Altınsoy et al. (2010) investigate time varying behaviour of beta for Turkish REICs from 2002 to 2009. Findings of the study are similar to other emerging and developed REITs markets that Turkish REICs have a declining beta. Empirical results suggest that REICs return more closely track stock market in high growth economic conditions. Türkmen and Demirel (2012) analyze the effects of macroeconomic conditions and financial ratios on performance of REICs. Results indicate that net profit after tax/equity ratio of REIC has significant effect by Dollar/Turkish Lira currency volatility, however consumer price ratio, benchmark interest rates and Euro/Turkish Lira currency do not have significant effects.

3 Data and Methodology

This study measures the performance of direct real estate investment and real estate investment companies by property types, using risk-adjustment return. The main stages are given as below to calculate return and to compare performance between direct real estate investment and real estate investment companies:

- Classification of REICs regarding their property portfolio by type (residential, retail and office) and selection of benchmark REICs for each property type,
- REICs return calculation; quarterly return of the selected REICs by property type is calculated based on ISE monthly return data,
- Direct real estate return calculation; quarterly return of commercial properties is calculated by using DTZ Pamir and Soyuer's investment transaction database (ITD) and property market indicators (PMI), while quarterly return of residential properties is calculated by using DTZ Pamir and Soyuer's property market indicators (PMI) and Central Bank of Turkish Republic (CBTR) new housing price index (TNHPI),
- Performance analysis of REICs and direct real estate by property types (residential, retail and office); the Sharpe Ratio is used in comparing the performance of REIC versus direct real estate for each property type.

4 Data Sources

Turkish REICs became publicly listed in ISE (re-named as Borsa İstanbul-BIST) since 1997 and monthly return data is available for all REICs and also for REIC sector index (XGMYO) since 2000. However, limited data is available for the historical return of unlisted or direct real estate vehicles in Turkey. No public data or indices (such as NCREIF, INREV) is available for direct real estate investments. Therefore, the available data by different source is used to calculate direct real estate return, which is described in the next section. Due to direct real estate data is available since 2002Q1 and also the latest available REIC return data is 2012Q3, all quarterly return series is obtained from 2002Q1 to 2012Q3. Similarly, residential price index is available since 2010, therefore direct residential real estate return series is obtained from 2010Q1 to 2012Q3. The data sources and measurement are summarized in Table 2.

Table 2 Data source and measurement

	Source	Measurement
REICs	ISE return index	Quarterly total return (2002Q1–2012Q3)[a]
Direct RE		
Office	DTZ Pamir and Soyuer investment transaction database (ITD) and property market indicators (PMI)	Quarterly total return using transaction based index (2002Q1–2012Q3)[b]
Retail	DTZ Pamir and Soyuer investment transaction database (ITD) and property market indicators (PMI)	Quarterly total return using transaction based index (2002Q1–2012Q3)[b]
Residential	Central Bank of Turkish Republic (CBTR) new housing price index (TNHPI) and DTZ Pamir and Soyuer property market indicators (PMI)	Quarterly total return using valuation based index (2010Q1–2012Q3)[c]
Risk-free rate	Central Bank of Turkish Republic (CBTR)	91-day government bond

Note [a]The latest available data cover 2012Q3
[b]DTZ Pamir and Soyuer ITD and PMI database is available since 2002Q1
[c]TNHPI is available since 2010

5 Return Calculation

Past researches have often used valuation based indices for the direct real estate market, but an extensive literature exists that highlights problems with such indices. These are criticised with regard to their perceived smoothing and lagging of market performance, due to issues in the valuation process, such as the availability of sales evidence, using past evidences when conducting a new valuation (Daveney et al. 2012). Transaction based indices for direct real estate markets have been researched for many years, but the most widely known is developed for the US by the MIT Centre for Real Estate in collaboration with National Council of Real Estate Index Fiduciaries (NCREIF). Since such indices are not available for direct real estate market in Turkey, total return is calculated using different sources; for direct commercial investments based on transaction indices while for direct residential investments based on valuation indices. In real estate investment industry, total return (r) accounts for two components of return: income (y) and capital appreciation (g). Income return component is more directly relevant to the income objective of investors, while the appreciation return component is more directly relevant to the growth objectives (Geltner et al. 2001). Total return is calculated as given below:

$$r_t = y_{t-1} + g_t$$

For retail and office properties, PMI data is used for income return (y), while ITD is used for appreciation return (g) based on DTZ Pamir and Soyuer database which is available since 2002Q1.

For residential properties, DTZ Pamir and Soyuer PMI database is used for income return (y), while CBRT new housing price index (TNHPI) is used for appreciation return (g). TNHPI is available since 2010Q1, therefore total return series of residential property is obtained for the period 2010Q1 to 2012Q3.

6 Performance Measure: The Sharpe Ratio

The Sharpe ratio is a risk-adjusted measure of return that is often used to evaluate the performance of a portfolio. The ratio helps to compare the performance of different portfolios by making an adjustment for risk. Since the Sharpe ratio was derived in 1966 by William Sharpe, it has been one of the most referenced risk/return measures used in finance, and much of this popularity can be attributed to its simplicity. William Forsyth Sharpe developed what is now known as the Sharpe ratio in 1966. Sharpe originally called it the "reward-to-variability" ratio (Sharpe 1966) before it began being called the Sharpe ratio by later academics and financial operators. Sharpe's 1994 revision acknowledged that the basis of comparison should be an applicable benchmark, such as the risk free rate of return or an index (S&P 500, etc.), which changes with time (Sharpe 1994). It is broken down into just three components: asset return, risk-free return and standard deviation of return. After calculating the excess return, it's divided by the standard deviation of the risky asset to get its Sharpe ratio. The ratio describes how much excess return you are receiving for the extra volatility of holding the risky asset over a risk-free asset. The Sharpe ratio is expressed mathematically as;

$$S(x) = \frac{r_x - R_f}{StdDev_{(x)}}$$

where;

X	is the investment
r_x	is the average rate of return of x
Rf	is the best available rate of return of a risk-free security (i.e. T-bills)
StdDex (x)	is the standard deviation of r_x.

The return (r_x) measured can be of any frequency (i.e. daily, weekly, monthly), as long as they are normally distributed, as the returns can always be annualized. However, because it is based on the mean-variance theory, it is valid only for either normally distributed returns or quadratic preferences. If returns are not normally distributed, the Sharpe ratio can lead to misleading conclusions and unsatisfactory paradoxes (Zakamulin and Koekebakker 2008; Ziemba 2005; Hodges 1998). Abnormalities like kurtosis, fatter tails and higher peaks, or skewness on the distribution can be a problematic for the ratio, as standard deviation doesn't have the same effectiveness, when these problems exist.

The risk-free rate of return (Rf) is used to see if you are being properly compensated for the additional risk you are taking on with the risky asset. Traditionally, the risk-free rate of return is the shortest dated government T-bill (i.e. U.S. T-Bill). In this study, 91-day Central Bank of Turkish Republic (CBTR) Bond return index is used to calculate risk free rate. While the Treynor ratio works only with systemic risk of a portfolio, the Sharpe ratio observes both systemic and idiosyncratic risks. The Sharpe ratio characterizes how well the return of an asset compensates the investor for the risk taken. When comparing two assets versus a common benchmark, the one with a higher Sharpe ratio provides better return for the same risk (or, equivalently, the same return for lower risk). In general, a higher number is better, since the higher number indicates a greater return for less risk.

7 Classification of REICs

24 REICs were listed on the ISE, as end of 2012. We focus on only office, retail and residential property types which are invested predominantly in one property type and other types (hotel, industrial) are excluded from the analysis. Of those we chose REICs that are 35 % or more invested in one property type and the remaining shares comprise of mostly land and/or on-going development projects. Regarding their predominant property types, REICs are classified and 11 REICs are selected, given as in Table 3.

Preliminary tests are applied for further analysis to compare return performance of REICs and direct real estate. The detailed information is given in the Sect. 9. Preliminary Tests.

8 Return Calculation

The total return of direct real estate and REICs are computed as explained in the previous section. Two different periods are used to calculate for direct real estate return regarding data availability. First period cover 43 quarters from 2002Q1 to 2012Q3, which retail and office return data is available. Second period runs for 11

Table 3 Classification of REICs

Property type	REICs
Residential	EKGYO, IDGYO, SNGYO,
Retail	AGYO, AKMGYO, PEGYO, TRGYO, TSGYO
Office	ISGYO, NUGYO, VKGYO

Note Abbreviations are explained in Appendix 1 Table 13

Table 4 Direct real estate and REIC return and standard deviation

	Direct real estate		REICs		
	Average return	Standard deviation		Average return	Standard deviation
2002–2012			XGMYO	0.046	0.201
2010–2012				0.039	0.150
Office			ISGYO	0.047	0.211
2002–2012	0.123	0.067	NUGYO	0.122	0.421
2010–2012	0.119	0.052	VKGYO	0.098	0.345
Retail			AGYO	0.063	0.239
2002–2012	0.110	0.076	AKMGY	0.052	0.260
2010–2012	0.113	0.063	PEGYO	0.051	0.462
			TSGYO	−0.033	0.120
			TRGYO	0.002	0.194
Residential			SNGYO	0.009	0.203
2010–2012	0.152	0.049	EKGYO	0.056	0.194
			IDGYO	0.344	0.997

quarters from 2010Q1 to 2012Q3, which is governed by availability of the residential return. REIC return series is computed for the time period which all series are available for both REIC and the related direct real estate type. In addition, REIC sector return (XGMYO) data is also included into the analysis. Return data is summarized in Table 4 and also historical returns for all property types are given in Fig. 3.

Regarding properties types, similar results are obtained for both direct real estate and REICs. In terms of average return, residential REICs and also direct residential properties, performed better compared to office and retail REICs, they provided higher return with less volatility. Direct office and retail properties have quite similar average return, 12.3 and 11 %, respectively. However, return of office REICs are better than retail REICs, ranged between 4.7 and 12.2 %, while return of retail REICs ranged between −3.3 and 6.3 %. As mentioned above, second return data series covering 2010Q1–2012Q3 period is calculated to obtain a comparable return series for all property types. Based on this period, residential properties provided the highest average return at 15.2 %, while office and retail assets provided relatively lower average returns, which were 11.9 and 11.3 %, respectively. For the same period, REIC sector average (XGMYO) was only 3.88 %. It should be noted that real estate market has started to recover in 2010, following the negative effect of the global credit crunch. Therefore, residential return data is not included the effect of the global crisis and declining residential price.

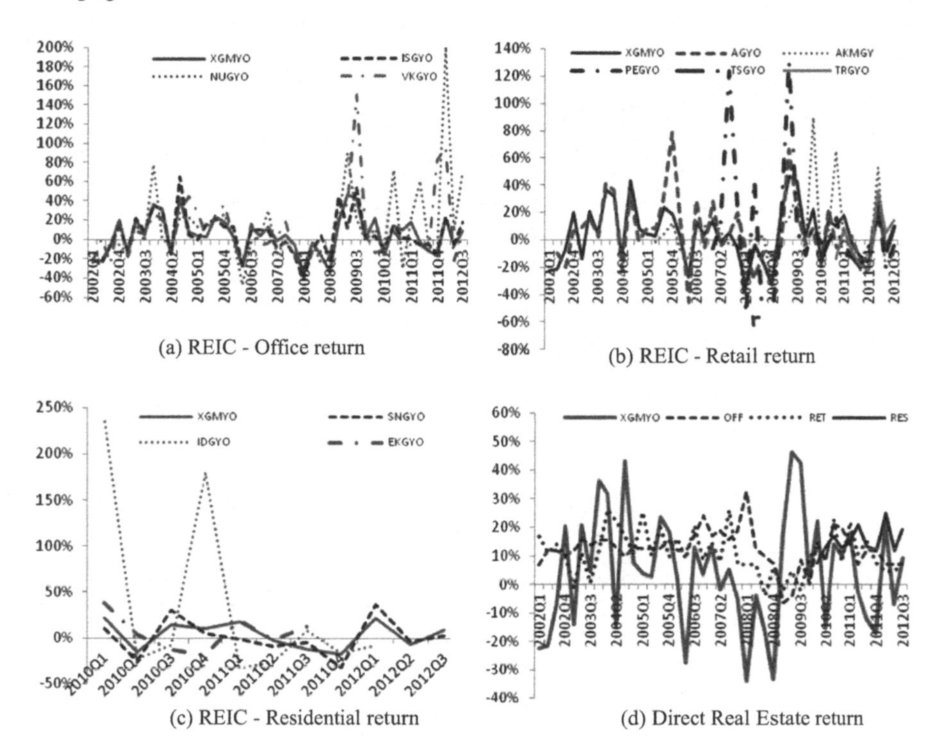

Fig. 3 Historical return of direct real estate and REIC

9 Preliminary Tests

Selected 11 real estate investment companies are analysed according to quarterly returns. In order to understand the characteristics of real estate investment companies performance, the distribution of returns analysed for normal distribution tests. In Table 5 shows the distribution of total rates of returns in quarterly and ranges from minimum −64 % to maximum 234 % with an average of 7 % return.

After analysis of normality in the returns of REICs for Sharpe ratio calculations EKGYO, SNGYO, TRGYO, TSGYO, ISGYO passed all normality tests. AGYO with a slightly on edge of rejection of normality also had taken into calculations which less than 1.14 skewness. In general the distribution of returns being skewed toward positive returns with 1.07 and degree of kurtosis of 2.01. Normal distribution test of Shapiro-Wilk and Kolomogorov-Smirnov detect problems in NUGYO, VKGYO, AKMGY, PEGYO, IDGYO. Consequently data analysis results in excluding some of them in further stages of model. Box-plot graph shows that returns of residential REICs are homogeneous and clustered around the median, office returns are more dispersed and retail has highest IQR (Fig. 4). These

Table 5 Descriptive analysis result of REICs

Stock	N	Min	Max	Mean	Std dev.	Skewness	Kurtosis	K-S	Sig	S-W	Sig
XGMYO	43	−0.339	0.464	0.046	0.201	0.113	−0.406	0.058	0.200[a]	0.983	0.762[a]
ISGYO	43	−0.407	0.646	0.047	0.211	0.615	0.950	0.092	0.200[a]	0.972	0.375[a]
NUGYO	43	−0.471	2.003	0.122	0.421	2.440	8.760	0.212	0.000	0.786	0.000
VKGYO	43	−0.325	1.506	0.098	0.345	2.155	6.310	0.164	0.005	0.811	0.000
AGYO	42	−0.460	0.783	0.063	0.239	0.819	1.528	0.120	0.134	0.954	0.090[a]
AKMGY	30	−0.331	0.895	0.052	0.260	1.692	3.492	0.209	0.002	0.840	0.000
PEGYO	23	−0.642	1.308	0.051	0.462	1.480	3.074	0.192	0.027	0.839	0.002
TSGYO	10	−0.200	0.183	−0.033	0.120	0.484	−0.472	0.127	0.200[a]	0.971	0.899[a]
TRGYO	8	−0.219	0.355	0.002	0.194	0.729	−0.063	0.180	0.200[a]	0.935	0.566[a]
SNGYO	11	−0.317	0.364	0.009	0.203	0.251	0.089	0.146	0.200[a]	0.952	0.666[a]
EKGYO	7	−0.188	0.379	0.056	0.194	0.492	−0.220	0.125	0.200[a]	0.975	0.930[a]
IDGYO	9	−0.334	2.346	0.344	0.997	1.638	1.160	0.381	0.000	0.673	0.001
Mean	26	−0.353	0.936	0.071	0.321	1.076	2.017	0.167	0.114	0.891	0.358[a]

Note [a]Normal distribution under null for Kolmogorov-Smirnov and Shapiro-Wilk tests. Normality assumption limit for skewness is +1.14 and kurtosis +3

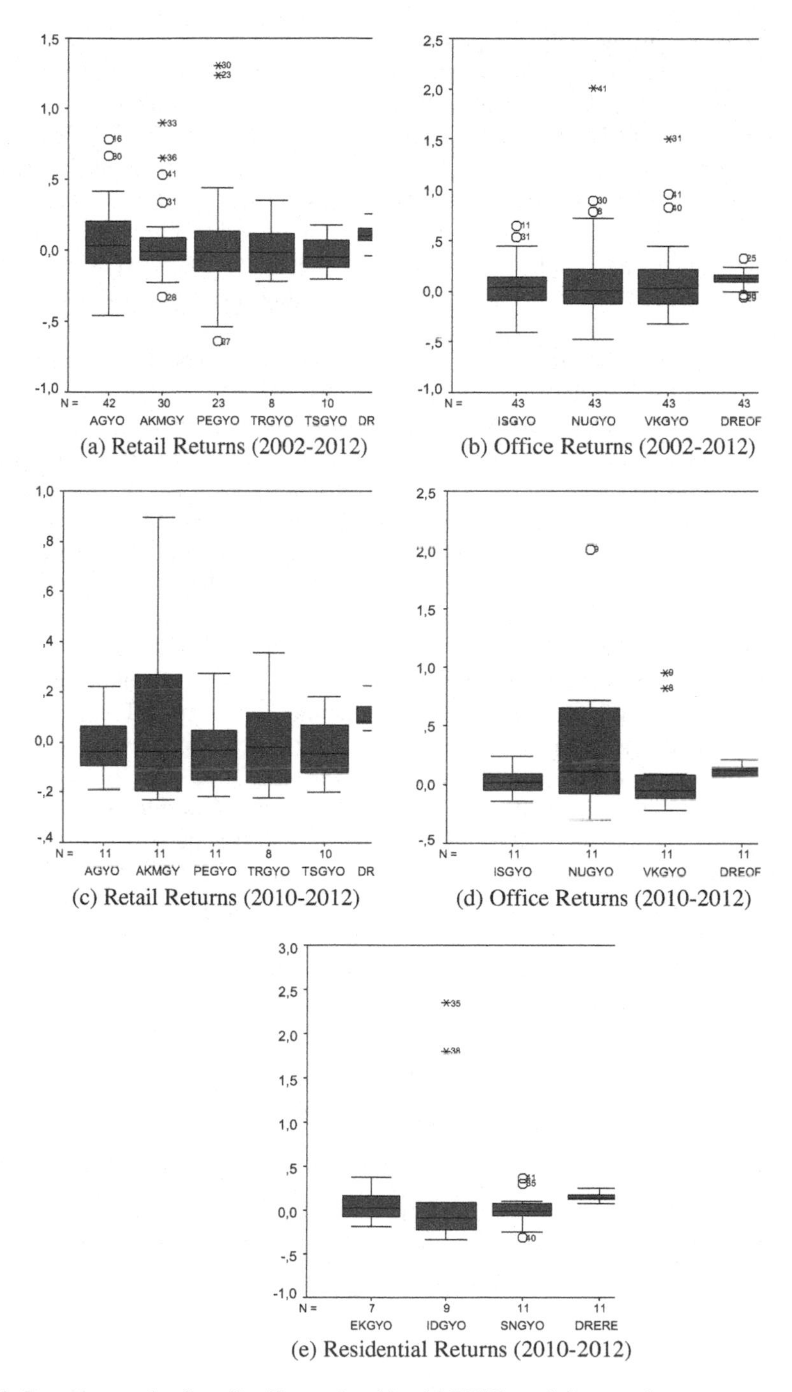

Fig. 4 Box-plot graph of retail, office and residential REICs and direct real estate returns

observations are in line with the general assumption that higher the variability in returns the riskier the REICs where retail displays higher volatility. Furthermore the outlier percentage is different from general assumptions that office REICs also has as much as the retail REICs. Some of the listed REICs traded in ISE with small real estate portfolio left outside of the analysis. Expected pattern of retail oriented REICs more volatile returns than office REICs and residential, particularly. Table 6 and Fig. 4 confirm mean and median total returns of residential are considerably lower than office and retail. Furthermore for the all property types of returns, standard deviation is the lowest for residential over time and higher for retail market returns.

Distribution of direct real estate returns in normality test for skewness and kurtosis passed for 1.14 and ±3. Shapiro-Wilk test passed for retail and residential with high level of significance. Where office distribution found to be problematic but skewness and kurtosis levels are inside confidence limits. For better comparison of different property type returns, time-span of quarterly data is divided into two sets. First set of data covering 2002–2012 and second is 2010–2012 which based on residential return availability. For the second set starting from 2010 gives more clear comparison for all property types, where the Turkish REICs for residential shows better performance compared to other property types. This type of unusual/unexpected behaviour where risk is low and return is higher could be described as aggressive pricing competition of REICs for residential properties. Volatility of retail and office properties are in line with expected returns performance. Correlation analysis is conduct to examine relationship between REICs and the benchmark direct real estate by property types and time periods. As given in Table 7, direct retail investment and retail REICs returns found to be uncorrelated and insignificant for both 2002–2012 and 2010–2012 periods. Direct office investment and office REIC returns are also found insignificant which correlation coefficients diverse from −0.259 to 0.061 between two time periods (Table 8). Only residential direct returns found to be correlated and significant with residential REICs (Table 9). This can be interpreted as residential REICs behave like direct real estate residential vehicle.

Table 6 Descriptive analysis result of direct real estate

	N	Min	Max	Mean	St. dev	Skewness	Kurtosis	Kol-S	Sig.	Shap-W	Sig.
Office	43	−0.063	0.325	0.123	0.067	−0.150	2.576	0.132	0.057	0.934	0.017*
Retail	43	−0.038	0.262	0.110	0.076	0.216	−0.299	0.130	0.064	0.965	0.213*
Res	11	0.075	0.252	0.152	0.050	0.681	0.314	0.202	0.200	0.936	0.480*

Table 7 Correlations of retail REICs returns to direct retail returns

2002–2012						
	X100	XGMYO	DRERET	AGYO	TRGYO	TSGYO
X100	1.000	0.868	0.030	0.737	0.899	0.626
XGMYO	0.868	1.000	0.141	0.731	0.772	0.789
DRERET	0.030	0.141	1.000	0.148	−0.210	−0.241
AGYO	0.737	0.731	0.148	1.000	0.888	0.689
TRGYO	0.899	0.772	−0.210	0.888	1.000	0.707
TSGYO	0.626	0.789	−0.241	0.689	0.707	1.000
2010–2012						
	X100	XGMYO	DRERET	AGYO	TRGYO	TSGYO
X100	1.000	0.715	0.099	0.904	0.899	0.626
XGMYO	0.715	1.000	0.312	0.785	0.772	0.789
DRERET	0.099	0.312	1.000	0.177	−0.210	−0.241
AGYO	0.904	0.785	0.177	1.000	0.888	0.689
TRGYO	0.899	0.772	−0.210	0.888	1.000	0.707
TSGYO	0.626	0.789	−0.241	0.689	0.707	1.000

Table 8 Correlations of office REICs returns to direct office returns 2002–2012

2002–2012				
	X100	XGMYO	DREOFF	ISGYO
X100	1.000	0.868	−0.161	0.734
XGMYO	0.868	1.000	−0.170	0.868
DREOFF	−0.161	−0.170	1.000	−0.259
ISGYO	0.734	0.868	−0.259	1.000
2010–2012				
	X100	XGMYO	DREOFF	ISGYO
X100	1.000	0.715	0.098	0.884
XGMYO	0.715	1.000	0.342	0.793
DREOFF	0.098	0.342	1.000	0.061
ISGYO	0.884	0.793	0.061	1.000

Table 9 Correlations of residential REICs returns to direct returns 2010–2012

	X100	XGMYO	DRERES	EKGYO	SNGYO
X100	1.000	0.715	0.513	0.535	0.923
XGMYO	0.715	1.000	0.254	0.932	0.821
DRERES	0.513	0.254	1.000	0.436	0.456
EKGYO	0.535	0.932	0.436	1.000	0.590
SNGYO	0.923	0.821	0.456	0.590	1.000

Table 10 The Sharpe ratio for direct real estate by asset types

	2002Q1–2012Q3			2010Q1–2012Q3		
	Mean	St. dev	Sharpe ratio	Mean	St. dev	Sharpe ratio
XGMYO	−0.01	0.208	−0.03	0.02	0.149	0.12
Office	0.07	0.078	0.90	0.10	0.054	1.82
Retail	0.06	0.086	0.68	0.09	0.065	1.43
Residential	–	–	–	0.13	0.048	2.75

10 Performance Analysis: Sharpe Ratio

The Sharpe Ratio, which is a risk-adjusted measure of return, is used to compare the performance of REICs versus direct real estate for each property types; residential, retail and office. The Sharpe ratio of direct real estate by property types is given in Table 10. Two different time period is used to compare across property types. First period cover 43 quarters from 2002Q1 to 2012Q3. Second period runs for 11 quarters from 2010Q1 to 2012Q3, which data for all three property types are available.

As stated in earlier section, higher Sharpe ratio indicates a greater return for less risk. According to results, residential properties provided higher return compared to office and retail properties during 2010Q1–2012Q3 period. Office properties provided relatively higher return compared to retail properties, although both of them performed well for two different periods. Besides, all three property types performed better compared to REIC sector average. Considering these two periods, it should be noted that real estate market started to recover in the beginning of 2010, after global credit crunch, which had a negative effect especially on commercial markets in Turkey. Both office and retail markets witnessed a slowdown in rentals and property values (DTZ Pamir and Soyuer 2010). Due to low housing loan ratio in Turkish market, the effect was quite limited on residential market and no sharp decline was observed in residential prices. Sharpe ratio is improved for office and retail sector for recovery period. It increased from 0.90 to 1.82 for office, while it increased from 0.68 to 1.43 for retail. To compare direct real estate and REIC performance by property types, the Sharpe ratios are computed for the same time period which both direct real estate and REIC data is available. The period start time and Sharpe ratio calculation are summarized in Table 11.

Table 11 The Sharpe ratio for direct real estate and REICs by property types

		Direct real estate			REICs			
	Period[1]	Mean	St. dev	Sharpe		Mean	St. dev	Sharpe
Office	2002Q1	0.07	0.078	0.90	ISGYO	−0.01	0.217	−0.03
Retail	2002Q2	0.06	0.086	0.69	AGYO	0.01	0.241	0.05
	2010Q2	0.09	0.068	1.33	TSGYO	−0.05	0.118	−0.46
	2010Q4	0.08	0.062	1.27	TRGYO	−0.03	0.123	−0.21
Residential	2010Q1	0.13	0.048	2.75	SNGYO	−0.01	0.201	−0.06
	2011Q1	0.15	0.049	2.99	EKGYO	0.04	0.194	0.18

Table 12 Turkish REICs characteristics

Characteristics	ISGYO	AGYO	TSGYO	TRGYO	EKGYO	SNGYO
Investor type	Developer/investor	Developer/investor	Developer/investor	Developer/investor	Developer	Developer
Investment strategy	Develop/hold	Develop/hold	Develop/hold	Develop/hold	Develop/sell	Develop/sell
Investment product	Income producing assets	Income producing assets	Income producing assets	Income producing assets	Residential sales	Residential sales
Core properties	Office	Retail	Retail	Retail	Residential	Residential
IPO date	1999	2002	2010	2010	2010	2007
Core business	Banking	Investment	Banking	Construc. company	Governm. institution	Construc. company

Direct real estate properties displayed quite well performance for all property types, compared to REICs performance for the same periods. The major characteristics of REICs may help to understand underlying reasons, which are summarized in Table 12. Turkish REICs can be categorized regarding their investor type, investment product and strategy (Pekdemir and Soyuer 2012). Among the selected REICs in this study, office and retail REICs focus on income producing property. They have conservative strategies and take less risk. In contrast, residential REICs are focused on residential sale, which force them to apply aggressive strategies with higher risk (Pekdemir 2013).

For office property, direct office properties provided good performance for 2002Q1–2012Q3 period. However, the negative Sharpe ratio for office REIC (ISGYO) indicates that its return is lower than the risk-free rate of return. As pointed above, ISGYO is a "developer/investor" REIC and has a "develop-hold" strategy. As one of the early period REICs, ISGYO developed its properties, mainly due to lack of investment grade product in Turkey. Considering long time period, the negative effect of the both Turkish banking crisis and also global credit crunch can be observed on its return. For retail properties, performance of direct retail properties is much better in all periods against retail REICs. Among retail REICs, only AGYO has positive Sharpe ratio, which means it generated an excess return for the holding period, although at a lower rate. The rest of them, TSGYO and TRGYO have negative Sharpe ratio. It can be explained that short time series data of TSGYO and TRGYO, which is not enough to explain risk adjusted return for the short term. Like office REIC, all selected retail REICs are also "developer/investor", and their returns include "developer risk". Compared to AGYO, the portfolio age of TSGYO and TRGYO are quite young and their portfolios have more development risk than AGYO. For residential properties, direct residential properties performed very well against residential REICs. Among residential REICs, only EKGYO can be able to produce excess return, although at a lower rate. It can be attributed to different characteristics and strategies of two companies. EKGYO is a government-oriented company and is the largest REIC, representing almost half of the total REIC market cap and NAV.

11 Concluding Remarks

As the government guarantees to subsidize on infrastructure REITs; it might generate a risk of moral hazard in future as stated for other emerging markets. The risk of Basel III requirements for emerging markets, as banks already hold most of their capital in form of equity and not ventured in hybrid forms of capital and transition directly to the more robust definition as Basel III makes the most sense for emerging countries. For construction and real estate industry; "high volatility commercial real estate" financing institutions will change construction loan lending which has to be lower Loan-to-Value ratios, increased down payment equity, and shorter durations and other factors. The changes may give an outcome of moving commercial mortgage lending out of banking system to a commercial mortgage backed securities market or another unregulated market.

The performance of the direct real estate investments and REICs by different property types, office, retail and residential, is analyzed measuring risk-adjusted return. The results confirmed that the performance of direct real estate investments in all property types is quite well compared to REICs in the examining periods. In other word, direct real estate provides great return for less risk in Turkish market, however REICs provided less return for the same risk.

The analysis conducts for two different periods, 2002–2012 and 2010–2012. Results reveal that direct real estate investments performed better than REICs even in the long term. In contrast with international literature (Boudry et al. 2012; Hoesli and Oikarinen 2012; Yunus et al. 2012), no evidence is found that they can be substituted each other in portfolio allocation.

The weak performance of REICs can be attributed to their portfolio allocation. The current asset composition of Turkish REICs emphasizes development of their own assets, due to the lack of investment grade property portfolios in Turkey. REICs become a "developer's vehicle" for construction companies and contractors. They act like "developer" instead of "investor" and also focus on "developer's profit" instead of "rental income and "capital gains". Their behaviours indicate unique characteristics of Turkish REICs, and therefore they may call as "Real Estate Development Companies—REDCs" (Pekdemir 2013).

In fact, the lack of investment grade product is a chronic problem in Turkish real estate market, both for direct investment and REICs. The investment grade office portfolio is limited with Grade A office properties in the primary areas. Besides, office investors also have started to adopt "develop and sell" strategy to take the advantage of creating source for finance by pre-sale of individual office units in the last a few years. Retail market looks more promising, due to high quality product availability and increasing interest of international investors that are looking for opportunities in emerging markets.

Among direct real estate investments, residential properties provide higher return compared to commercial properties. It should be noted a unique investment characteristic in Turkey, residential assets are considered for sale instead of income producing asset, compared with mature markets. No large companies invest in residential portfolios for leasing purposes, because residential properties provide much more return from capital gains and less income yield. Therefore, both individual and corporate investors prefer to take advantage of capital gain by residential sale. The aforementioned characteristics of Turkish REICs may be attributed to the three major reasons; the lack of investment product, difficulties in financing and benefiting tax advantage of REIC structure (Pekdemir 2013). First, the lack of investment grade products becomes a chronic problem for especially corporate investors. For this reason, REICs prefer to develop their own assets which force them to act like a developer instead of investor. Second reason is difficulties to provide financing. Especially residential properties are able to create source for finance by pre-sale of residential units. Thus, developer/investor can finance the remaining developments or buy land for new projects. Third, REICs provide tax advantage and especially construction companies can benefit from this advantage. The major challenge to conduct this study is data availability, because no return indices for direct real estate investment are available for Turkish real estate market. The total return is calculated using transaction based data series for capital appreciation and income return components. This is the first attempt to use such transaction based series which is well documented in the international literature (Boudry et al. 2012; Hoesli and Oikarinen 2012; Daveney et al. 2012). Although the results of this study should be viewed with caution due to data availability, they still provide strong evidence on good performance of direct real estate investments against REICs, for both residential and commercial properties. The performance of the direct and indirect real estate investments are analyzed by only risk-adjusted return. Other factors, which may help to measure portfolio performance of the investments, should be incorporated into the analysis and also more sophisticated models should be used in the further studies. Besides, other specialized REICs, hotel and industrial, will be included in the analysis in the forthcoming research, when adequate time series data will be available. Finally, the size of REICs and also investment markets in Turkey is quite low compared to mature markets. However, this study may provide a contribution to international literature by analyzing an emerging real estate market. For further steps, more research is required to understand market characteristics and underlying reasons.

Appendix 1

See Table 13.

Table 13 Market capitalization and NAV, 2012Q3 (CMBT)

	REIC	NAV (million USD)	Market cap (million USD)	Premium/discount (%)
AFGYO	AKFEN	611.3	164.3	−73.1
AKMGY	AKMERKEZ	527.1	384.4	308.8
ALGYO	ALARKO	128.1	110.4	−13.8
AGYO	ATAKULE	126.6	54.8	−56.7
AVGYO	AVRASYA	44.8	22.5	−49.8
DGGYO	DOĞUŞ-GE	110.9	82.7	−25.5
EGYO	EGS	75.8	6.1	−91.9
EKGYO	EMLAK KONUT	4266.5	3,459.3	−18.9
IDGYO	İDEALİST	5.2	16.2	213.4
ISGYO	İŞ	725.7	448.6	−38.2
KLGYO	KİLER	265.3	121.1	−54.4
MRGYO	MARTI	119.9	32.5	−72.9
NUGYO	NUROL	276.5	184.6	−33.3
OZGYO	ÖZAK	450.5	197.1	−56.2
OZKGY	ÖZDERİCİ	72.9	42.4	−41.8

(continued)

Table 13 (continued)

	REIC	NAV (million USD)	Market cap (million USD)	Premium/discount (%)
PEGYO	PERA	122.4	27.9	−77.3
RYGYO	REYSAŞ	245.7	79.6	−67.6
SAFGY	SAF	402.9	450.2	11.7
SNGYO	SİNPAŞ	1,087.7	425.2	−60.9
TRGYO	TORUNLAR	2,372.1	783.9	−67.0
TSGYO	TSKB	193.9	57.8	−70.2
VKGYO	VAKIF	107.6	212.7	97.6
YGYO	YEŞİL	619.5	89.2	−85.6
YKGYO	YAPI KREDİ KORAY	62.7	28.8	−54.1

Appendix 2

Endnotes

1. BIST Companies Monthly Price and Return Data: Monthly and compounded returns of equities were calculated by using the closing prices on the last trading day of each month. Compounded returns are calculated with the following assumptions:

    - the dividend received during the month is reinvested to buy back the concerning equity at the closing price at the end of the month,
    - pre-emptive rights are exercised in case the price of the equity exceeds its subscription price.

    The prices of new shares (shares that are not entitled to dividends from previous year's net profits) were not taken into account in the calculation of returns.
    The abbreviations used in the tables and the definitions of the terms are indicated below.
    HAF/İİF: Initial public offering or first trading price.
    PRICE: The closing price of a equity with a nominal value of TL 1,000/TRY 1 on the last trading day of the month unless stated otherwise. If the equity is not traded during the month, it is the last closing price of the equity.
    MONTHLY RETURN: The monthly return of a equity is calculated according to the following formula. US Dollar based monthly returns are calculated by adjusting the TL/TRY based returns according to monthly devaluation rate of US Dollar.

$$G_i = \frac{F_i \cdot (BDL + BDZ + 1) - R \cdot BDL + T - F_{i-1}}{F_{i-1}}$$

G_i	Return for the month "i"
F_i	The closing price the equity on the last trading day of the month "i"
BDL	The number of rights issues received during the month
BDZ	The number of bonus issues received during the month
R	The price for exercising rights (i.e. subscription price)
T	The amount of net dividends received during the month for a equity with a nominal value of TL 1,000/TRY 1
F_{i-1}	The closing price of a equity on the last trading day of the month "i − 1".

Compounded Return: This shows the value of a equity that is sold and bought at the end of each month relative to its value at the beginning period and is calculated according to the following formula. In calculation of US Dollar based compounded returns, US Dollar based monthly returns are used.

$$BG_n = (1 + G_1)(1 + G_2)\ldots(1 + G_n) = \prod_{i=1}^{n} 1 + G_i$$

BG_n	The compounded return for the month "n"
BG_i	The compounded return for the month "i"
G_i	Return for the month "i"
n	The number of periods (months)
Source	http://borsaistanbul.com/en/data/data/price-and-return-data.

2. The HPI, which covers the whole country, is constructed for the purpose of monitoring price movements in the Turkish housing market. Price data related to all houses subject to sale, regardless of the construction year are used to develop the HPI. In the housing market, as the prices of properties become available when they are actually sold, house prices indicated in valuation reports prepared at the time of approval of individual housing loans are used as a proxy for price. The actual sale of the property and utilization of the loan is not required and all houses appraised are included in the scope. To construct the House Price Index for Turkey (THPI) representing the whole country, all valuation reports are used, whereas, to construct the New Housing Price Index for Turkey (TNHPI) again representing the entire country, valuation reports for houses built in the current and previous years are used.
New Housing Price Index for Turkey (TNHPI): Price index for houses constructed in the current year and the previous year, which covers the whole country. The HPI series which starts in January 2010 is issued within 40 days following the 3-month reference period it covers (For index release dates, please refer to:
http://www3.tcmb.gov.tr/veritakvim/calendar.php) Indices produced are accessible under the Data/Periodic Data/House Price Index menu on the CBRT website.

References

Aktan, B., & Özturk, M. (2009). Empirical examination of REITs in Turkey: An emerging market perspective. *Journal of Property Investment & Finance, 27*(4), 373–403.

Altınsoy, G., Erol, I., & Yıldırak, S. (2010). Time-varying beta risk of turkish real estate investment trusts. *Middle East Technical University Studies in Development, 37*(2), 83–114.

Arslanlı, K., Pekdemir, D., & Lee, S. (2011). Initial return performance of turkish REIC IPOs. In *Paper presented at the 18th ERES Conference*, Eindhoven, 15–18 June 2011.

Baum, A. (2006). *Real estate investment through indirect vehicles: An initial view of risk and return characteristics*. Retrieved December 2012 from http://www.inspen.gov.my/inspen/v2/wp-content/uploads/2009/08/Plenary-IRERS-2006-International.pdf

Boudry, W., Coulson, N. E., Kallberg, J. G., & Liu, C. H. (2012). On the hybrid nature of REITs. *Journal of Real Estate Finance and Economics, 44*, 230–349.

Chou, R. K., Ho, K., & Lu, C. (2013). The diversification effects of real estate investment trusts. *A Global Perspective Journal Of Financial Studies, 21*(1), 1–27.

Daveney, S., Xiao, Q., & Clacy-Jones, M. (2012). *Listed and direct real estate investment: A European analysis*, EPRA Research Brussels.

DTZ Pamir & Soyuer. (2010). *Property times 2010 Q1*. Investment Transaction Database: Property Market Indicators.

Erol, I., & Tırtıroğlu, D. (2008). The inflation-hedging properties of Turkish REITs. *Applied Economics, 40*(19–21), 2671–2696.

Erol, I., & Tırtıroğlu, D. (2010). Concentrated ownership, no dividend payout requirement and capital structure of REITs: Evidence from Turkey. *Journal of Real Estate Finance and Economics, 43*(1–2), 174–204.

Geltner, D., Miller, N. G., Clayton, J., Eichholtz, P. (2001). *Commercial real estate analysis and investments* (2nd Ed.). South-Western Educational Publication.

Gyamfi-Yeboah, F., Ling, D. C., & Naranjo, A. (2012). Information, uncertainty, and behavioral effects: Evidence from abnormal returns around real estate investment trust earnings announcements. *Journal of International Money and Finance, 31*(7), 1930–1952.

Hamzah, A., Mohammad Badri, R., & Tahir, I. (2010). Empirical Investigation on the performance of the malaysian real estate investment trusts in pre-crisis, during crisis and post-crisis period. *International Journal Of Economics and Finance, 2*(2), 62–69.

Hodges, S. (1998). *A generalization of the sharpe ratio and its applications to valuation bounds and risk measures*, Working Paper, Financial Options Research Centre, University of Warwick.

Hoesli, M., & Oikarinen, E. (2012). Are REITs real estate? Evidence from international sector level data. *Journal of International Money and Finance, 31*, 1823–1850.

Ling, D., & Naranjo, A. (2003). The dynamics of REIT capital flows and returns. *Real Estate Economics, 31*(3), 405–434.

Pavlov, A. & Wachter, S. (2010). REITs and underlying real estate markets: Is there a link: Working Paper. Retrieved December 2012 from http://realestate.wharton.upenn.edu/research/papers/full/693.pdf

Pavlov, A., & Wachter, S. (2011). Subprime lending and real estate prices. *Real Estate Economics, 39*(1), 1–17.

Pekdemir, D. (2013). Turkish REICs: Real estate investment or real estate development companies? In S. McGreal & R. Sotelo (Eds.), *Real estate investment trusts in Europe—Evolution, regulation, and opportunities for growth*. Springer,—forthcoming.

Pekdemir, D., & Soyuer, F. (2012). *Turkish REICs or REDCs: Analysis of the effect of asset allocation on return*. Paper presented at the 19th ERES Conference, Edinburgh, 15–18 June 2012.

Pham, A. (2012). The dynamics of returns and volatility in the emerging and developed asian REIT markets. *Journal Of Real Estate Literature, 20*(1), 79–96.

Sharpe, W. F. (1966). Mutual Fund Performance. *Journal of Business, 39*(1), 119–138.

Sharpe, W. F. (1994), The Sharpe ratio. *The Journal of Portfolio Management, 21*(1):49–58. Retrieved December 3, 2012 from http://www.stanford.edu/~wfsharpe/art/sr/sr.htm

Springer, T. M. & Cheng, P. (2006). Real estate property portfolio risk: Evidence from REIT portfolios, real estate research institute, Annual Conference Chicago, Illinois April 27–28 2006.

Tuluca, S., Myer, F. C. N., & Webb, J. R. (2000). Dynamics of private and public real estate markets. *Jornal of Real Estate Finance and Economics, 21*(3), 279–296.

Türkmen, S. Y., & Demirel, E. (2012). Economic factors affecting financial ratios: Real estate investment trusts case on ISE. *European Journal of Scientific Research, 69*(1), 42–51.

Yunus, N., Hansz, J. A., & Kennedy, P. J. (2012). Dynamic Interactions between private and public real estate markets: Some international evidence. *Journal of Real Estate Finance and Economics, 45*, 1021–1040.

Zakamulin, V. & Koekebakker, S. (2008). Portfolio performance evaluation with generalized Sharpe ratios: Beyond the mean and variance (February 8, 2008). SSRN:http://ssrn.com/abstract=1028715 or http://dx.doi.org/10.2139/ssrn.1028715

Zhou, J. (2012). Multiscale analysis of international linkages of REIT returns and volatilities. *Journal of Real Estate Finance and Economics, 45*, 1062–1087.

Ziemba, W. (2005). The symmetric downside—risk Sharpe ratio. *Journal of Portfolio Management, 32*(1), 108–122.

An Application of Short Tab MCA to Podgorica

Maurizio d'Amato, Vladimir Cvorovich and Paola Amoruso

Abstract The work is based on the application of particular version of an Italian approach to Sales Comparison Approach (Simonotti 1997) to define an "hedonic" relationship between property price and characteristics. This version is also known as Short Tab Market Comparison Approach (d'Amato 2015a, b, c). The method was originally proposed to provide a forecast for the price of residential properties to be built on underdeveloped land. In this case will be used to provide an estimation of an appraisal function and relative hedonic prices based on few data. The proposed methodology may be useful to try to create an appraisal "function" in emerging real estate market where few comparables are available. The model will be applied to a small group of comparables in the city of Podgorica in Montenegro.

Keywords Market comparison approach · Appraisal system · Automated valuation methods

1 Introduction

The interest in developing techniques for automated valuation methodologies is increasing because of their wider application in property taxation, insurance and mortgage management. The contribution propose the application of an AVM methodology for an emerging market like Montenegro. The method is based on the

The paper has been written in strict cooperation between the three authors. Therefore the credit of the article should be equally divided among them.

M. d'Amato (✉)
DICATECh, Technical University-politecnico di Bari, Bari, Italy
e-mail: madamato@fastwebnet.it

V. Cvorovich
Montenegro Business School, Podgorica, Montenegro

P. Amoruso
University Lum Jean Monnet, Casamassima, BA, Italy

M. d'Amato and T. Kauko (eds.), *Advances in Automated Valuation Modeling*,
Studies in Systems, Decision and Control 86, DOI 10.1007/978-3-319-49746-4_7

determination of an appraisal function based on a specific theory of marginal prices (Simonotti 1997). In the application of the method it is possible to determine marginal prices having few data. The idea is the definition of an appraisal function and therefore the valuation is derived by the product between marginal prices and the characteristics of the property to be estimated. The method may be useful in those countries whose data are not (Kauko and d'Amato 2008) precise and well organized. The contribution is organized as follows. In the next paragraph Short Tab Sales Comparison Approach is presented. In the following paragraph an application of this method is proposed to the residential real estate market of Podgorica in Montenegro. Final remarks will be offered at the end.

2 Short Tab Sales Comparison Approach

Short Tab Market Comparison Approach is based on the theory of marginal prices determination (Simonotti 1997; Ciuna 2010, 2011; Ciuna and Simonotti 2014; Ciuna et al. 2014a, b, 2015a, b, 2016). The most important concept of this theory are illustrated in a contribution in this book (Ciuna et al. 2016). Normally this theory is used in Italy to determine a specific Sales Comparison Approach (also defined by the professional document Market Comparison Approach). In this context the appraisal function is built to estimate a property having few and not well organized data. Scarce and not ordered data can be considered a problem recurring mainly in the emerging market but can be also a problem in specific urban context with several specific different market segments (Biłozor and Renigier-Biłozor 2016; Renigier-Biłozor et al. 2014a, b; Renigier-Biłozor and Biłozor 2016a, b). The Short Tab Sales Comparison Approach starts from the definition of hedonic prices. In this case each comparable allow the appraiser to define an appraisal function. Normally, in the first part of the method are determined only the quantitative characteristics of the property. Therefore in a second method the difference between the actual price and the price expressed by quantitative characteristics is related to qualitative variables using simple matrix calculation (Sistema Integrativo di Stima). In this second part the location variable is also defined. The marginal prices determined in the two different parts form the appraisal function.

Therefore in the former phase of the method is calculated a partial equation having only the quantitative variable like square meters or data.

$$Vpfs = p'_{DAT} \cdot DAT + p'_{SUI} \cdot SUI + p'_{SUB} \cdot SUB \tag{1}$$

In the Formula 1 the partial appraisal function Vpfs for quantitative variables. The dependent variable is the final result summing up the marginal prices for the quantitative characteristics. The latter phase will consider the remaining difference between the price and the partial appraisal function to determine the marginal prices of the remaining qualitative variables:

$$\begin{bmatrix} P_j - Vpfs_j \\ \\ P_m - Vpfs_m \end{bmatrix} = \begin{bmatrix} 1 & \ldots & \ldots \\ 1 & \ldots & \ldots \\ 1 & \ldots & \ldots \end{bmatrix} \begin{bmatrix} LOC \\ \ldots \\ \ldots \end{bmatrix} \quad (2)$$

A practical example will clarify the concept. The proposed model have been tested on a small sample of property provided by one author of this work. All the data are in a specific area of Podgorica called Zabjelo In Montenegro there are problems to have the exact price in transaction because of tax burden The data obtained have been divided in two different samples the first one is indicated in the Table 1.

In the Table 1 the first column identifies the comparable with a latin letter (A, B, C, D) the second column identifies the date measured in month from the moment of transaction to the moment of valuation. The third column indicates the measure in square meters of the apartments, the fourth column is the square meters of balcony. Finally it is possible to observe both the presence or the absence of a elevator and the presence or the absence of a park. All these are the elements of comparison taken into account to determine the appraisal function. The last two columns indicates both the property price and the are of the city in Podgorica. A further small sample of three properties sold in the same area is reported in the Table 2 below.

Table 1 A first small sample of four real estate transactions in Podgorica

Acronym	Date	SUI	Sub	ELEV	PARK	Price	Area of the city
	Month	Square meters	Square meters	Dummy	Dummy		
A	11	48	3	0	0	€52,000.00	Podgorica, Zabjelo
B	24	60	3	1	1	€65,589.30	Podgorica, Zabjelo
C	36	70	5	1	1	€77,000.00	Podgorica, Zabjelo
D	27	75	6	1	1	€82,332.80	Podgorica, Zabjelo

Table 2 A second small sample of three real estate transactions in Podgorica

Acronym	Date	SUI	SUB	ELEV	PARK	Price	AREA of the city
	Month	Square meters	Square meters	Dummy	Dummy		
E	4	31	0	0	1	€39,000.00	Podgorica Zabjelo
F	6	68	4	1	1	€82,000.00	Podgorica, Zabjelo
G	7	45	7	1	1	€57,000.00	Podgorica, Zabjelo

Table 3 Sales summary grid of short table sales comparison approach

	A	B	C	D
Price	52000	65589,3	77000	82332,8
DAT (Month)	5	18	30	21
SUI (Sq.Meter)	45	60	70	75
SUB (Sq.Meter Balcony)	3	0	5	6
ELEV (Elevator)	0	1	1	0
PARK (Parking)	1	0	0	0

Table 4 Marginal price of internal area

Comparable	Medium price calculated	Lowest medium price	Highest medium price
A	1154.78 €/sqm	1080.48 €/sqm	1154.78 €/sqm
B	1093.15 €/sqm		
C	1084.5 €/sqm		
D	1080.48 €/sqm		

The method proposed will be applied to the fourth properties in order to determine the appraisal function, therefore the calculated appraisal function will be applied to the three properties of the second small sample in the same area (Table 2) in order to calculate the percentage error between the actual prices and the estimated prices. The application of Short Tab Sales Comparison Approach starts with the specific sales summary grid indicated in the Table 3. In the following Table 4 the marginal price of real estate surface is calculated.

It is worth to notice that in this sales summary grid there is not the column of subject. In fact the purpose of the method is not determining the value of a specific property but defining an appraisal function in those context with limited in formation. In the second phase the calculation of hedonic price will follow the theoretical background of the Italian Market Comparison Approach (Salvo et al. 2013a, b; Salvo and De Ruggiero 2011, 2013). Therefore the marginal price of the date will be calculated as an adjustment percentage per each months:

$$p'(\mathrm{DAT}) = -\frac{0.01}{12} = -0.00083 \quad (3)$$

The marginal price of internal area is indicated below (Simonotti 1997). Assuming the other characteristics constant the appraisal function should be equal to:

$$p = \overline{p_{SUI}}\mathrm{SUI} + \overline{p_{SUB}}\mathrm{SUB} \quad (4)$$

The following ratio (market ratio) is obtained by the market:

$$\pi = \frac{\overline{p_{SUB}}}{\overline{p_{SUI}}} = 0.20 \quad (5)$$

Knowing the marginal prices the relations 4 and 5 becomes.

$$P = p'_{SUI}SUI + p'_{SUB}SUB \Rightarrow \pi = \frac{p'_{SUI}}{p'_{SUB}} \Rightarrow p'_{SUB} = \pi p'_{SUI} \tag{6}$$

Therefore the part of the appraisal function indicated in the Formula 4 will be rewritten as follows:

$$P = \overline{p_{SUI}}SUI + \overline{p_{SUI}} \cdot \pi \cdot SUB \tag{7}$$

Therefore it will be possible to write:

$$\bar{p}_{SUI} = \frac{P}{SUI + \cdot \pi \cdot SUB} \tag{8}$$

It is possible to transform a medium price in a marginal price through the product between the Formula 8 and the following ratio:

$$\sigma = \frac{p'_{SUI}}{p_{SUI}} \tag{9}$$

In order to calculate the following ratio it is necessary define a criteria for the calculation of marginal price and medium price. In this case we have four properties and the calculation will have the following results showing four different medium prices.

In the Table 4 it is possible to observe the lowest and the highest medium price. In the following graphic 1 it is possible to observe the shape of the mathematical relationship between the square meters and the property price (Di Pasquale and Wheaton 1996).

As a consequence if we imagine the medium price of four comparables there will be a correspondent graphic relationship between price and square meters in the curve b. Therefore there will be four different dots in the curve c representing the correspondent marginal prices (Simonotti 1997; d'Amato 2004, 2010, 2015a, b).

Graphic 1 Price and square meter relation, marginal price and medium price of internal area

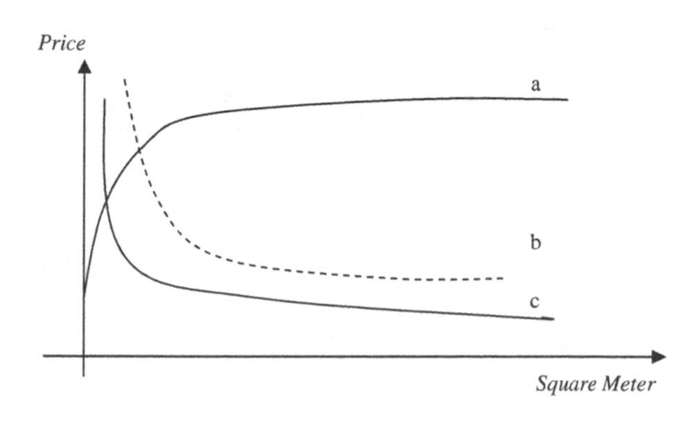

In order to determine the marginal prices it is possible to select the minimum of medium prices because it will be closer to the curve of the marginal prices (curve c) in the specific market segment. On the other hand the maximum of medium prices will be closer to the curve b and will represent the medium price in the market segment. Therefore the ratio indicated in the Formula 9 can be calculated as follows:

$$\sigma = \frac{p'_{SUI}}{p_{SUI}} = \frac{\min(\overline{p_j})}{\max(\overline{p_j})} = \frac{1{,}080.48€/mq}{1{,}154.78€/mq} = 0.9356 \tag{10}$$

This ratio allow the appraiser to determine the marginal price in fact:

$$p'_{SUI} = \overline{p_{SUI}}\sigma = \overline{p_{SUI}}\frac{p'_{SUI}}{\overline{p_{SUI}}} = \frac{P}{SUI + \cdot\pi\cdot SUB}\sigma \tag{11}$$

According to the theory expressed in the Graphic 1 a possible determination of marginal price of square meters will be possible. In the following Table 5 it is possible to observe the product between the product between the medium prices calculated in the second column of Table 4 and the ratio between the third and the fourth column of the Table 4 indicated in the Formula 9:

Therefore the marginal price will be the lower among the different marginal prices calculated for each comparable property. Once the marginal price of the SUI or internal area is calculated the marginal price of SUB balcony will be easily calculated using the Formula 6, therefore:

$$P = p'_{SUI}SUI + p'_{SUB}SUB \Rightarrow \pi = \frac{p'_{SUI}}{p'_{SUB}} \Rightarrow p'_{SUB} = \pi p'_{SUI} = 0.20 \cdot 1010.96€ = 216.09€ \tag{12}$$

Now it is possible to calculate the value using the marginal price of Market Comparison Approach previously calculated. Table 6 indicated the value of the property A:

In a similar way it is possible to calculate the value of comparable B, C and D.

There is a difference between the actual price and the value estimated using the marginal prices of DATA (date) SUI (internal area) SUB (internal area balcony) (d'Amato 2015c; d'Amato and Kauko 2008, 2012; Kauko and d'Amato 2008, 2011). This difference is motivated by the location variable (LOC) and two further elements of comparison the presence or the absence of parking (PARK) and the

Table 5 Marginal price of internal area

Comparables	Marginal prices	Marginal price
A	€1,080.48	€1,010.96
B	€1,022.82	
C	€1,014.73	
D	€1,010.96	

Table 6 Determination of marginal prices for comparable A

Description	Acronym	A	
Price	PRC	€52,000.00	
Month	DAT	−€216.67	−0,000833*5*52,000 €
Sq.Meter	SUI	€45,493.25	45*1,010.96 €
Sq.Meter Balcony	SUB	€648.29	3*216.09 €
		€45,924.87	

presence or the absence of elevator (ELEV). For this reason the difference between the value obtained using the marginal prices and the actual price is considered directly related to the three missing variables. Therefore it is possible to write the following relationship:

$$\begin{bmatrix} \mathrm{PRICE_A - VALUE_A} \\ \mathrm{PRICE_B - VALUE_B} \\ \mathrm{PRICE_C - VALUE_C} \\ \mathrm{PRICE_D - VALUE_D} \end{bmatrix} = \begin{bmatrix} 1 & 1 & 0 \\ 1 & 0 & 1 \\ 1 & 0 & 0 \end{bmatrix} \begin{bmatrix} p'_{LOC} \\ p'_{PARK} \\ p'_{ELEV} \end{bmatrix} \tag{13}$$

In the mathematical relationship provided by the Formula 13 the difference of each price of the comparables and the presence or the absence of parking (second column) elevator (third column). The first column indicates the location variable always constant in all the comparables. In a similar way it is possible to calculate the value of comparable B, C and D as in the Tables 6, 7, 8 and 9 below:

$$\begin{bmatrix} 52{,}000\,€ - 45{,}924.87\,€ \\ 65{,}589.30\,€ - 59{,}673.82\,€ \\ 77{,}000\,€ - 74{,}789.29\,€ \\ 82{,}332.80\,€ - 80{,}891.98\,€ \end{bmatrix} = \begin{bmatrix} 1 & 1 & 0 \\ 1 & 0 & 1 \\ 1 & 0 & 0 \end{bmatrix} \begin{bmatrix} p'_{LOC} \\ p'_{PARK} \\ p'_{ELEV} \end{bmatrix} \tag{14}$$

The final result will be obtained as follows:

$$\begin{bmatrix} 1440.82\,€ \\ 4634.30\,€ \\ 2622.26\,€ \end{bmatrix} = \begin{bmatrix} p'_{LOC} \\ p'_{PARE} \\ p'_{ELEV} \end{bmatrix} \tag{15}$$

In the Table 10 it is possible to observe the marginal price determination based on the theory previously indicated. These marginal prices has been tested on three further real transactions observed in Podgorica. This second small sample is composed by the transactions indicated in the Table 11.

In order to test the validity of the marginal prices obtained and exposed in the Table 10 they have been applied to determine the price of the three properties E, F and G in order to calculate the percentage error between the actual price and the

Table 7 Determination of marginal prices for comparable B

Acronym	B	
PRC	€65,589.30	
DAT	−€983.84	−0.000833*18*65,589.30 €
SUI	€60,658	60*1,010.96 €
SUB	0	3*216.09 €
	€59,673.82	

Table 8 Determination of marginal prices for comparable C

Acronym	C	
PRC	€77,000,00	
DAT	−€1,925.00	−0.000833*30*77,000.00 €
SUI	€75,633.81	70*1,010.96 €
SUB	€1,080.48	5*216.09 €
	€74,789.29	

Table 9 Determination of marginal prices for comparable D

Acronym	D	
PRC	€82,332.80	
DAT	−€1,440.82	−0.000833*21*82,332.80 €
SUI	€81,036.22	75*1,010.96 €
SUB	€1,296.58	6*216.09 €
	€80,891.98	

Table 10 Determination of valuation function

Variable	Marginal price
DAT	−0.00083*P
SUI	€1,080.48
SUB	€216.10
LOC	€1,440.82
PARK	€4,634.30
ELEV	€2,622.27

estimated price. Therefore we will have for the comparable E the following calculation indicated in the Table 12.

In the Table 12 the first column list the elements of comparisons the second column the product between the marginal prices and the characteristics of the property E. Summing up the product the final value will be 39,570.10 €. The real price is 37,000 € therefore it is possible to calculate the percentage error considering also the DAT adjustment. The appraised value considers also the variable DAT. In fact similar adjustment has been done taking into account the difference between the date of transaction and the date of the valuation therefore (Table 13):

Table 11 Three observations E, F, and G

	E	F	G
PRICE	39000	82000	57000
DAT (Month)	4	6	7
SUI (Sq.Meter)	31	68	45
SUB (Sq.Meter Balcony)	0	4	7
ELEV (Elevator)	0	1	1
PARK (Parking)	1	1	1

Table 12 Opinion of value of property E based on the short tab MCA

Property E		
LOC	€1,440.82	€1,440.82
SUI	€33,497.97	31*1,080.48 €
SUB	€-	0*216.10 €
ELEV	€-	0*2,622.27 €
PARK	€4,634.30	1*4,634.30 €
	€39,570.10	

Table 13 Property E percentage error between actual price and appraised value

Property E percentage error		
Appraised value	€39,702.00	0.0177
Actual price	€39,000.00	

$$\begin{aligned} P &= 39{,}570.10\,€ + DAT \cdot 39{,}570.10\,€ \\ P &= 39{,}570.10\,€(1 + DAT) \end{aligned} \tag{16}$$

For the comparable F the following calculation indicated in the Table 14.

In the Table 14 the first column list the elements of comparisons the second column the product between the marginal prices and the characteristics of the property F. Summing up the product the final value will be 80,412.35 €. The real price is 82,000 € therefore it is possible to calculate the percentage error considering also the DAT adjustment. The appraised value considers also the variable DAT. In fact similar adjustment has been done taking into account the difference between the date of transaction and the date of the valuation therefore (Table 15).

$$\begin{aligned} P &= 80{,}412.35\,€ + DAT \cdot 80{,}412.35\,€ \\ P &= 80{,}412.35\,€(1 + DAT) \end{aligned} \tag{17}$$

As a consequence it will be possible to write:

Table 14 Property F percentage error between actual price and appraised value

Property F		
LOC	€1,440.82	€1,440.82
SUI	€73,472.84	68*1,080.48 €
SUB	€864.39	4*216.10 €
ELEV	€-	0*2,622.27 €
PARK	€4,634.30	1*4,634.30 €
	€80,412.35	

Table 15 Property F percentage error between actual price and appraised value

Property F percentage error		
Appraised value	€80,814.41	0.0147
Actual price	€82,000.00	

Table 16 Property G percentage error between actual price and appraised value

Property G		
LOC	€1,440.82	€1,440.82
SUI	€48,621.73	45*1,080.48€
SUB	€1,512.68	7*216.10 €
ELEV	€2,622.27	1*2,622.27 €
PARK	€4,634.30	1*4,634.30 €
	€58,831.80	

To test the validity of the marginal prices obtained and exposed in the Table 10 they have been applied to determine the price of the three property G in order to calculate the percentage error between the actual price and the estimated price. Therefore we will have for the comparable G the following calculation indicated in the Table 16.

In the table the first column list the elements of comparisons the second column the product between the marginal prices and the characteristics of the property G. Summing up the product the final value will be 58,831.80 €. The real price is 57,000 € therefore it is possible to calculate the percentage error considering also the DAT adjustment. The appraised value considers also the variable DAT. In fact similar adjustment has been done taking into account the difference between the date of transaction and the date of the valuation therefore:

$$
\begin{aligned}
P &= 58{,}831.80\,€ + DAT \cdot 58{,}831.80\,€ \\
P &= 58{,}831.80\,€(1 + DAT)
\end{aligned}
\tag{18}
$$

In all the three cases the percentage error is less than 0.05 therefore the mathematical method proposed seems to be interesting for future analysis and further

Table 17 Property G percentage error between actual price and appraised value

Property G percentage error		
Appraised value	€ 59,174.99	0,0368
Actual price	€ 57,000.00	

studies especially in those cases with real estate market contexts with high variability prices or specific property markets without a significant number of property data (Table 17).

3 Conclusions

The paper tested a valuation methodology to deliver an opinion of value on properties using an appraisal function based on the early studies of Italian literature on marginal prices determination (Simonotti 1997). The use of Short Tab MCA can be useful in those real estate market without a precise information on comparables or real estate data. In these cases it may difficult a direct comparison among properties without an accurate description of data. In context without a precise information the construction of a appraisal function based on the theory of marginal prices exposed previously may replace the traditional methods of Automated Valuation Methodology based on the Sales Comparison Approach and the multiple regression analysis. Short Tab MCA has been applied several times to real cases in Italy and demonstrates a primary use in providing a forecast for the price of the property to be built on underdeveloped land (d'Amato 2015a, b). In this case the application of Sales Comparison Approach appears to be problematic because of the lack of a subject. In this case Short Tab MCA allow the appraiser to use comparable to determine the appraisal function. Once the appraisal function has been calculated each marginal price will be multiplied for the characteristics of the property to be built which may be also generic. Further studies may improve the analysis on a greater sample analysing the difference between the actual prices and the predicted prices.

References

Biłozor, A., & Renigier-Biłozor, M. (2016). The procedure of assessing usefulness of the land in the process of optimal investment location for multi-family housing function. In *World Multidisciplinary Civil Engineering-Architecture-Urban Planning Symposium*. WMCAUS 2016 Praga.

Ciuna, M., Salvo, F., & Simonotti, M. (2016). An estimative model of automated valuation method. In M. d'Amato (Ed.), *Automated valuation methodologies*. AVM after the non agency mortgage crisis, Springer Verlag in print.

Ciuna, M. (2010). L'Allocation Method per la stima delle aree edificabili. *AESTIMUM, 57*, 171–181.
Ciuna, M. (2011). The valuation error in the compound values. AESTIMUM [S.l.], 569–583, Aug. 2013. ISSN 1724-2118.
Ciuna, M., Salvo, F., & Simonotti, M. (2015a). Appraisal value and assessed value in Italy. *International Journal of Economics and Statistics, 24–31*. ISSN:2309-0685.
Ciuna, M., Salvo, F., & Simonotti, M. (2015b). Compensation appraisal processes for the realization of hydraulic works in an agricultural area. In *Proceedings of XLIV INCONTRO DI STUDI Ce.S.E.T. Il danno*. Elementi giuridici, urbanistici e economico-estimativi. Bologna, Italy. November 27–28, 2014, pp. 69–82. ISBN:978-88-99459-21-5.
Ciuna, M., & Simonotti, M. (2014). Real estate surfaces appraisal. AESTIMUM 64, Giugno 2014, 1–13.
Ciuna, M., Salvo, F., & Simonotti, M. (2014a). The expertise in the real estate appraisal in Italy. In *Proceedings of the 5th European Conference of Civil Engineering* (ECCIE '14). Mathematics and Computers in Science and Engineering Series (pp. 120–129), North Atlantic University Union, ISSN: 2227-4588, Firenze, 22/24 November 2014.
Ciuna, M., Salvo, F., & Simonotti, M. (2014b). Multilevel methodology approach for the construction of real estate monthly index numbers. *Journal of Real Estate Literature, 22*(2), 281–302.
d'Amato, M. (2015a). Stima del valore di trasformazione utilizzando la funzione di stima. Il MCA a tabella dei dati ridotta, Territorio Italia, June, pp. 97–106.
d'Amato, M. (2015b). MCA a Tabella Ridotta e Sistema Integrativo di Stima. Un secondo caso a Bari, Territorio Italia, December, pp. 97–109.
d'Amato, M. (2015c). Income approach and property market cycle. *International Journal of Strategic Property Management, 29*(3), 207–219.
d'Amato. (2010). A location value response surface model for mass appraising: An "Iterative" location adjustment factor in Bari, Italy. *International Journal of Strategic Property Management, 14*, 231–244.
d'Amato, M. (2004). A comparison between RST and MRA for mass appraisal purposes. A case in Bari. *International Journal of Strategic Property Management, 8*, 205–217.
d'Amato, M., & Kauko, T. (2012). Sustainability and risk premium estimation in property valuation and assessment of worth. *Building Research and Information, 40*(2), 174–185.
d'Amato, M., & Kauko, T. (2008). Property market classification and mass appraisal methodology. In M. d'Amato & T. Kauko (Eds.), *Mass appraisal methods. An international perspective for property valuers* (pp. 280–298) RICS Real Estate Issue, Wiley Blackwell Publishers.
Di Pasquale, D., & Wheaton, W. C. (1996). *Urban economics and real estate markets*. Prentice Hall.
Kauko, T., & d'Amato, M. (2008). *Mass appraising. An international perspective for property valuers*. London: Wiley Blackwell.
Kauko, T., & d'Amato, M. (2011). Neighbourhood effect. In *International encyclopedia of housing and home*. Elsevier Publisher.
Renigier-Biłozor, M., Wiśniewski, R., Biłozor A., & Kaklauskas, A. (2014a). Rating methodology for real estate markets – Poland case study. Pub. *International Journal of Strategic Property Management, 18*(2), 198–212. ISNN:1648-715X.
Renigier-Biłozor, M., Dawidowicz, A., & Radzewicz, A. (2014b). An algorithm for the purposes of determining the real estate markets efficiency in land administration system. Pub. *Survey Review, 46*(336), 189–204.
Renigier-Biłozor, M., & Biłozor, A. (2016a). Proximity and propinquity of residential market area - Polish and Italian case study. In *16th International Multidisciplinary Scientific GeoConferences SGEM*. Bułgaria. (web of science).
Renigier-Biłozor, M., & Biłozor, A. (2016b). The use of geoinformation in the process of shaping a safe space. In 16th *International Multidisciplinary Scientific GeoConferences SGEM Bułgaria*.

Simonotti. (1997). La Stima Immobiliare, Utetlibreria, Torino Italy.

Salvo, F., Ciuna, M., & d'Amato, M. (2013a). Appraising building area's index numbers using repeat values model. A case study in Paternò (CT). In *Dynamics of land values and agricultural policies* (pp. 63–71). Bologna: Medimond International Proceedings, Editografica. ISBN:978-88-7587-690-6, Palermo, 22–23/11/2012.

Salvo, F., Ciuna, M., & d'Amato, M. (2013b). The appraisal smoothing in the real estate idices. In (a cura di): Maria Crescimanno, Leonardo Casini and Antonino Galati, Dynamics of land values agricultural policies (pp. 99–111). Bologna: Medimond Monduzzi Editore International Proceeding Division, ISBN:978-88-7587-690-6, Palermo, 22–23/11/2012.

Salvo, F., & De Ruggiero, M. (2013). Market Comparison Approach between tradition and innovation. A simplifying approach. *AESTIMUM, 62*, 585–594, ISSN:1592-6117.

Salvo, F., & De Ruggiero, M. (2011). Misure di similarità negli Adjustment Grid Methods. *AESTIMUM, 58*, 47–58. ISSN:1592-6117.

Part III
AVM Methodological Challenges: Dealing with the Spatial Issue

Spatial Analysis of Residential Real Estate Rental Market with Geoadditive Models

Vincenzo Del Giudice and Pierfrancesco De Paola

Abstract A study of geographical variability of real estate rents in the central urban area of Naples (Italy) benefits from geostatistical mapping or kriging. Often, some of the observed variables can have non-linear relationships with the response variable. To account for such effects properly we combine kriging techniques with additive models to obtain the geoadditive models, expressing both as linear mixed models. The resulting mixed model representation for the geoadditive model allows for fitting and analysis using standard methodology and software. In effect, the geoadditive models represent efficient and flexible tools, useful in modeling realistically complex situations, often based on semi-parametric regressions integrated by Kriging techniques for the spatial interpolation. In this paper a geoadditive model based on penalized spline functions has been applied, in order to obtain improvements respect to usual Kriging techniques, an analysis of rents values and their spatial distribution for the neighborhoods of Chiaia and Santa Lucia in Naples.

Keywords Geoadditive models · Kriging · Mixed models · Penalized spline · Real estate rental market

1 Introduction

In this study a geoadditive model characterized by penalized spline functions has been implemented, in order to analyze the spatial distribution of unitary real estate rents in the urban area central of Naples (neighborhoods of Chiaia and Santa Lucia). In particular, geoadditive models have many advantages, even in small local real estate markets, being able to analyze the variation of rents values in an area of interest (Kammann et al. 2003; Ruppert et al. 2003; Del Giudice et al. 2014). In

---

V. Del Giudice (✉) · P. De Paola
Department of Industrial Engineering, University of Naples "Federico II", Naples, Italy
e-mail: vincenzo.delgiudice@unina.it

P. De Paola
e-mail: pfdepaola@libero.it

M. d'Amato and T. Kauko (eds.), *Advances in Automated Valuation Modeling*,
Studies in Systems, Decision and Control 86, DOI 10.1007/978-3-319-49746-4_8

addition, these models allow to predict, quantify and locate in real time where and how rents values vary in urban context, with possibility to correlate these variations with any phenomenon or economic effect (for example, delimitation of micro-zones, modeling of locational variables, delimitation of areas with homogeneous values) (Manganelli et al. 2014; Morano et al. 2014; Del Giudice et al. 2014) The combined use of penalized splines with techniques of spatial statistics (Kriging) allows to obtain spatial maps with high reliability on which to base any decisions related to urban investments. Kriging is a regressive technique used for geostatistic spatial analysis that allows to interpolate a variable in the space minimizing the mean square error. Generally, knowing the variable value in some points of space, it is possible to determine the variable value in other points for which there are no measures, this through a weighted average of known values. The weights that are assigned to known measures (real estate rents) depend on the spatial relationship of measured values in the range of unknown point (for which no information is available). For the calculation of weights a semi-variogram is often used, it is a graph capable of putting in relation the distance between two points and the semi-variance value among the measurements made respect to these two points. Substantially, a semi-variogram exposes in a qualitative and quantitative mode the spatial autocorrelation between two points (Kammann et al. 2003; Wand 2003; Ruppert 2003). In Kriging the spatial interpolation is based on the autocorrelation of variable, considering that a variable vary continuously in space. Currently Kriging is often used in the implementation of Geographic Information Systems (G. I.S.), which represent computer systems able to produce, manage and analyze the spatial data associating one or more alphanumeric descriptions to each geographical element (Manganelli et al. 2014; Morano et al. 2014) In particular, the determination of interpolation areas can currently be done through the use of exponential, gaussian, linear, rational or spherical functions. As alternative to these functions the use of geoadditive models with penalized spline functions has been used in this paper in order to achieve significant improvements in forecasting for surfaces interpolation (Wand 2003; Ruppert 2003).

2 Penalized Spline Additive Models and Geostatistical Extension

The complexity of relationship between real estate rents and explanatory variables has conducted to implementation of a geoadditive model. Generally, geoadditive models are composed by a semi-parametric additive component, which it serves to express the relationship between model's non-linear response and explanatory variables, and a model with linear mixed effects that expresses the spatial correlation of observed values (Kammann et al. 2003). The first component involves a low rank mixed model representation of additive models. For simplicity we shall present the case of two additive components. Suppose that (s_i, t_i, y_i), $1 \leq i \leq n$, represent measurements on two predictors s and t for the response variable y, in this case the additive model is:

$$y_i = \beta_0 + f(s_i) + g(t_i) + \varepsilon_i \quad (1)$$

where f and g are unspecified smooth functions of s and t respectively. Therefore, if we define u_+ to equal u for $u > 0$ and 0 otherwise, a penalized spline version of model (1) involves the following functional form (Manganelli et al. 2014):

$$y_i = \beta_0 + \beta_s \cdot s_i + \sum_{k=1}^{Ks} u_k^s \left(s_i - \kappa_k^s\right)_+ + \beta_t \cdot t_i + \sum_{k=1}^{Kt} u_k^t \left(t_i - \kappa_k^t\right)_+ + \varepsilon_i \quad (2)$$

In Eq. (2) there is the penalization of the knot coefficients u_k^s and u_k^t, where κ_1^s, …, κ_{ks}^s and κ_1^t, …, κ_{kt}^t are knots in the s and t directions respectively. The penalization of the u_k^s and u_k^t is equivalent to treating them as random effects in a mixed model (Kammann et al. 2003). Setting $\beta = (\beta_0, \beta_s, \beta_t)^T$, $u = (u_1^s, \ldots, u_{ks}^s, u_1^t, \ldots, u_{kt}^t)^T$, $X = (1\ s_i\ t_i)$ with $1 \leq i \leq n$, $Z = (Z_s \mid Z_t)$,
with:

$$Z_s = [(s_i - \kappa_k^s)_+]_{1 \leq i \leq n,\ 1 \leq k \leq Ks},\ Z_t = [(t_i - \kappa_k^t)_+]_{1 \leq i \leq n,\ 1 \leq k \leq Kt} \quad (3)$$

penalized least squares is equivalent to best linear unbiased prediction in the mixed model:

$$y = X\beta + Zu + \varepsilon \quad E\begin{pmatrix} u \\ \varepsilon \end{pmatrix} = 0 \quad \text{cov}\begin{pmatrix} u \\ \varepsilon \end{pmatrix} = \begin{bmatrix} \sigma_s^2 \cdot I & 0 & 0 \\ 0 & \sigma_x^2 \cdot I & 0 \\ 0 & 0 & \sigma_\varepsilon^2 \cdot I \end{bmatrix} \quad (4)$$

Model (4) is a variance components model since the covariance matrix of $(u^T \varepsilon^T)^T$ is diagonal. The variance ratio $\sigma_\varepsilon^2/\sigma_s^2$ acts as a smoothing parameter in s direction. Penalized spline additive models are based on low rank smoothers, as defined by (Hastie 1996), considering that linear terms are easily incorporated into the model through the $X\beta$ component. At this point we can incorporate a geographical component by expressing kriging as a linear mixed model and merging it with an additive model such as model (4) to obtain a single mixed model (defined as geoadditive model).

Universal kriging model for (x_i, y_i), $1 \leq i \leq n$ (y_i are scalar and x_i represent geographical location included in R^2 domain) is (d'Amato, 2010):

$$y_i = \beta_0 + \beta_1^T x_i + S(x_i) + \varepsilon_i \quad (5)$$

where S(x) is a stationary zero-mean stochastic process and ε_i are assumed to be independent zero-mean random variables with common variance σ_ε^2 and distributed independently of S. Prediction at an arbitrary location x_0 is done through the following expression:

$$\widehat{y}(x_0) = \widehat{\beta}_0 + \widehat{\beta}_0^T x_0 + \widehat{S}(x_0)$$

Then for a know covariance structure of S the resulting equation is:

$$\widehat{y}(x_0) = \widehat{\beta}_0 + \widehat{\beta}_1^T x_0 + c_0^T (C + \sigma_\varepsilon^2 I)^{-1} \left(y - \widehat{\beta}_0 - \widehat{\beta}_1^T x \right) \tag{6}$$

where:

$$C = \left(\operatorname{cov}\{s(x_i), S(x_j)\} \right)_{1 \le i,j \le n}$$
$$c_0^T = \left(\operatorname{cov}\{s(x_0), S(x_i)\} \right)_{1 \le i \le n}$$

For the implementation of Eq. (6) we can use:

$$\operatorname{cov} = \{s(x),\ S(x')\} = C_\theta(\|x - x'\|) \tag{7}$$

where $\| v \| = \sqrt{(v^T v)}$ and C_θ is a term of a Matérn covariance function. The complete formulation of C_θ term corresponds to:

$$C\theta(r) = \sigma_x^2 (1 + |r|/\rho) \exp(-(r)/\rho) \tag{8}$$

Equation (8) is the simplest member of the Matérn family and ρ term can be choose with following rule to ensure scale invariance and numerical stability (Kammann and Wand 2003):

$$\widehat{\rho} = \max_{1 \le i,j \le n} \|x_i - x_j\| \tag{9}$$

For all aspects and matters above reported, a geoadditive model can be described, substantially, as a single linear mixed model as follow:

$$y_i = \beta_0 + f(s_i) + g(t_i) + \beta^T{}_1 \cdot x_i + S(x_i) + \varepsilon_i \tag{10}$$

It we put $X = (1\ s_i\ t_i\ x_i^T)$ with $1 \le i \le n$ and $Z = (Z_s \mid Z_t \mid Z_x)$, where Z_s and Z_t are defined by Eq. (3) and $Z_x = Z\Omega^{-1/2}$ with:

$X = (1\ x_i^T)_{1 \le i \le n}$

$$Z = \left[C_0 \left(\|x_i - \kappa_k\| / \rho \right)_{1 \le k \le K} \right]_{1 \le i \le n}$$
$$\Omega = \left[C_0 \left(\|\kappa_k - \kappa_{k'}\| / \rho \right)_{1 \le k, k' \le K} \right]$$
$$C_0(r) = (1 + |r|) \exp(-|r|)$$

The model has representation:

$$y = X\beta + Zu + \varepsilon \tag{11}$$

where:

$$E\begin{pmatrix} u^s \\ u^t \\ \tilde{u} \end{pmatrix} = 0 \qquad \mathrm{cov}\begin{pmatrix} u \\ \varepsilon \end{pmatrix} = \begin{bmatrix} \sigma_s^2 I & 0 & 0 & 0 \\ 0 & \sigma_t^2 I & 0 & 0 \\ 0 & 0 & \sigma_x^2 I & 0 \\ 0 & 0 & 0 & \sigma_\varepsilon^2 I \end{bmatrix} \tag{12}$$

Model (10) can be extended to incorporate linear covariates through the Xβ term. The extension to more than two additive components is straightforward.

3 Empirical Analysis

In this section are presented the results of theoretical model described in previous section for a sample of real estate data. In neighborhoods of Chiaia and Santa Lucia, n. 64 rents of residential units located in a limited geographical area have been observed in the last 3 months. Obviously here are presented the results for a test designed to verify the validity of the model proposed. The residential units have the same building types and are included into a homogeneous urban area in terms of services and infrastructural qualification. For each property the real estate rents and the amounts of some real estate characteristics are known, as shown in Table 1. On the basis of data the following semi-parametric model was implemented:

$$rent = man + f(lev) + f(xcoord,\ ycoord) + \varepsilon_i \tag{13}$$

Results and main indices of model verification are presented in tables and graphics that follow. The determination of knots for the spatial component and its geographical coordinates are identified by the space filling algorithm, implemented in default knots. 2D function library of R Software (Wand et al. 2005; Wand et al. 2003) (Table 2).

The model (13) was therefore estimated by the Re.M.L. method using the spm library of R software. The estimates of effects in the non-linear model have been

Table 1 Variables description

Variables		Description
Rent for residential unit	(rent)	Expressed in Euro
Maintenance status	(man)	Expressed by scale score
Floor level	(lev)	Expressed in number of floor level
Geographic coordinates	(xcoord, ycoord)	Expressed in degrees for lat. and long

Table 2 Statistic description of data

Variables	Std. dev.	Median	Mean	Min	Max
Rent	3.46	13.66	14.41	9.23	28.26
Man	0.59	2.00	1.50	0.00	2.00
Lev	1.68	2.00	2.63	0.00	7.00

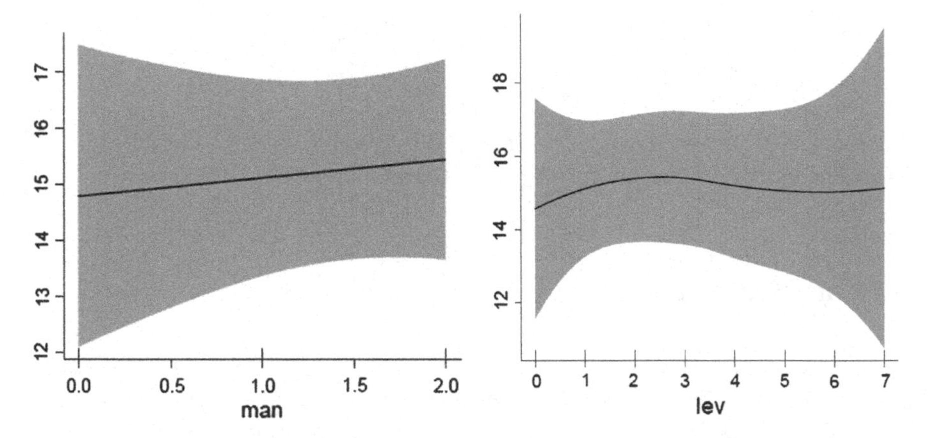

Fig. 1 Values of the predictions obtained for variables "man" and "lev"

significant, as shown by values of freedom degrees (df) and smoothing parameters (spar). The values of obtained predictions are consistent with observed data, also analysis of residuals has not shown any abnormality in its structure. In examined area the spatial distribution of unitary real estate rents clearly shows how the geographical component affect the rents of sampled properties. Main result of interpolation is a thematic map depicting the unitary real estate rents values in the urban context considered, in which blue and red colors represent unitary rents, respectively, lowest and highest (see Figs. 1 and 2) (Table 3).

4 Closing Remarks

The geoadditive models are an effective vehicle for the analysis of spatial real estate data and other applications where geographic point data are accompanied by covariate measurements.

The low rank mixed model formulation allows a straightforward implementation and fast processing of large databases, thus facilitating the use of the model in real estate property and rent markets. In this paper the spatial distribution of unitary real estate rents was analyzed for a central urban area of Naples (neighborhoods of Chiaia and Santa Lucia), such experimentation has allowed to verify the reliability

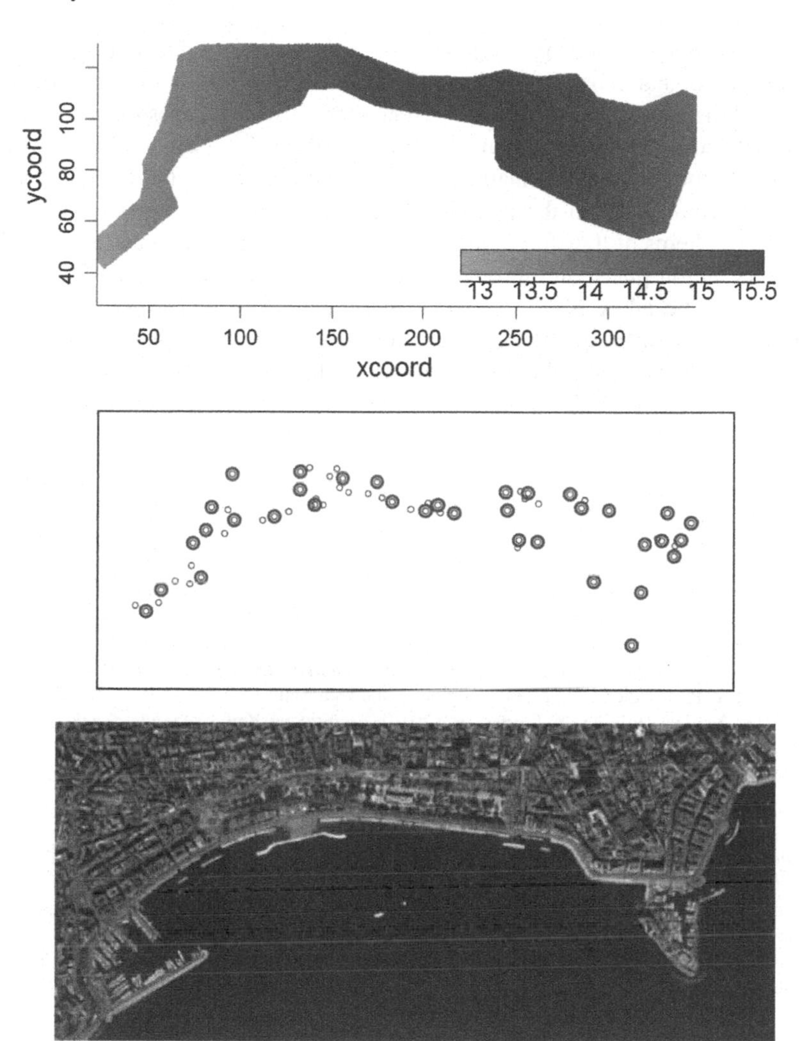

Fig. 2 Spatial distribution of rents values per unit, knots placement and location of housing units

Table 3 Main model results

Non-linear components		df		spar		knots
f (lev)	2242		2979		1	
f (xcoord, ycoord)		3048		635.200		35
Linear components	coef		se		ratio	p-value
Intercept	13.91		5677		2.45	0.017
Man	0.3688		0777		0.4746	0.6367

of proposed model. The positive results obtained by the application of proposed model suggests that geoadditive models can be successfully used for the prediction and spatial distribution of unitary real estate rents. The objectives pursued with the theoretical model proposed are many and varied, such as the study of different segments of local real estate markets, or even the prediction and interpretation of the phenomena related to the genesis of rewards of position, with particular reference to problems of transformation and investments for urban areas affected from projects or action plans, and in order to optimize the choices of use of goods and resources. In the analysis of real estate markets fundamental importance has the proposed model for to define and delineate the scenarios of real estate markets in urbanized contexts, being able to identify any "alarm status" in relation to the data used as, for example, in assessing the presence of a speculative real estate bubble.

References

d'Amato. (2010). A location value response surface model for mass appraising: An "iterative" location adjustment factor in bari, Italy. *International Journal of Strategic Property Management, 14,* 231–244.

Del Giudice, V., & De Paola, P. (2014). *Geoadditive models for property market* (pp. 584–586). Trans Tech Publications: Applied Mechanics and Materials.

Del Giudice, V., De Paola, P., & Torrieri, F. (2014). *An integrated choice model for the evaluation of urban sustainable renewal scenarios* (pp. 1030–1032). Trans Tech Publications, Vols: Advanced Materials Research.

Hastie, T. J.(1996). Pseudosplines. *Journal of the Royal Statistic Society, Series B 58.*

Kammann, E. E., & Wand M. P. (2003.). Geoadditive models. *Applied Statistics*, 52.

Manganelli, B., Morano, P., & Tajani, F. (2014). House prices and rents. *The Italian Experience, WSEAS Transactions on Business and Economics, 11.*

Morano, P., & Manganelli, B. (2014). Estimating the market value of the building sites for homogeneous areas. *Advanced Materials Research*, 869–870.

Ruppert, D., Wand, M. P., & Carroll, R. J. (2003). *Semiparametric regressions*, Cambridge University Press.

Wand, M. P. (2003). Smoothing and mixed models. *Computational Statistics, 18.*

Wand, M. P., French, J. L., Ganguli, B., Kammann, E. E., Stuadenmayer, J., & Zanobetti, A. (2005). SemiPar 1.0 R package. http://cran.r-project.org

A Spatial Analysis for the Real Estate Market Applications

Rocco Curto, Elena Fregonara and Patrizia Semeraro

Abstract The importance of location effect on prices is recognized by real estate literature. This paper proposes a spatial analysis of prices, but also of asset liquidity. In particular we introduce a new approach to measure the relative improvement in price and asset liquidity prediction when the location is known, as opposed to when the location is unknown. As a first application we considered a simplified model, where the location is represented through geographical submarkets. We applied the model to an Italian case study, where real estate markets are divided into geographical submarkets according to the law. We investigated location effect on selling price and asking price, that in Italy is often used for appraisal purposes. We find empirical evidence of the Italian submarket impact on house prices. By contrast, we show that the liquidity of the market, proxied by the time on the market and the discount ratio, is not associated with geographical submarkets.

Keywords Spatial models · Housing submarkets · Housing prices · Liquidity measures

R. Curto (✉) · E. Fregonara
Department of Architecture and Design, Politecnico di Torino,
Viale Mattioli 39, Turin, Italy
e-mail: rocco.curto@polito.it

E. Fregonara
e-mail: elena.fregonara@polito.it

P. Semeraro
Department of Mathematical Sciences G.L. Lagrange, Politecnico di Torino,
C.so Duca degli Abruzzi 24, Torino, Italy
e-mail: patrizia.semeraro@polito.it

M. d'Amato and T. Kauko (eds.), *Advances in Automated Valuation Modeling*,
Studies in Systems, Decision and Control 86, DOI 10.1007/978-3-319-49746-4_9

1 Introduction

Spatial statistics has been introduced into real estate literature to incorporate location effect into house prices. In fact, the empirical analysis on spatial autocorrelation of prices, discussed, among others, in Basu and Thibodeau (1998), Diggle and Ribeiro (2007) and Dubin (1998) supported the importance of location in price prediction. As a consequence, many papers have dealt with the spatial analysis of house prices. For example, Dubin (1998) compared the OLS technique with the geostatistical approach, Pace and Gilley (1997) developed lattice models, and Case et al. (2004) compared different methods, such as Bourassa et al. (2010). An overview of the relationship between spatial statistics and the real estate market can be found in Pace et al. (1998), where the following are considered: lattice models, e.g. geographically weighted regression (see Cleveland and Devlin (1988) and Fortheringham et al. (2002)) and geostatistical models, such as Kriging (see Diggle and Ribeiro (2007) as a standard reference on Kriging). The Bourassa et al. (2010) analysis supports the conclusion that lattice models are not well-suited for mass appraisal purposes. According to their findings this article proposes a geostatistical approach that is well-suited to analyzing real estate markets. On a theoretical level the present article proposes a new approach to perform spatial analysis: to measure the reduction of the response variation when the location is known. By so doing, we measure the relative improvement of predicting the response when the location is known, as opposed to when it is unknown. Within the background of the real estate market, we first measured the association between location and house prices. Secondly, we extended our investigation and measured the location influence on the market liquidity. The liquidity measures we used are time on the market and spread between asking price and transaction price divided by the asking price for any asset, as discussed in Jud et al. (1996) and references therein. We use the approach introduced to analyse the impact of location on the real estate market of Turin, a city in the North of Italy. The importance of location in the Italian market was evidenced in d'Amato (2010) and also in Fregonara and Semeraro (2013). The Italian real estate market is characterized by the segmentation of urban house markets—according to the Italian law—in geographical areas that are defined on the basis of location, type of buildings, green, house values and so on. Recent real estate research deals with the comparison of the simple approach to incorporate spatial dependence—based on geographical submarket definitions—with the geostatistical approach (see Goodman and Tibodeau (1998)). In particular Bourassa et al. (2005, 2008) conclude that geographical housing submarkets improve price prediction accuracy. On account of these results we departed from a continuous model and performed the empirical analysis using a geographical segmentation. By doing so we tested if there is empirical evidence that the Italian segmentation matters. First we investigated whether sellers and agents incorporate location into asking prices: asking prices in Italy are often used by real estate agents for appraisal purposes. In addition, the importance of asking prices in prediction accuracy has also been considered in recent literature. Some references on this topic are Horowitz (1992), Anglin et al. (2003), Knight (2002) Secondly we considered the improvement of selling price prediction when the location is known, and finally we discussed the association between the location and liquidity of the assets. The paper is structured as follows: Sect. 2 introduces the statistical model. The empirical investigation is presented in Sect. 3. Section 4 introduces the application to the Real Estate market. It discusses the case study and provides the empirical results. Lastly, we present our conclusions.

2 The Model

This section introduces a spatial process Y (see Diggle and Ribeiro (2007) as a standard reference for geostatistics). Let $A \subseteq R^2$ a geographical area. A local response is a family of random variables parameterized by the position Z:

$$Y = \{Y(z),\, z \in A\} \tag{1}$$

where $z \in A$ represents the position. The sampled data is $\{(y_1, z_1) : i = 1, \ldots, n\}$, where z_1 represents the spatial location and y_1 is the corresponding realization of Y. Let Z be a random variable representing location, independent from the process Y.[1] Let $f_z(z)$ be the probability density of Z.

Roughly speaking, if $B \subseteq A$, $\int_B f_z(z)ds$ represents the probability that a measurement occurs in the area B. We do not make any assumptions on the distribution f_z. Let us introduce the random variable Y as a mixture Y(Z), defined by $Y(z) := [Y(Z)|Z = z]$, whose distribution, say F_Y, is the following:

$$F_y(x) = \int_A f(z)F_z(x)dz \tag{2}$$

where f is the density of Z and $F_Z(x)$ is the conditional distribution of Y(z).

We investigated the spatial variability of the response Y, taking inspiration from a measure of association named proportional reduction in variance (PRV) and introduced by Kendall and Stuart (1979) to study the association between categorical data. The PRV measure for a pair of random variables (Y, X) was defined as:

$$\text{PRVX(Y)} = \frac{Y[Y] - E(V[Y|X])}{V[Y]} \tag{3}$$

[1]We assume that Z and $Y(z)$ are independent random variables for each $z \in A$.

where V[P] is a measure of variation for the marginal distribution of the response Y, and E[V [Y|X]] is the expectation of the conditional variation taken with respect to the distribution of X. (Tomizawa and Yukawa 2004). If Y is a quantitative variable we use the variance as a measure of variation. Note that Eq. (4) defines the family of R2-like measures, including the square of the famous Pearson correlation coefficient, to assess the strength of linear correlation. Also note that if (X, Y) is a pair of standardized bivariate Normal random variables Eq. (4) becomes the correlation ratio defined by Renyi (1959).

Let (Y, Z) now be a pair of random variables, where Y is a response variable and Z is the location. We name Spatial proportional reduction in variance (Spatial PRV), the measure $PRV_Z(Y)$ defined by Eq. (4). The strength of association measured by $PRV_Z(Y)$ is the relative improvement (in variation) in predicting the response Y when the location is known, as opposed to when it is unknown.

Remark 1. We recall the conditional variance formula that, applied to the pair (Y,Z), becomes:

$$V[Y] = E[V[Y\ |Z]] + V[E[Y|Z]]. \tag{4}$$

According to (4) it is possible to separate the variability of Y into two parts:

- the first one (E[V[Y|Z]]) is the mean of conditional variances of Y given its location: the conditional variance given the location is not due to the spatial variability;
- the second component, V[E[Y|Z]] is the variability of the conditional mean: the conditional mean E[Y|Z] = h(Z) is a function of the location Z, thus V[E[Y| Z]] = V[h(Z)] is a function of the random location Z. We name this component spatial component of variation, SCV.

In this research we consider a discrete version of the above model, by partitioning the geographical area A. Let $A = \{A_1, \ldots, A_n\}$ be a finite partition of A and let it be fine enough to assume that the SCV is negligible within each $A_i \in A$. Then we can assume $V[Y(z)] = \sigma_i^2, \ldots, \forall z \in A_i$. We name $\sigma_i^2, i = 1, \ldots, n$ local variances. We also assume that $E[Y(z)] = \mu_i, \forall z \in A_i$, and we name $\mu_i, i = 1, \ldots, n$, local means. By introducing the spatial finite partition A the location Z is a discrete variable. Let $p_i = P(Z \in A_i), A_i \in A$ be the probability that a measurement occurs in A_i.

Remark 2. The assumptions $E[Y(z)] = \mu_i$ and $V[Y(z)] = \sigma_i^2, \ldots \forall z \in A_i$, are weaker than $Y(z) = Y_i$ on A_i.

According to Eq. (5), Eq. (4) becomes:

$$PRV_X(Y) = \frac{\sum_i p_i(\mu_i - \mu)^2}{\sigma^2} = \frac{\sum_i p_i \sigma_i^2}{\sigma^2} \tag{5}$$

where $E[Y(z)] = \mu_i$ and $V[Y(z)] = \sigma_i^2, \ldots \forall z \in A_i$. Note that $V(E[Y|Z]) = \sum_i p_i(\mu_i - \mu)^2$, therefore $\sum_i p_i(\mu_i - \mu)^2$ is the spatial component of total variance, SCV.

Note that the measure of association PRV is a function of the variance σ^2, of the local variances σ_i^2, the local means μ_i and the location discrete distribution.

We conclude this section presenting the estimation procedure adopted. According to Eq. (6) we have to estimate the response variance σ, the conditional variances of Y given the location σ_i, $i = 1, \ldots, n$ and the probabilities p_i.

The conditional variances σ_i, $i = 1, \ldots, n$ are estimated using the conditional sample variances, defined by:

$$S_i^2 = \frac{\sum_{j=1}^{n_i}(X_{ij} - \bar{X}_i^2)}{n_i - 1}, i = 1, \ldots, n. \tag{6}$$

where for each i = 1, …, n, $\overline{X_i}$ is the conditional sample mean.

The probabilities pi estimates are the transaction relative frequencies, i.e. $\hat{p}_i = \frac{n_i}{n}$ where n is the total number of measurements and n_i is the number of measurements that occurred in the area Ai.

We estimate E[V(Y|Z)], using:

$$S_W^2 := \sum_{i-1}^{40} \hat{p}_i S_i^2 \tag{7}$$

The estimator for the SCV, S_{SCV}^2 follows straightforward:

$$S_{SCV}^2 = S^2 - S_W^2 \tag{8}$$

where S^2 is the sample variance of Y.

Finally the estimate P $\hat{R}$VZ(Y) of PRV_Z (Y) is:

$$P\hat{R}V_Z(Y) = \frac{S_{SCV}^2}{S^2} \tag{9}$$

The above analysis revokes the classical analysis of variance (called ANOVA). We recall that the ANOVA is a procedure introduced and developed by Fisher (1958) at the beginning of 20th century. The presentation of ANOVA is out of the aim of the present paper, as a standard reference see Stuart et al. (1994).

3 The Empirical Investigation

This section introduces the spatial analysis of the real estate market. First we formally introduce the local response variables considered, which are the house prices and liquidity measures of the market. Secondly, we address the question of partitioning the geographical area studied.

3.1 The Response: House Prices, Price Spreads and Time on the Market

This section formally introduces the response variables considered in order to discuss the influence of location on real estate market. First of all we discuss the impact of location on prices. We considered both asking prices Pa and selling prices Ps; by so doing we investigated if sellers and agents incorporate submarket values into listing prices and if the sale prices at the end of the negotiation also show the same dependence on location. Second, we considered the measures of market liquidity, represented by the time on the market (TOM) and the price spreads, i.e. the spread between listing and selling price (see Jud et al. (1996)). Note that the listing price influences how long it takes to find a buyer, and TOM affects the final price, see Yavas and Yang (1995). Therefore the variables under consideration are closely related to house prices. Henceforth we investigated their relationship with location. Since both TOM and price spreads are continuous responses associated to a sale, they can be modelled as spatial processes. Let T represent the time on the market and ΔP the price spread, i.e.:

$$\Delta P = \frac{P_a - P_s}{P_a} \tag{10}$$

where Pa is the asking price and Ps the corresponding selling price. In order to perform a spatial analysis of price and asset liquidity, we define the local processes corresponding to prices, T and ΔP. By so doing, Eq. (5) can be applied to T and ΔP to identify the spatial component of TOM and the spatial component of price spreads: $V[E[T|Z]]$ and $V[E[\Delta P|Z]]$.

A local price is a family of random variables parameterized by the position z:

$$P = \{P(z), z \in A\} \tag{11}$$

where P(z) m is the price of a house in the position $z \in A$. Similarly let

$$\Delta P = \{\Delta P(z), z \in A\} \tag{12}$$

$$\Theta = \{T(z), z \in A\} \tag{13}$$

be respectively the local spreads and the local TOM processes. In this framework, the random variable Z represents the location of a transaction. Roughly speaking, if B is any subset of A, i.e. any subarea, $\int_B f_z(z)ds$ represents the probability that a sale occurs in the area B.

According to the Eq. (4) we define the PRV of prices, as follows:

$$PRV_Z(P) = \frac{V(E[P|Z])}{V(P)} \tag{14}$$

Note that the local price could be a selling price or the corresponding asking price.

The PRV of location on T and ΔP is defined respectively as:

$$PRV_Z(T) = \frac{V(E[T|Z])}{V(T)} \tag{15}$$

and

$$PRV_Z(\Delta P) = \frac{V(E[\Delta P|Z])}{V(\Delta P)} \tag{16}$$

The following section addresses the definition of a geographical segmentation to model the random location.

3.2 *The Partition: Housing Submarkets*

Traditionally housing submarkets—such as those defined by agents—are geographical areas. Consequently they are closely related to the spatial analysis of markets. In particular Bourassa et al. (2003) conclude that housing submarkets matter, and location plays the major role in explaining why they matter. Moreover their empirical evidence suggests that geographical submarkets are more important in predicting house prices than the more fluid approach which permits "submarkets" to vary from house to hous. In this framework we estimated the impact of geographical submarkets on house market analysis in a case study. Formally the partition *A* is a geographical segmentation, whose elements Ai, i = 1, …, n represent geographical submarkets. In this case we empirically computed $PRV_Z(Y)$, representing the improvement in predicting Y obtained using a geographical segmentation. In the sequel, by abuse of notation we write $PRV_A(Y)$ instead of $PRV_Z(Y)$ to underscore the geographical area analyzed. The paper Bostic et al. (2007) deals with the general question of submarket definition. In particular, concerning our case study, we considered a geographical partition into cadastral zones that are typical of Italian cities and that we are going to discuss when introducing the case study, in order to find some empirical evidence that supports its relevance in price prediction accuracy and Italian market analysis.

4 Case Study: The Turin Real Estate Market

We performed a first empirical analysis on an Italian case study: the Turin real estate market. Before attempting the empirical analysis we describe the Italian real estate market, focusing on the relationship between location and selling process. Thus we focus on listing prices, selling prices, TOM and discount ratio. Usually properties are listed by real estate agents in public advertisements. The real estate agents estimate the property value and agree with the seller to define the asking price. Therefore the seller commits to selling the property at the established price. The selling price is the result of a negotiation between the agent and a possible buyer. Once a buyer makes a bid for the house, usually lower than the list price, the seller decides whether or not to close the transaction.

In order to incorporate the location effect on asking price, agents consider a cadastral and geographical segmentation that is typical of the Italian market. According to the Presidential Decree 138/1998 (shortly DPR 138/1998) and subsequent Regulation of the Ministry of Finance, each city should be divided into homogeneous cadastral zones, named Microzones. In particular the buildings belonging to the same Microzone are similar in terms of accessibility, structural characteristics, green, etc. Furthermore they should represent different sub-markets for the level of prices. It is worthwhile mentioning that Italian real estate observatories provide statistical indices of unit prices for each Microzone—mean value, standard deviation, maximum and minimum of unit prices. Therefore we suppose that sellers and buyers define house value taking local indices into account. This fact could increase the association between Microzones and listing prices, but also between Microzones and selling prices. On the basis of this background we decided to use the spatial partition defined by Microzones. This choice is in line with Bourassa et al. (2003), who conclude that it is probably unhelpful to employ elaborate statistical methods to define submarkets, emphasizing the value of the practical knowledge of appraisers. We mention the relevance of the unit price, i.e. the price/size ratio within the Italian house market, in fact real estate observatories provide indices of unit prices. For this reason we performed the analysis using unit prices. A further issue to be addressed is the role played by listing prices within the Italian real estate market. In fact selling prices are not public and are difficult to obtain. Therefore appraisers and analysts use asking prices to study the market and for appraisal purposes. The key role of asking prices to improve prediction accuracy is also supported by empirical studies, such as Horowitz (1992), Anglin et al. (2003), Knight (2002). As a consequence, in order to investigate the Italian market we can not disregard the analysis of asking prices. In particular, we consider two different frameworks: first asking prices of houses on sale, second asking prices and selling prices of sold houses. The first analysis investigates the seller and agent listing behaviour, the second analysis allows the comparison of the location effect on listing prices and corresponding selling prices. Finally, we investigated the relationship between location and asset liquidity. To this purpose, we computed the Microzones PRV on discount ratio and time on the market. Regarding our case study, Turin is divided into forty Microzones. The Microzones are numbered from the center of the city to the suburbs. The number is not related to the Microzone amenities and disamenities, whose attractiveness depend on subjective taste. On the occasion of the market segmentation, the Turtin Real Estate Market Observatory started—as part of an agreement between the Politecnico di Torino, the Chamber of Commerce and the Municipality—with the aim of analyzing the residential real estate market of the city. The residential buildings in Turin are mainly apartment buildings. There are also detached houses concentrated in the hill side zone. The more dynamic Microzones are the semi-central areas, built mainly after 1960. The central Microzones are characterized by the presence of historical buildings and of particular assets that are often not listed in real estate advertisements. In particular Microzone 16 corresponds to a central pedestrian zone with desirable properties and house sales do not take place on the market. In fact the number of transactions or asking prices sampled in this area are often not sufficient to be statistically significant. The hill zone, Microzone 24, is characterized by the presence of detached houses as well as apartment buildings. Summing up, the partition $A = \{A_1, \ldots, A_{40}\}$—considered to empirically compute the Spatial PRV on asking prices, selling prices spreads and TOM—is composed of the 40 cadastral Microzones of Turin (see Fig. 1). Remark 3. Note that the Microzones define a nominal statistical variable. In fact the Microzones differ for environmental amentias and disamenities, whose appreciation depends on subjective taste.

4.1 Data Choice

The databases used to perform the empirical analysis, is the property of the Turin Real Estate Market Observatory (TREMO). TREMO was implemented in 2000 following an agreement between the Politecnico di Torino, the Chamber of Commerce and the Municipality, with the aim of monitoring the Turin real estate market. TREMO collects data and organize them in a in a territorial information system. The sample are cleaned by removing outliers. The samples considered to perform the empirical analysis are property of the TREMO, which collects real estate data and after removing the outliers. The two samples considered have are structured in a territorial information system. However it is not possible to establish a correspondence between the two databases because we considered apartments in apartment buildings, so two different transactions with the same address could occur.

4.2 Sample 1

The first sample, which we name Agents, consists of listing prices and market prices of houses sold in 2008, 2009, 2010. The database has been collected by real

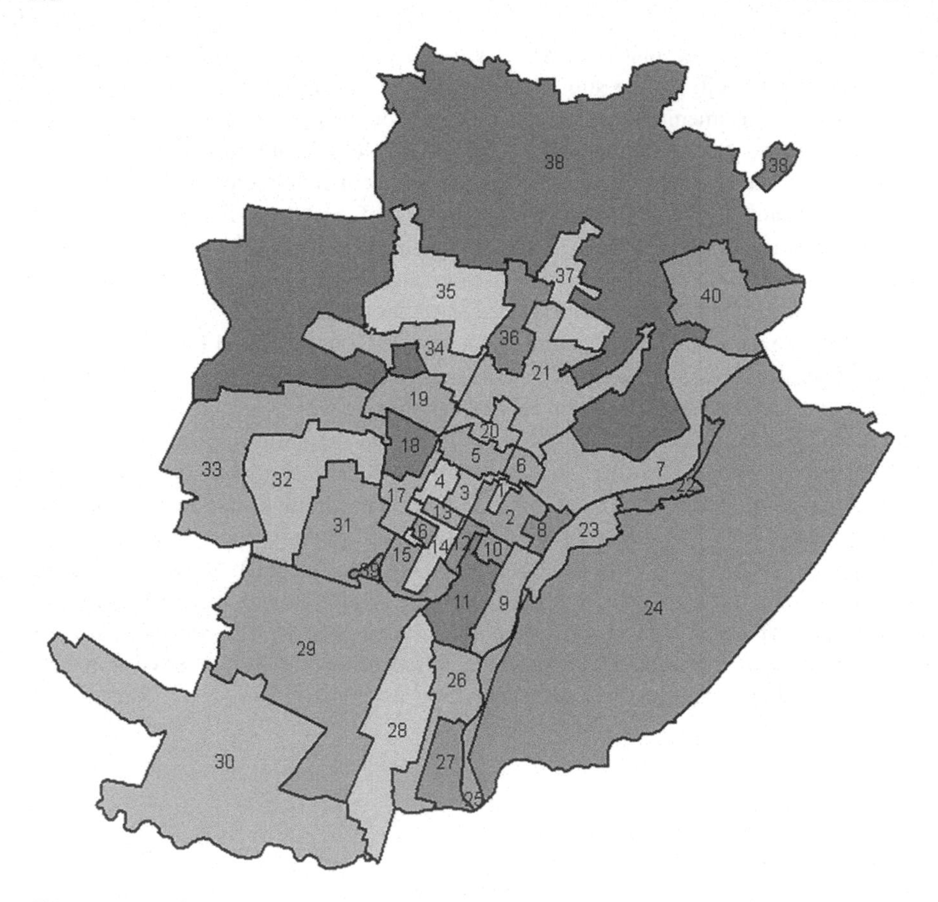

Fig. 1 Turin Microzones

estate agents and sent to TREMO within an agreement between Politecnico di Torino, a real estate agencies association and the Chamber of Commerce (after the validation of the Chamber of Commerce). We have a total of 509 housing sales. For each of them we consider: the offer price, selling price, size (m2) and sale data. Table 1 shows the data unit mean and unit variance, expressed respectively in euro/m2.

It emerges from Table 1 that the sample does not allow a spatial analysis of the whole city. A* is the area defined by the Microzones listed in Table 1 (see also Fig. 2).

Table 1 Sample Agents descriptive statistics

Microzone	Mean asking prices	St. dev.	Mean selling prices	St. dev. sell
7	2842.45	1016.42	2623.16	994.09
10	2625.00	356.00	2450.00	391.93
11	2306.26	614.68	2061.46	586.78
12	2339.88	648.93	1914.88	624.889
14	3623.68	346.71	3328.39	272.65
15	2905.78	583.08	2715.04	541.57
18	2120.32	338.85	1974.32	355.13
19	1764.98	411.07	1646.90	396.16
21	1804.88	259.70	1553.96	276.32
22	2926.85	448.20	2630.66	531.45
23	5886.90	1318.59	5547.06	1291.29
26	2121.52	257.52	1965.04	248.53
27	2136.464	180.39	1971.11	171.44
28	2486.28	419.85	2237.16	387.25
29	2471.61	464.32	2264.99	453.484
31	2220.38	373.93	1972.24	318.88
32	2448.87	548.90	2189.72	519.26
33	2375.01	583.78	2158.353	557.42
35	1834.91	300.94	1660.71	280.77
36	1751.06	382.42	1589.46	326.46
37	2018.00	299.74	1746.59	281.14
38	2041.14	308.53	1798.10	270.64

4.3 *Sample 2*

The second sample—named Residential Used (RU)—consists of asking prices collected in 2008, 2009, 2010. The RU database does not include new built property. The asking prices are randomly collected from real estate advertisements, by so doing, unsold houses are also included in the database. The whole sample consists of 3179 houses.[2] Table 2 shows that the number of dwellings for every Microzone is variable. Table 3 presents the unit sample means and the unit standard deviations, expressed respectively in euro/m2 both for the whole city and for each Microzone.

[2]We note that the sample size is much bigger that Agents sample size. In fact asking prices are publicized on advertisements. In contrast, we recall that sales data is not public in Italy.

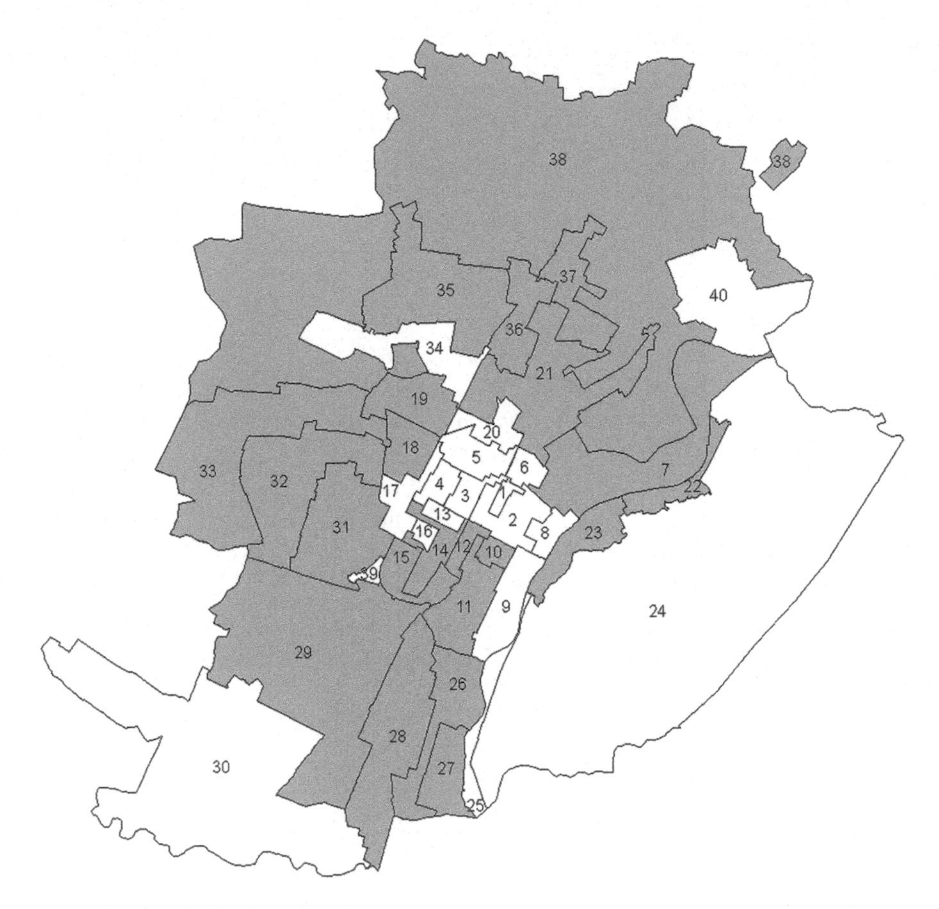

Fig. 2 Spatial distribution of data

4.4 Empirical Results Spatial PRV of Housing Prices

This section presents the results of price analysis. The studied area A is the whole city and the partition *A* is its segmentation into 40 Microzones. However the Agents data belongs to a smaller area: the union of the Microzones listed in Table 3 and colored in Fig. 2, for a total of 22 Microzones (A*). Area A* does not correspond to the whole city, therefore we can not compare the resulting empirical Spatial PRV with the PRV computed on the RU database. In order to compare the results obtained from the two samples analysis (houses on sale and sold houses) we selected a RU subsample, that we name RU-A, considering the data extracted from the Microzones listed in Table 3. The empirical $PRV_A(P)$, corresponding to each sample, are listed in Table 4, where Pa and Ps represent asking and selling prices respectively:

Table 2 Sample RU. Number of dwellings across Microzones

Microzone	Name	Sample size	Microzone	Name	Sample size
01	Roma	20	02	Carlo Emanuele II	111
03	Solferino	72	04	Vinzaglio	52
05	Garibaldi	104	06	Castello	25
07	Vanchiglia	120	08	Rocca	57
09	Valentino	59	10	San Salvario	40
11	Dante	108	12	San Secondo	72
13	Stati Uniti	7	14	Galileo Ferraris	45
15	De Gasperi	75	16	Duca D'Aosta	3
17	Spina 2—Politecnico	13	18	Duchessa Jolanda	91
19	S. Donato	113	20	Porta Palazzo	61
21	Palermo	190	22	Michelotti	79
23	Crimea	71	24	Collina	132
25	Zara	34	26	Carducci	81
27	Unità d'Italia	33	28	Lingotto	81
29	Santa Rita—Mirafori	150	30	Mirafori Sud	96
31	San Paolo	147	32	Pozzo Strada	113
33	Aeronautica-Parella	149	34	Spina 3—Eurotorino	101
35	Madonna Di Campagna	135	36	Spina 4—Docks Dora	58
37	Rebaudengo	84	38	Corona Nord Ovest	106
39	Spina 1—Marmolada	25	40	Barca Bertolla	66

We find empirical evidence that the Microzones explain more than 50 % of price variation, for all samples considered. This fact supports the idea that real estate agents and sellers incorporate Microzones value into asking price. In addition, the empirical evidence highlights a perfect correspondence between the Spatial PRV of the asking and corresponding selling prices of sold houses—Agents sample. However, the Spatial PRV of asking prices computed on unsold houses—RU and RU-A samples—is lower. This fact indicates that location could affect the transaction closure. A three-year period was studied as the Agent database size does not allow for single year analysis. In contrast, the RU sample size allows single year analysis: the empirical Spatial PRV of asking prices is confirmed at about 56 % for each year. A further analysis is performed on RU data, whose sample size also enables the analysis of different urban subareas. We selected three urban areas to fit the Spatial PRV of a geographical segmentation corresponding respectively to a homogeneous central zone, a homogeneous but bigger and peripheral zone and a heterogeneous zone. The first area, composed of central, neighboring and small

Table 3 Sample RU descriptive statistics

Microzone	Mean	St. dev.	Microzone	Mean	St. dev.
1	4625.28	1155.67	2	4394.78	1500.19
3	4463.35	953.19	4	3912.37	1209.72
5	3696.48	829.22	6	4250.61	761.98
7	3135.61	809.41	8	4621.39	1580.16
9	3568.87	1273.99	10	2245.36	498.57
11	3431.37	1359.79	12	3088.19	663.42
13	4222.10	552.36	14	3693.19	928.31
15	3349.02	654.63	16	5500.00	163.89
17	2901.28	702.16	18	3252.52	647.58
19	2643.61	831.29	20	2563.19	840.56
21	2378.61	602.41	22	3377.46	791.02
23	5093.82	1558.33	24	3816.78	1079.74
25	2772.31	755.28	26	2507.79	831.99
27	2444.79	697.21	28	2329.66	544.02
29	2693.54	567.03	30	2137.18	398.71
31	2779.36	599.43	32	3137.67	836.10
33	2793.16	696.71	34	2587.99	593.94
35	2372.39	616.42	36	1915.41	421.71
37	2162.73	420.95	38	2208.99	540.91
39	3762.76	802.24	40	2399.88	574.60

Table 4 Empirical PRV_A for each sample

Sample	$\hat{PRV}_A(P)$
RU Pa	56 %
Sample	$\hat{PRV}_A(P)$
Agents Pa	66 %
Agents Ps	66 %
RU Ps	52 %

Microzones, indicates a low PRV. Area A is the union of Microzones 1, 2, 3, 4 and 5. The estimated PRV_A(Pa) is in Table 5.

The second selected area is a semi-central/peripheral family of Microzones, i.e. from Microzone 34 to Microzone 40. The selected Microzones are neighboring, but they cover a wide, heterogeneous zone of the city area. The empirical evidence indicates a bigger PRV. The estimated PRV is in Table 6.

The third area is a union of Microzones belonging to locations with different characteristics: 1, 23, 24, 29, 38. The first one is a central submarket, while the second and third ones are hilly areas. The last two Microzones are peripheral, but one is in the North and the other in the south of the city, moreover they have different dimensions. The estimated $PRV_A(P_a)$ is in Table 7.

Table 5 Empirical PRV_A. Central area

Sample	$P\hat{R}V_A(P_a)$
RU Pa	7 %

Table 6 Empirical PRV_A. Semi-central/peripheral area

Sample	$P\hat{R}V_A(P_a)$
RU Pa	13 %

Table 7 Empirical PRV_A. Third heterogeneous area

Sample	$P\hat{R}V_A(P_a)$
RU Pa	50 %

The last subarea shows a bigger PRV of submarkets on housing prices. The empirical evidence suggests that, the more the urban area is heterogeneous, the more the segmentation matters.

4.5 Spatial PRV of Market Liquidity

We now discuss the submarkets PRV of market liquidity. We considered the Agents sample, whose data includes the time on the market, the asking prices and the transaction prices. The empirical spatial PRV in the following table suggests that the Microzones do not influence the time on the market.

Table 8 indicates that Microzones do not explain the variation of selling time, Table 9 indicates a low PRV of price spreads.

The empirical results suggest that the Microzones are more associated with discount ratio than with TOM, at least for the data we consider.[3] We find empirical evidence that Microzones are important factors in negotiation as concerns prices and discount ratio. In contrast they do not affect the time to close a transaction.

Table 8 Sample agents. Selling time. Empirical PRV_A

Sample	$P\hat{R}V_A(P_a)$
Agents T	2 %

Table 9 Sample agents. Selling time. Empirical PRV_A

Sample	$P\hat{R}V_A(P_a)$
Agents ΔP	12 %

[3]This result is in line with the findings in Fregonara and Semeraro (2013) where they empirically measured the impact of characteristics and location on the selling process on a sample of transactions occurred in Turin.

5 Conclusions

This paper presents a spatial analysis of variance aiming at identifying the impact of location on a quantitative location-driven response. It provides a first empirical application to the Italian real estate market. On a theoretical level, we start from a spatial process and introduce a spatial analysis of variance. We measure the relative improvement in predicting a quantitative response when the location is known and perform a first analysis on an Italian case study. We apply the model in order to investigate the importance of the typical Italian submarkets—which are defined according to the Italian law—in order to improve price prediction accuracy and market analysis. In particular, we consider both house prices and market liquidity. Concerning house prices, we analyze selling prices, but also asking prices. In fact the latter are often used by agents for appraisal purposes, since selling price is not public in Italy. We proxy the market liquidity through time on the market and discount ratio. We have found empirical evidence that Microzones explain more than 50 % of price variation, supporting their usefulness in improving the price prediction accuracy. Furthermore the empirical results have highlighted the importance of location to close a transaction and its influence on discount ratio, and therefore on negotiation. In contrast, their association with market liquidity is negligible, which confirms that Microzones do not affect the selling time.

References

Anglin, P., Rutherford, R., & Springer, T. (2003). The trade-off between the selling prices of residential properties and time-on-the-market: The impact of price setting. *Journal of Real Estate Finance and Econometrics, 26*(1), 95–111.

Basu, S., & Thibodeau, T. (1998). Analysis of spatial autocorrelation in house prices. *Journal of real Estate Finance and Economics, 17*(1), 61–85.

Bostic, R., Longhofer, S., & Redfearn, C. (2007). Land leverage: Decomposing home price dynamics. *Real Estate Economics, 35*, 183–208.

Bourassa, S., Cantoni, E. & Hoesli, M. (2005). Spatial dependence, housing submarkets, and house prices, s.l.: International Center for Financial Asset Management and engineering, Research paper 151.

Bourassa, S., Cantoni, E., & Hoesli, M. (2010). Predicting house prices with spatial dependence: a comparison of alternative methods. *Journal of Real Estate Research, 32*(2), 139–159.

Bourassa, S. C., Cantoni, E. & Hoesli, M. (2008). *Predicting house prices with spatial dependance: Impacts of alternative submarket definition* (Vol. 08-01). Swiss Finance Institute.

Bourassa, S., Hoesli, M., & Peng, V. S. (2003). Do housing submarkets really matter? *Journal of Housing Economics, 12*(1), 12–28.

Case, B., Clapp, J., Dubin, R., & Rodriguez, M. (2004). Modelling spatial and temporal house price patterns: A comparison of four models. *Journal of Real Estate Finance and Economics, 29*(2), 167–191.

Cleveland, W. S., & Devlin, S. (1988). Locally weighted regression: An approach to regression analysis by local fitting. *Journal of the American Statistical Association, 83*(403), 596–610.

D'Amato, M. (2010). A location value response surface model for mass appraising: An "iterative" location adjustment factor in Bari, Italy. *International Journal of Strategic Property Management, 14*(3), 231–244.

Diggle, P. J., & Ribeiro, P. J. J. (2007). *Model-based Geostatistics*. New York: Springer.

Dubin, R. A. (1998). Estimation of regression coefficients in the presence of spatially autocorrelated error terms. *Review of Economics and Statistics, 70*(3), 466–474.

Fisher, R. A. (1958). *Statistical Methods for Research Workers* (13th ed.). Edinburgh: Oliver & Boyd.

Fotheringham, S., Brunsdon, C. & Charlton, M. E. (2002). Geographical weighted regression: The analysis of spatially varying relationships. London: Wiley.

Fregonara, E., & Semeraro, P. (2013). The impact of house characteristics on the bargaining outcome. *Journal of European Real Estate Research, Emerald Group Publishing Limited, 6*(3), 262–278.

Goodman, A. C., & Thibodeau, T. G. (1998). Housing market segmentation. *Journal of Housing Economics, 7*, 121–143.

Horowitz, J. (1992). The role of the list price in housing markets: Theory and an econometric model. *Journal of Applied Econometrics, 7*(2), 115–129.

Jud, G. D., Winkler, D. T., & Kissling, G. (1996). Price spreads and residential housing market liquidity. *Journal of Real Estate Finance and Economics, 12*(3), 447–458.

Kendall, M., & Stuart, A. (1979). *The advanced theory of statistics 2, Inference and relationship*. New York: Macmillan.

Knight, J. R. (2002). Listing price, time on market, and ultimate selling causes and effects of listing price changes. *Real Estate Economics, American Real Estate and Urban Economics Association, 30*(2), 213–237.

Pace, R. K., Barry, R., & Sirmans, C. F. (1998). Spatial Statistics and Real Estate. *Journal of Real Estate Finance and Economics, 17*(1), 5–13.

Pace, R. K., & Gilley, O. W. (1997). Using the spatial configuration of data to improve estimation. *Journal of Real Estate Finance and Economics, 14*(3), 333–340.

Renyi, A. (1959). On measures of dependence. *Acta Mathematica Academiae Scientiarium Hungarica, 10*, 441–451.

Stuart, A., Ord, K., & Arnold, S. (1994). *Kendall's advanced theory of statistics*. London: Edward Arnold.

Tomizawa, Y., & Yukawa, T. (2004). Proportional reduction in variation measure for two-way contingency tables with ordered categories. *Journal of Statistical Research, 38*(1), 45–59.

Yavas, A., & Yang, S. (1995). The strategic role of listing price in marketing real estate: Theory and evidence. *Real estate Economics, 23*, 347–368.

Location Value Response Surface Model as Automated Valuation Methodology a Case in Bari

Maurizio d'Amato

Abstract The paper is focused on the third application in the Italian context of Location Value Response Surface Modelling for Automated Valuation Modelling. LVRS (Connor in Locational Valuation Derived Directly from the Real Estate Market with the Assistance of Response Surface Techniques 1982) modelling is a procedure normally applied for automated valuation method purposes. In this context has been tested for a group of property prices exploring its use as automatic valuation methodology. The results showed that this method may have a potential role in those automated valuation model dealing with spatial autocorrelation.

Keywords Spatial models · Location value response surface · Housing prices

1 Introduction

Location value response surface modelling is an automated valuation method to appraise the estimate value of a property. The application of this method provides an integration between the tradition Multiple Regression Analysis with a location adjustment factor defined in this work as LAF. These factors can be used to include the distance as a measure of proximity effect of a specific place increasing or decreasing the property value in a specific urban context. The article shows a successful application of this kind of method to the Italian urban context. The article is organized as follows: the next paragraph will offer a brief literature review of Location Value Response Surface models, the second paragraph will show a second application of this model to an italian sample in Bari an italian south eastern city. Final remark and future direction of research will be offered at the end.

M. d'Amato (✉)
DICATECh Technical University,
Politecnico di Bari, Bari, Italy
e-mail: madamato@fastwebnet.it

M. d'Amato and T. Kauko (eds.), *Advances in Automated Valuation Modeling*,
Studies in Systems, Decision and Control 86, DOI 10.1007/978-3-319-49746-4_10

2 Location Value Response Surface Models: A Literature Review

Among different AVM modeling, LVRS model has been recently applied to Italian real estate market (d'Amato 2010). This works represents the third application of this automated valuation model to Italian real estate market. Location Value Response Surface (LVRS) Models (O'Connor 1982) were introduced to appraise single family houses in Lucas County (USA) without referring to fixed neighbourhoods or composite submarkets analysis. The method has been applied in the U.S. (Eichenbaum 1989, 1995; Ward et al. 1999), in England (Gallimore et al. 1996), Northern Ireland (McCluskey et al. 2000) and in Italy (d'Amato 2010, 2011). There are three approaches to LVRS. The first one (McCluskey et al. 2000) calculate a location adjustment factor referred to the spatial distribution of the selling prices. In this case a contour plot based on the ratio between prices and square meters will be originated overlying the area. The map will show the area and the point with higher and lower value of property values which are also called value influence centres (VICs). The distance from each VIC indicated in the contour map is calculated for each property of the sample. The impact of each VIC on any property is determined using different possible measures of the distance from the property to the VIC (Eckert 1990; Eckert et al. 1993). In fact these models are strongly dependent on the VIC positions and the adopted distance measure and transformation. The local adjustment factor will vary from −1 to +1 measuring the impact of location in the final regression model. Location adjustment factor does not indicate the value of a certain location, but only the relative location values for the property analysed. A further approach to LVRS consists in measuring the variance between actual prices and predicted prices using a MRA location blind model. Considering the error ratio related to under valuation or over valuation together with the coordinates of each observation an error map will be generated. The coefficient is included in the original location insensitive regression model. The third approach starts from an interpolation grid modelled to reflect the influence on each property of the location ratio factors within its proximity. In this work it is proposed an application to residential flats in the Italian real estate market. Spatial interpolation needs that the surface of the z variable (selling price or error term) would be continuous and the data value at any location can be estimated if sufficient information about the surface is given using the sample. The variable (selling price or error term) must be spatially correlated and the value at any specific location should be related to the values of surrounding locations.

3 LVRS Modelling

The sample is composed by 114 observations residential properties in an urban area called Carrassi near the downtown of Bari a city located in the south east of Italy. The sample of properties has been partially used for a previous application. The data are referred to residential flats in condominium in a temporal range between 20/05/1992 and 01/04/1997. Statistics on the selected observations are indicated in Table 1.

The dependent variable is the PRICE while DATE; SQM; SQM_BAL and PARK are the selected independent variables whose explanation is indicated in the Table 2.

Hedonic modelling have been used as a method to appraise the market value of a house for several decades (Palmquist 1980; Rosen 1974; Ciuna et al. 2014a, b). Selecting an appropriate functional form has been a recurring problem in the literature (Halvorsen and Pollakowski 1981). Five outliers has been detected. The problem is caused by the absence of theoretical justification for the appropriate functional relationship between housing price and its attributes (d'Amato and Siniak 2008; d'Amato 2008; d'Amato and Kauko 2012; d'Amato 2015). The functional form which better fit of the data in term of mean absolute percentage error and Box Cox test (for box cox test equal to 1), is the linear multiple regression model. The model location blind (or location insensitive) is indicated in the following Formula (1).

$$\begin{aligned} PRICE = {} & 101397.87 + 1391.762\,SQM + 482.614\,BAL \\ & + 38654.23\,PARK - 1132.63\,DATE \end{aligned} \tag{1}$$

The output of this regression model is indicated in the Table 3.

The variables are not correlated and there is not multicollinearity as the VIF (Variance Inflation Factor) index is always less than the threshold. In the model are

Table 1 Descriptive statistics of the sample

Minimum	€50,850.00
Maximum	€293,000.00
Mean	€150,698.14
Standard deviation	€60,066.02

Table 2 Description of dependent and independent variables

Variables	Explication	Type of variable	Measure
DATE	Date of sales	Cardinal	Month
SQM	Square meters of flat	Cardinal	SQM
SQM_BAL	Square meters of balcony	Cardinal	SQM
PARK	Presence of parking	Dummy	Dichotomic
PRICE	Price of the property	Cardinal	Euro

Table 3 Linear MRA model n. 1 location blind

Variable	Acronym	Coefficient	T-stat	F-ratio	Adj. R2	MAPE	N. Obs
Location	LOC	101397	5.343	91.524	78.32	16.71	109
Square meters	SQM	1391.767	12.149				
Balcony	BAL	482.614	1.022				
Date	DAT	−1132.63	−6.9452				
Parking	PARK	38654.23	4.297				

included observation belonging to three fixed neighbourhood (Carrassi, Poggiofranco, S. Pasquale). All of these "neighbourhood" belongs to the same residential area called Picone. The new linear multiple regression model was applied dividing the observation between two fixed neighborhoods as indicated in the Formula (2) avoiding the dummy variable trap (Green 2003). The new model will be

$$\begin{aligned} \text{PRZ} = & \ 84.570 + 1{,}410.22 \cdot \text{SQM} + 511.252 \cdot \text{BALCONY} \\ & - \text{DATE} \cdot 982.63 + \text{PARK} \cdot 42{,}243.32 + \\ & - \text{NG1}{,}16849.324 + \text{NG2} \cdot 43725.623 \end{aligned} \tag{2}$$

These neighborhoods are coincident with two different zones in the area, the third one is included in the constant term. The output of regression can be read in the Table 4.

The variables are not correlated and there is not multicollinearity as the VIF (Variance Inflation Factor) index is always less than the threshold. The role of spatially modeled variable in hedonic pricing has been highlighted in several different works (Des Rosiers et al. 2003, 2005). In order to analyse the opportunity of the application of spatial related model, it is necessary to consider the degree of spatial autocorrelation of a fundamental variable as the ratio between price and square meters. The most commonly used and robust indicator was proposed by the statistician Moran (1948, 1950) and it is normally indicated as Moran I test indicated in the Formula (3).

Table 4 Linear MRA model n. 2 using 3 fixed neighbour groups

Variable	Acronym	Coefficient	T-stat	F-ratio	Adj. R2	MAPE	N. Obs
Location	LOC	84570	5.0012	77.422	79.57	16.19	109
Square meters	SQM	1410.22	12.4272				
Balcony	BAL	511.252	−7.0752				
Date	DAT	−982.63	1.284				
Parking	PARK	42243.32	4.924				
Neighbour group 1	NG_1	−16849.324	−1.254				
Neighbour group 2	NG_2	43725.623	2.822				

$$I = \frac{N \sum_{i=1}^{n} \sum_{j=1}^{n} w_{ij}(x_i - \bar{x})(x_j - \bar{x})}{\sum_{i=1}^{n} \sum_{j=1}^{n} w_{ij}(x_i - \bar{x})^2} \tag{3}$$

where x is the variable (, and wij represents the set of neighbours j for observation i). In this case the inverse squared distance among the observations has been considered according to previous works (Des Rosiers et al. 1999; Shiller 1993; Wyatt 1997; Kauko and d'Amato 2008a; Kauko and d'Amato 2008b; Kauko and d'Amato 2008c; Kauko and d'Amato 2011; Kaklauskas et al. 2012). The Moran's I ranges from −1 to +1 and each observation is only compared with its relevant neighbourhood as a consequence positive Moran's I means positive autocorrelation. In this case high values for x value should be located near other high values while lower market basket values should be located near other lower market basket values. In this work the Moran I test showed a high positive autocorrelation assuming a value of 0.7274. There is a theoretical premise to integrate traditional multiple regression analysis with spatial analysis. A location adjustment factor will be added to the model location blind indicated in the Formula (1). This location

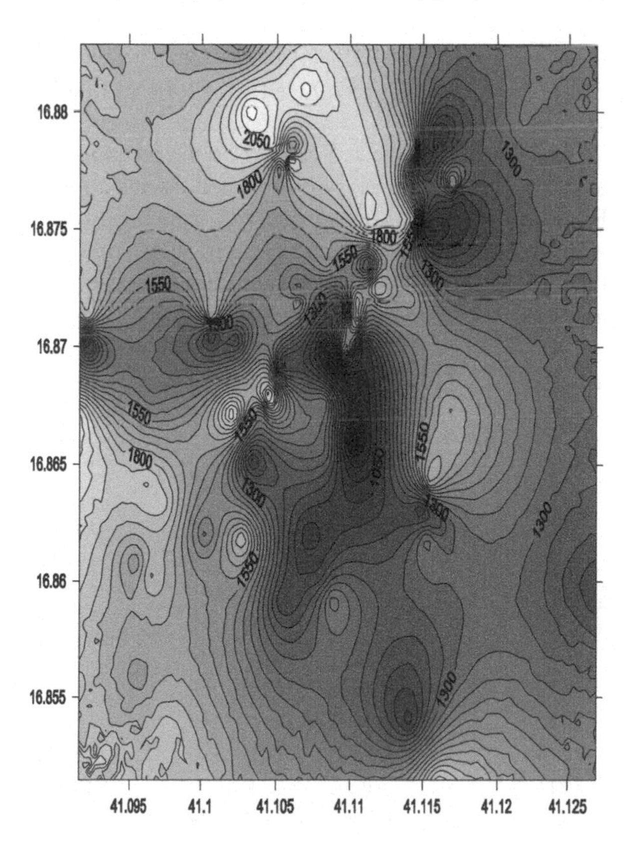

Fig. 1 Contour plot of market basket value carrassi san pasquale area using a linear variogram

adjustment factor will be based on a contour map developed on the ratio price per square meters. The Fig. 1 indicate the contour map originated by a linear variogram in the context calculated using SURFER 8.

In the Fig. 1 there are 12 value influence centers located having the coordinates indicated in the Table 5.

In the Table 5 each VIC is individuated with the decimal coordinates (longitude and latitude). The final column of Table 6 indicates the nature of the value influence center, H means higher values while the L indicates lower values. It is worth to notice that the value influence centers n. 2; 4; 7; 10; 11; 12 are closer to an important urban park in the area called *Parco Largo 2 Giugno* therefore higher values can be observed. This confirms the cause of spatial autocorrelation indicated in previous scientific work (Gillen and Thibodeau et al. 2001). A location adjustment factor was calculated regressing the market basket value as dependent variable on the location and the distance between each point and the value influence center. Including VIC in the regression model require a method to transform physical distance such as half Gaussian or gravity models. They transform the physical measure in a proximity variable. In this case the transformation occurred using a half Gaussian function normally indicated in literature (O'Connor 1982). The model including the location adjustment factor is indicated in the Formula (4).

$$\begin{aligned}\text{PRZ} = & -15,194.30 + 1,384.12 \cdot \text{SQM} + 307.91 \cdot \text{BALCONY} \\ & - \text{DATE} \cdot 857.93 + \text{PARK} \cdot 19,243.84 + \text{LAF} \cdot 107,249.80 + \varepsilon\end{aligned} \tag{4}$$

The last term indicated as LAF is the location adjustment factors. The constant term of the Formula (4) is diminished compared to the other models, the adjusted R^2 of the model n. 3 is improved growing to 0.836 and the t-student test are significant for all the variable except for Balcony. There is also an improvement in term of mean absolute percentage error. Table 6 indicates the output of regression analysis.

Table 5 Value influence centers individuated in the contour map of Fig. 1

Value influence center	Latitude	Longitude	Nature VIC
VIC1	41.1100000	16.8693216	L
VIC2	41.1038779	16.8800000	H
VIC3	41.1153174	16.8543407	L
VIC4	41.0999831	16.8752060	H
VIC5	41.1177992	16.8782044	L
VIC6	41.1156999	16.8786358	L
VIC7	41.1111999	16.8771798	H
VIC8	41.1029999	16.8716999	L
VIC9	41.1087545	16.8713759	L
VIC10	41.1132996	16.8578977	H
VIC11	41.1191000	16.8526769	H
VIC12	41.0936229	16.8810161	H

Table 6 Location value response surface model

Variable	Acronym	Coefficient	T-stat	F-ratio	Adj. R2	MAPE	N. Obs
Location	LOC	−15,194.30	1.9342	116.29	83.6	13.21	109
Square meters	SQM	1384.12	12.3729				
Balcony	BAL	307.91	1.5329				
Date	DAT	−857.93	−4.7983				
Parking	PARK	19,243.84	4.924				
Location adjust factor	LAF	107,249.80	7.0002				

Table 7 Comparing the three model's main findings

	Model 1	Model 2	Model 3
	Location blind	*Model with neighbour groups*	*Location value response surface*
MAPE	16.71	16.19	13.21
Adj R2	78.32	79.57	83.6

The variables are not correlated and there is not multicollinearity as the VIF (Variance Inflation Factor) index is always less than the threshold. Finally the Table 7 compares the mean absolute percentage errors and the adjusted R^2 of the four models presented.

The model 1 is the location blind indicated in the Formula (1), the model number 2 is the fixed neighbor model indicated in the Formula (2) while the third is an application of LVRS model presented in the previous paragraph.

4 Conclusions and Future Directions of Research

Empirical findings demonstrates that LVRS may be an interesting option for automated valuation methodologies, too. This application in Italian context and residential segment showed an evident increase of the quality of the model. Empirical studies are required to analyse the difference between property valuation carried out by valuers, mass appraisal and price. This would be helpful to understand the relation between the price and the valuation (Renigier-Biłozor 2014a, b) and how spatial context affect both. In particular further works may be required testing also the error correction model in the Italian context. The application showed an evident vagueness of boundaries among different VIC, a typical situation determined by the nature of Italian context with several different urban areas. As one can see the number of VIC used in italian application is much higher than in other experience because the complex nature of Italian urban context.

References

Box, G., & Cox, D. (1964). An analysis of transformations. *Journal of the Royal Statistical Society, 26*, 211–252.
Ciuna, M., Salvo, F., & De Ruggiero, M. (2014a). Property prices index numbers and derived indices. *Property management, 32*(2), 139–153. DOI(Permanent URL):10.1108/PM-03-2013-0021.
Ciuna, M., Salvo, F., & Simonotti, M. (2014b). Multilevel methodology approach for the construction of real estate monthly index numbers. *Journal of Real Estate Literature, 22*(2), 281–302.
d'Amato, M. (2015). Income approach and property market cycle. *International Journal of Strategic Property Management, 29*(3), 207–219.
d'Amato, M. & Kauko, T. (2012). Sustainability and risk premium estimation in property valuation and assessment of worth. *Building Research and Information, 40*(2), 174–185 (March–April 2012).
d'Amato, M. (2008). Rough set theory as property valuation methodology: The whole story (Chap. 11). In T. Kauko & M. d'Amato (Eds.), *Mass appraisal an international perspective for property valuers* (pp. 220–258). Wiley Blackwell.
d'Amato, M. (2010). A location value response surface model for mass appraising. An "iterative" location adjustment factor in Bari, Italy. *International Journal of Strategic Property Management, 14*, 231–244.
d'Amato, M. & Kauko, T. (2008). Property market classification and mass appraisal methodology (Chap. 13). In T. Kauko & M. d'Amato (Eds.), *Mass appraisal an international perspective for property valuers* (pp. 280–303). Wiley Blackwell.
d'Amato, M. & Siniak, N. (2008). Using fuzzy numbers in mass appraisal: The case of belarusian property market (Chap. 5). In T. Kauko & M. d'Amato (Eds.), *Mass appraisal an international perspective for property valuers* (pp. 91–107). Wiley Blackwell.
Des, Rosiers F., Theriault, M., & Menetrier, L. (2005). Spatial versus non-spatial determinants of shopping center rents: Modeling location and neighborhood-related factors. *Journal of Real Estate Research, 27*(3), 293–319.
Des Rosiers, F., Theriault, M., Villeneuve, P., & Kestens, Y. (1999). *House price and spatial dependence: Towards an integrated procedure to model neighborhood dynamics.* Paper presented at 1999 AREUEA annual meeting.
Des, Rosiers F., Theriault, M., Villeneuve, P., & Kestens, Y. (2003). Modelling interactions of location with specific value of housing attributes. *Property Management, 21*(1), 25–62.
Eckert, J., Ed. (1990). Property appraisal and assessment administration. *International Association of Assessing Officers.*
Eckert, J., O'Connor, P., & Chamberlain, C. (1993). Computer-assisted real estate appraisal: a california savings and loan case study. *The Appraisal Journal, LXI*(4), 524–532.
Eichenbaum, J. (1989). Incorporating location into computer-assisted valuation. *Property Tax Journal, 8*(2), 151–169.
Eichenbaum, J. (1995). The location variable in world class cities: Lessons from CAMA valuation in New York City. *Journal of Property Tax Assessment & Administration, 1*(3), 46–60.
Gallimore, P., Fletcher, M., & Carter, M. (1996). Modeling the influence of location on value. *Journal of Property Valuation and Investment, 14*(1), 6–19.
Gillen, K., Thibodeau, T., & Wachter, S. (2001). Anisotropic autocorrelation in house prices. *Journal of Real Estate Finance and Economics, 23*(1), 5–30.
Green, W. H. (2003). *Econometric analysis*. New Jersey: Prentice Hall.
Halvorsen, R., & Pollakowski, H. (1981). Choice of functional form for hedonic price equations. *Journal of Urban Economics, 10*, 37–49.

Kaklauskas, A., Daniūnas, A., Dilanthi, A., Vilius, U., Lill, I., Gudauskas, R., et al. (2012). Life cycle process model of a market-oriented and student centered higher education. *International Journal of Strategic Property Management, 16* (4), 414–430.

Kauko, T., & d'Amato, M. (2011). *Neighbourhood effect, international encyclopedia of housing and home.* Edited by Elsevier Publisher.

Kauko, T., & d'Amato, M. (2008a). *Mass appraising. An international perspective for property valuers.* London: Wiley Blackwell.

Kauko, T., & d'Amato, M. (2008b). Introduction: Suitability issues in mass appraisal methodology. In T. Kauko & M. d'Amato (Eds.), *Mass appraisal an international perspective for property valuers* (pp. 1–24). Wiley Blackwell.

Kauko, T., & d'Amato, M. (2008c). Preface. In T. Kauko & M. d'Amato. (2008). (Eds.), *Mass appraisal an international perspective for property valuers* (p. 1). Wiley Blackwell.

McCluskey, W. J., Deddis, W. G., Lamont, I. G., & Borst, R. A. (2000). The application of surface generated interpolation 3a models for the prediction of residential property values. *Journal of Property Investment and Finance, 18*(2), pp. 162–176. MCB University Press.

Moran, P. A. P. (1948). The interpretation of statistical maps. *Journal of the Royal Statistical Society B, 10*, 243–251.

Moran, P. A. P. (1950). Notes on continuous stochastic phenomena. *Biometrika, 37*, 17–23.

O'Connor, P. (1982). *Locational valuation derived directly from the real estate market with the assistance of response surface techniques*, Lincoln Institute of Land Policy.

Palmquist, R. (1980). Alternative techniques for developing real estate price indexes. *Review of Economics and Statistics, 62*, 442–448.

Rasmussen, D., & Zuehlke, T. (1990). On the choice of functional form for hedonic price functions. *Applied Economics, 22*, 431–438.

Renigier-Biłozor, M., Wiśniewski, R., Biłozor, A., & Kaklauskas, A. (2014a). Rating methodology for real estate markets—Poland case study. *Pub. International Journal of Strategic Property Management. 18*(2), 198–212. ISNN. 1648-715X.

Renigier-Biłozor, M., Dawidowicz, A., & Radzewicz, A. (2014b). An algorithm for the purposes of determining the real estate markets efficiency in Land Administration System. *Pub. Survey Review, 46*(336), 189–204.

Rosen, S. (1974). Hedonic prices and implicit markets: Product differentiation in pure competition. *Journal of Political Economy, 82*, 34–55.

Shiller, R. (1993). Measuring asset value for cash settlement in derivative markets: Hedonic repeated measures indices and perpetual futures. *Journal of Finance, 48*, 911–931.

Ward, R. D., Weaver, J. R., & German, J. C. (1999). Improving models using geographic information systems/response surface analysis location factors. *Assessment Journal, 6*(1), 30–38 (Jan/Feb 1999).

Wyatt, P. J. (1997). The development of a GIS-based property information system for real estate valuation. *International Journal of Geographical Information Science, 11*(5), 435–450.

Further Evaluating the Impact of Kernel and Bandwidth Specifications of Geographically Weighted Regression on the Equity and Uniformity of Mass Appraisal Models

Paul E. Bidanset, John R. Lombard, Peadar Davis, Michael McCord and William J. McCluskey

Abstract Research has consistently demonstrated that geographically weighted regression (GWR) models significantly improve upon accuracy of ordinary least squares (OLS)-based computer-assisted mass appraisal (CAMA) models by more accurately accounting for the effects of location (Fotheringham et al. 2002; LeSage 2004; Huang et al. 2010). Bidanset and Lombard (2014a, 2017) previously studied the impacts of various kernel and bandwidth combinations employed in building residual (i.e. sale price less land value) GWR CAMA models and found that the specification of each does bear significant effect on valuation equity attainment. This paper builds upon the previous research by comparing performance of weighting specifications of non-building residual (i.e. full sale price) GWR CAMA models using new data of a different geographic real estate market. We find that the exponential kernel and fixed bandwidth together achieve a superior COD for our data, and that COD does fluctuate depending on the GWR weighting specification.

Keywords Geographic weighted regression · Automated valuation methods

P.E. Bidanset (✉) · P. Davis · M. McCord
School of the Built Environment, Ulster University, Newtownabbey, UK
e-mail: pbidanset@gmail.com

J.R. Lombard
Old Dominion University, Norfolk, VA, USA

W.J. McCluskey
African Tax Institute, University of Pretoria, Pretoria, South Africa

M. d'Amato and T. Kauko (eds.), *Advances in Automated Valuation Modeling*,
Studies in Systems, Decision and Control 86, DOI 10.1007/978-3-319-49746-4_11

1 Introduction and Background

Governments throughout the world have the responsibility of taxing the properties within their jurisdictions. In some countries, taxing authorities incur costs from potentially lengthy tax payer appeal processes that may even result in court time and, if unable to defend and justify their valuations, legal damages. Coupled with an ethical obligation to the tax payers to provide "fair" estimates of value, it is paramount that property assessments be accurate, equitable and defensible. With the advancement of methodologies and computational capabilities, computer-assisted mass appraisal (CAMA) models have improved appraisals with respect to both accuracy and defensibility (Moore and Myers 2010).

While helpful in revealing and quantifying drivers of price, ordinarily least squares (OLS)-based CAMA models may fail to account for spatial heterogeneity and spatial autocorrelation, resulting in misleading coefficient estimates (Ball 1973; Berry and Bednarz 1975; Anselin and Griffith 1988), and while location-based binary variables may help improve value estimates, estimates oftentimes still suffer from errors and biases (Berry and Bednarz 1975; Fotheringham et al. 2002; McMillen and Redfearn 2010). Locally weighted regression (LWR) has been shown to significantly improve the predictability power of OLS-based CAMA models by allocating more weight to similar observations during regression estimation (e.g. Brunson et al. 1996; McMillen 1996; Brunsdon 1998). Geographically weighted regression (GWR) is one such LWR technique that allows coefficient estimates to vary over geographic space (Fotheringham et al. 2002; LeSage 2004; Huang et al. 2010). With an ability to reflect particular behaviors of submarkets, GWR has been shown to significantly reduce spatial autocorrelation and spatial heterogeneity in CAMA models (Borst and McCluskey 2008; Moore 2009; Moore and Myers 2010; Lockwood and Rossini 2011; McCluskey et al. 2013; Bidanset and Lombard 2014b). For these reasons, it is increasing becoming an applied modeling technique among mass appraisal researchers and practitioners.

Previous research has found that the specification of kernel and bandwidth combinations of GWR models can have varying effects of overall model performance with respect to predictability power (e.g. Guo et al. 2008; Cho et al. 2010). Specifically in the field of mass appraisal, GWR kernel and bandwidth specifications have also been shown to have fluctuating impacts on equity and uniformity. Bidanset and Lombard (2014a, 2017) find that equity and uniformity attainment of a building residual[1] GWR model can vary significantly depending on the kernel and bandwidth employed (ceteris paribus). This paper will expand upon the understanding of GWR's capabilities in mass appraisal in two new ways: first by comparing performance of spatial weighting functions of non-residual CAMA models

[1]Building residual techniques subtract an a priori land value from the sale price of a valuation model (IAAO 2013). The resulting value is the theoretical price of the building (improvement) only, and the independent variables used in a building residual CAMA model are used to isolate price determinants of the physical structure(s) only (age, living area, condition, etc.).

(i.e. no a priori values are subtracted from sales prices prior to analyses), and secondly, by applying models to a geographic real estate market not previously analyzed.

2 Model Descriptions and Estimation Details

Ordinary least squares (OLS) is a regression technique represented by the following formula:

$$y_i = \beta_0 + \sum\nolimits_k \beta_k x_{ik} + \varepsilon_i \tag{1}$$

β_0 = intercept
β_k = kth coefficient
x_{ik} = kth variable for the ith sale
ε_i = error term of the ith sale

Geographically weighted regression is a modification of OLS represented by the following formula:

$$y_i = \beta_0(x_i, y_i) + \sum\nolimits_k \beta_k(x_i, y_i) x_{ik} + \varepsilon_i \tag{2}$$

where (x_i, y_i) = xy coordinates of the ith regression point.

GWR allows for a distance-decay spatial weighting component, whereby properties closer to a regression point to receive a higher weight than properties further away. Weighting specifications of GWR consist of a kernel and bandwidth, with the former specifying how weights will be calculated and the latter specifying the magnitude of the weighting assignment.

This paper will evaluate the Gaussian and exponential kernels. Neither of these kernels is discontinuous, meaning even observations outside of the specified bandwidth have an impact, albeit relatively small, on the local regression estimate. Figure 1 reveals how the weights of each kernels are calculated. Figure 2 depicts visually how the weights are assigned. Bandwidths can be either fixed (based on distance) or adaptive (based on a number of nearest neighbors to include). In the case of CAMA GWR models, a fixed bandwidth will include all sales within a

Fig. 1 Kernel functions

Gaussian $\quad w_{ij} = \exp\left(-\frac{1}{2}\left(\frac{d_{ij}}{b}\right)^2\right)$

Exponential $\quad w_{ij} = \exp\left(-\frac{|d_{ij}|}{b}\right)$

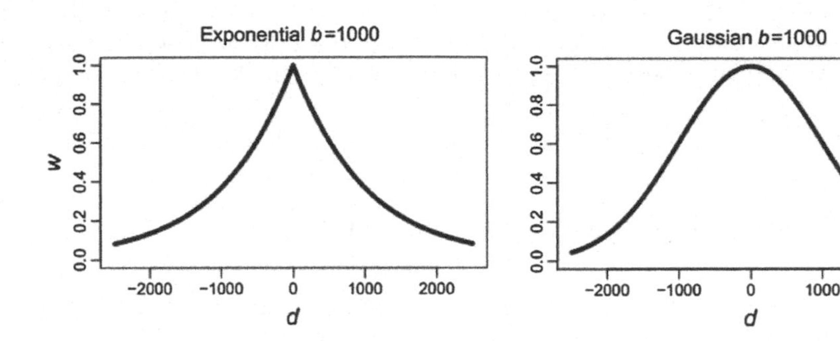

Fig. 2 Weight distribution of kernel functions

specified distance, while an adaptive bandwidth will include a specified number of sales nearest the subject property.

(Gollini et al. 2013, p. 5. Used with permission.)

where: w_{ij} is the weight applied to the *j*th property at regression point *i*

b is the bandwidth

d_{ij} is the geographic distance between regression point *i* and property *j*

(Gollini et al. 2013, p. 6. Used with permission.)

Whether fixed or adaptive, the optimal bandwidth is identified as that which corresponds to the lowest Akaike information criterion corrected (AICc) score[2] (Lu et al. 2014).

3 Model Valuation Performance

The International Association of Assessing Officers (IAAO) maintains standards to property valuation performance (IAAO 2003). They designate the coefficient of dispersion (COD) and price-related differential (PRD) to evaluate assessment uniformity and vertical equity, respectively. The COD is a statistic that measures the uniformity of a set of valuations. It is the average percent of dispersion around the median assessment-to-sale ratio:

$$COD = \frac{100}{R_m}\left[\sum_{1}^{N}\frac{(|R_i - R_m|)}{N}\right] \tag{3}$$

[2]The Akaike Information Criterion (AIC) is a goodness-of-fit measurement. AIC corrected (AICc) is a goodness-of-fit measurement that penalizes for irrelevant variables (Sugiura 1978).

R_m = median assessment-to-sale ratio
R_i = observed assessment-to-sale ratio for each sale
N = number of properties sampled.

For single-family homes, the IAAO designates 5.0–15.0 as an acceptable COD range; values below suggest sampling error or sales chasing (Gloudemans and Almy 2011).

The price-related differential (PRD) measures vertical equity of a set of valuations:

$$PRD = \frac{\sum_i \frac{\left[\frac{\hat{Y}_i}{Y_i}\right]}{n}}{\sum_i \frac{\left[Y_i * \left(\frac{\hat{Y}_i}{Y_i}\right)\right]}{\sum_i Y_i}} \tag{4}$$

The IAAO suggest a PRD should lie between 0.98 and 1.03; values above (below) this range suggest evidence of regressivity (progressivity) (Gloudemans 1999).

4 The Data

The data consists of 1925 arm's length, single-family home sales in Bloomington, IL, USA from 2011 to 2013. Sales that did not reflect market value—such as those between family members, foreclosures or other distressed sales—were omitted from the analysis. To safeguard against the inclusion of erroneously entered data that could impede model performance, sales with a price less than or equal to zero were filtered out, as were any with other counter-intuitive or impossible values (e.g. negative total living area). The natural logarithm of the sale price (LnSalePrice) is the dependent variable. The independent variables, along with their descriptions, are displayed in Table 1.

Table 1 Independent variables

Variable name	Description
TOTAL_SF	Total square feet of home
SH_LOT_SF	Total square feet of lot
GAR_SF	Total square feet of garage
FIN_BS_SF	Total square feet of finished basement
EXTR_COND	Linearized exterior condition variable
GRADE	Linearized grade variable
RM15	Reverse month time spline variable—15th month
RM24	Reverse month time spline variable—24th month

Table 2 Results by spatial weighting function

Model number	Kernel	Band-width	COD	PRD
1	Gaussian	Adaptive	14.57	1.04
2	Gaussian	Fixed	14.40	1.04
3	Exponential	Adaptive	12.93	1.03
4	Exponential	Fixed	11.67	1.03

TOTAL_SF is the size of the home, measured in square feet. *SH_LOT_SF* is the size of the lot, measured in square feet, on which the home sits. *GAR_SF* is the size of the garage—either attached, detached, or both—associated with the property. *FIN_BS_SF* is the total square feet of finished, livable basement space. *EXTR_COND* and *GRADE* are linearized variables variables representing the home's assigned exterior condition and grade, respectively.

To help account for fluctuations over time, temporal spline variables were constructed and assigned based on each transaction's reverse month of sale value[3]. The only splines that offered improved explanatory power (and were subsequently added to the model) were *RM15* and *RM24*. These spline variables have been shown add more explanatory power to CAMA models than monthly indicator variables (Borst 2013) Table 2.

5 Results

The most uniform valuations are produced by model 4 (exponential fixed), with a COD of 11.67, followed by model 3 (exponential adaptive [12.93]). Model 1 yields the least uniform valuations (COD = 14.57), with perhaps only a negligible improvement over model 2 (Gaussian fixed [14.40]). The COD range across all models is 2.9. With respect to IAAO standards, none of the models achieves an unacceptable level of COD, though models 1 and 2 do approach the 15.0 limit. The PRD remains relatively stable, ranging from 1.03 for models with an exponential kernel, to 1.04 for models with a Gaussian kernel. While this .01 change does cross over the upper limit of the IAAO PRD threshold (1.03), suggesting evidence of regressivity for models 1 and 2, it should be noted that the change is however, very small, and may not be statistically significant.

Figure 3 demonstrates the spatial variation in the assessment-to-sale price ratios of each of the four models.

[3]Sales in the most recent month of the dataset receive a reverse month of sale value of 1. Sales in the second most recent month of the dataset receive a reverse month of sale value of 2 (and so on). A dataset consisting of three full years of sales will have 36 reverse month of sale values (1–36) (Borst 2013).

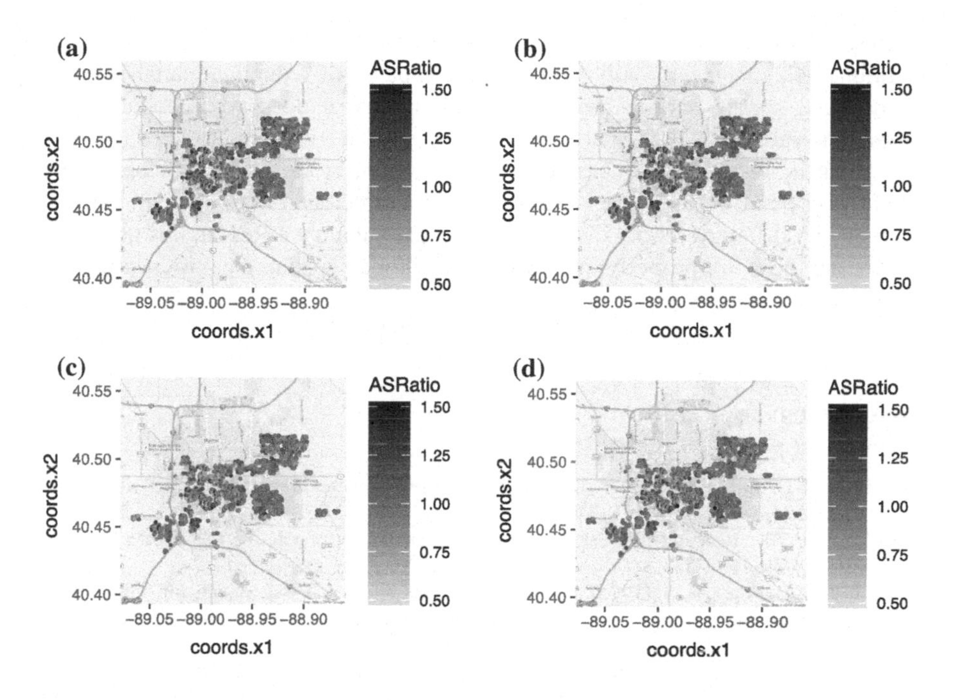

Fig. 3 Appraisal-to-sale price ratio maps by model **a** model 1—gaussian adaptive **b** model 2—gaussian fixed **c** model 3—exponential adaptive **d** model 4—exponential fixed

6 Conclusion

This paper expanded the understanding of GWR as a CAMA modeling technique by examining equity and uniformity attainment variability due to kernel and bandwidth specifications of non-residual property valuation models. The models were applied to arm's-length residential sales in Bloomington, IL, USA that transferred between 2011 to 2013.

Four models were compared with the only variations being the kernel (Gaussian or exponential) and bandwidth (fixed or adaptive) of each. The model combining an exponential kernel with a fixed bandwidth yielded the most uniform valuations (i.e. lowest COD). Exponential kernels outperformed Gaussian kernels with respect to uniformity regardless of the bandwidth employed (within this market). Building upon previous research, these findings suggest that kernel and bandwidth manipulation are beneficial to the optimization of both residual and non-residual CAMA models. A kernel or bandwidth combination deemed "optimal" in one market may not prove so in another (as exemplified by geographically varying assessment-to-sale price ratios). Combinations should be evaluated with respect to model performance as data change and/or update.

Research on the subject of GWR's role within the CAMA realm can benefit from additional advancements, still. GWR, while capable of reducing spatial autocorrelation and spatial heterogeneity, is limited in that it does not account for all geographic variations, particularly those experienced at a more micro-level (e.g. one sale being more closely situated to noise pollution, crime, or other externalities (negative or positive) than other sales on the same street; two sales in vicinity of one another with differing school districts or zoning regulations). Research is needed to understand how to properly incorporate additional spatial data to supplement GWR CAMA models.

References

Anselin, L., & Griffith, D. A. (1988). Do spatial effects really matter in regression analysis? *Papers in Regional Science, 65*(1), 11–34.

Ball, M. J. (1973). Recent empirical work on the determinants of relative house prices. *Urban Studies, 10*(2), 213—233.

Berry, B. J., & Bednarz, R. S. (1975). A hedonic model of prices and assessments for single-family homes: Does the assessor follow the market or the market follow the assessor? *Land Economics, 51*(1), 21–40.

Bidanset, P. E., & Lombard, J. R. (2014a). The effect of kernel and bandwidth specification in geographically weighted regression models on the accuracy and uniformity of mass real estate appraisal. *Journal of Property Tax Assessment & Administration, 11*(3).

Bidanset, P. E., & Lombard, J. R. (2014b). Evaluating spatial model accuracy in mass real estate appraisal: a comparison of geographically weighted regression and the spatial lag model. *Cityscape: A Journal of Policy Development and Research, 16*(3), 169.

Bidanset, P. E., & Lombard, J. R. (2017). Optimal kernel and bandwidth specifications for geographically weighted regression. *Applied Spatial Modelling and Planning*. J. R. Lombard, E. Stern, G. Clarke (Ed.). Abingdon, Oxon; New York, NY: Routledge.

Borst, R. (2013). Optimal market segmentation and temporal methods. *spatio-temporal methods in mass appraisal*. Fairfax, VA: International Property Tax Institute. Mason Inn Conference Center.

Borst, R. A., & McCluskey, W. J. (2008). Using geographically weighted regression to detect housing submarkets: Modeling large-scale spatial variations in value. *Journal of Property Tax Assessment and Administration, 5*(1), 21–51.

Brunsdon, C. (1998). Geographically weighted regression: A natural evolution of the expansion method for spatial data analysis. *Environment and planning A, 30*, 1905–1927.

Brunsdon, C., Fotheringham, A. S., & Charlton, M. E. (1996). Geographically weighted regression: A method for exploring spatial nonstationarity. *Geographical Analysis, 28*(4), 281–298.

Cho, S. H., Lambert, D. M., & Chen, Z. (2010). Geographically weighted regression bandwidth selection and spatial autocorrelation: An empirical example using Chinese agriculture data. *Applied Economics Letters, 17*(8), 767–772.

Cleveland, W. S., & Devlin, S. J. (1988). Locally weighted regression: An approach to regression analysis by local fitting. *Journal of the American Statistical Association, 83*(403), 596–610.

Fotheringham, A. S., Brunsdon, C., & Charlton, M. (2002). *Geographically weighted regression: The analysis of spatially varying relationships*. Chichester, West Sussex, England: Wiley.

Gloudemans, R. J. (1999). *Mass appraisal of real property*. Chicago, IL: International Association of Assessing Officers.

Gloudemans, R., & Almy, R. (2011). *Fundamentals of mass appraisal*. Kansas City, MO: International Association of Assessing Officers.

Gollini, I., Lu, B., Charlton, M., Brunsdon, C., & Harris, P. (2013). GWmodel: An R package for exploring spatial heterogeneity using geographically weighted models. arXiv preprint: arXiv: 1306.0413.

Guo, L., Ma, Z., & Zhang, L. (2008). Comparison of bandwidth selection in application of geographically weighted regression: A case study. *Canadian Journal of Forest Research, 38* (9), 2526–2534.

Huang, B., Wu, B., & Barry, M. (2010). Geographically and temporally weighted regression for modeling spatio-temporal variation in house prices. *International Journal of Geographical Information Science, 24*(3), 383–401.

IAAO (2003). Standard on automated valuation models (AVM's). *Assessment Journal* (Fall 2003), 109–154.

International Association of Assessing Officers. (2013). *Standard on ratio studies*. Chicago, IL: IAAO.

LeSage, J. P. (2004). A family of geographically weighted regression models. In *Advances in Spatial Econometrics* (pp. 241–264). Springer.

Lockwood, T., & Rossini, P. (2011). Efficacy in modelling location within the mass appraisal process. *Pacific Rim Property Research Journal, 17*(3), 418–442.

Lu, B., Harris, P., Charlton, M., & Brunsdon, C. (2014). The GWmodel R package: Further topics for exploring spatial heterogeneity using geographically weighted models. *Geo-spatial Information Science, 17*(2), 85–101.

McCluskey, W. J., McCord, M., Davis, P. T., Haran, M., & McIlhatton, D. (2013). Prediction accuracy in mass appraisal: A comparison of modern approaches. *Journal of Property Research, 30*(4), 239–265.

McMillen, D. P. (1996). One hundred fifty years of land values in Chicago: A nonparametric approach. *Journal of Urban Economics, 40*(1), 100–124.

McMillen, D. P., & Redfearn, C. L. (2010). Estimation and hypothesis testing for nonparametric hedonic house price functions. *Journal of Regional Science, 50*(3), 712–733.

Moore, J. W. (2009). A History of appraisal theory and practice looking back from IAAO's 75th year. *Journal of Property Tax Assessment & Administration, 6*(3), 23.

Moore, J. W., & Myers, J. (2010). Using geographic-attribute weighted regression for CAMA modeling. *Journal of Property Tax Assessment & Administration, 7*(3), 5–28.

Sugiura, N. (1978). Further analysts of the data by Akaike's information criterion and the finite corrections. *Communications in Statistics-Theory and Methods, 7*(1), 13–26.

Dealing with Spatial Modelling in Minsk

Maurizio d'Amato, Nikolaj Siniak and Paola Amoruso

Abstract The contribution is focused on comparing different automated valuation models dealing in different ways with location variable. Using a sample of 290 observations in Minsk, several different AVM models will be compared. A linear and log linear model of AVM with constant location variable will be applied. Therefore, after the determination of spatial correlation using the Moran I test the application of mixed regressive model integrating the geographic variable with the specific technical characteristics of the property. The results confirm an increasing quality of the model. Further works can be required to include also temporal variable in this class of models (Borst 2015).

Keywords Automated valuation modelling · Spatial lag models · Mixed autoregressive models

1 Introduction

Location and externalities play an important role in in property values (Krantz et al. 1982; Des Rosiers et al. 1996). The location may affect property value either determining the presence of excessive multicollinearity among attributes or spatial autocorrelation among residuals (Dubin 1988; Anselin and Rey 1991). In the specific case starting from a sample of 290 observations in Minsk several automated

---

The paper has been written in strict cooperation among the three authors. Therefore the credit of the article should be equally divided among them.

M. d'Amato (✉)
DICATECh, Technical University, Politecnico di Bari, Bari, Italy
e-mail: madamato@fastwebnet.it

N. Siniak
Belorussia State Technological University, Bari, Italy

P. Amoruso
LUM Jean Monnet University, Casamassima, Ba, Italy

M. d'Amato and T. Kauko (eds.), *Advances in Automated Valuation Modeling*,
Studies in Systems, Decision and Control 86, DOI 10.1007/978-3-319-49746-4_12

valuation techniques will be applied. The starting point will be the hedonic modelling linear and loglinear. Previously several contributions tried a comparison among different techniques dealing with spatial analysis. In particular Wilhelmsson (2002) compared the traditional multiple regression analysis using dichotomic variables with Spatial Autoregressive Models and Spatial Extension Method. Borst and McCluskey (2007) analysed Geographic Weighted Regression and Comparable Sales Method. Comparable Sales Method used in a spatial lag model diminish the Coefficient of Deviation more than Geographic Weighted Regression. Conway et al. (2010) demonstrates that Spatial Lag Model improves traditional multiple regression analysis performance considering spatial autocorrelation. Quintos (2013) use spatial lag models to create location adjustment factors therefore included them in traditional OLS. In this contribution we compare spatial lag models, spatial error models with the traditional linear hedonic model and the Cobb Douglas model with and without a dummy variable in order to take into account of spatial heterogeneity.

2 An Application of AVM Modelling in Minsk

The variables used in the first linear model are the following three (Table 1).

In the following Table 2 it is possible to observe the statistics of the sample of 290 residential properties in Minsk.

In order to analyse the quality of the results three indicators have been taken into account. The Mean Absolute Percentage Error whose formula is indicated below:

$$MAPE = \sum_{i=1}^{n} \frac{\left| \frac{PS_i - AS_i}{AS_i} \right| \cdot 100}{n}$$

In the formula PS are the predicting price with the application of the model, while AS are the actual selling, n is the number of the observations of the sample. In the IAAO standards two indicators are considered relevant in the application of AVM methodology. The former is the Coefficient of Dispersion also indicated as COD. It is indicated in the formula below

Table 1 Variables considered in the AVM model

Variable	
AR	Square meter of property
LEV	Level of floor
DAT	Date of sale
LAT	Latitude
LON	Longitude

Table 2 Descriptive statistics of the sample

	PRZ	DAT	LEV	AR
Minimum	41500.00	2.00	1.00	25.62
Maximum	74734.00	9.00	18.00	53.00
Mean	52483.82	8.11	6.28	38.21
St.dev	7962.21	0.85	4.09	6.54

$$\mathrm{COD} = \frac{100}{\mathrm{n}} \frac{\sum_{i=1}^{n} \left| \frac{\mathrm{PS_t}}{\mathrm{AS_t}} - \mathrm{Median}\left(\frac{\mathrm{PS_t}}{\mathrm{AS_t}}\right) \right|}{\mathrm{Median}\left(\frac{\mathrm{PS_t}}{\mathrm{AS_t}}\right)}$$

In the formula the term PS indicates the predicted selling price after the application of thc mcthod while the term AS means the actual or the real price analysed. A second important measure provided by the standards IAAO to detect vertical equity is the price related differential or PRD

$$\mathrm{PRD} = \frac{\mathrm{Mean}\left(\frac{\mathrm{PS_t}}{\mathrm{AS_t}}\right)}{\sum_{t=1}^{n} \frac{\mathrm{PS_t}}{\mathrm{AS_t}}}$$

All the output of the models will be tested using all the three ratios. The first application is a normal linear regression model having the following formula.

$$P = b_0 + b_1 x_1 + b_2 x_2 + \ldots + b_n x_n$$

In the first formula P means the dependent variable or prices (Ciuna 2010; Ciuna et al. 2015; Ciuna et al. 2014), while b are the marginal prices of the characteristics and x are the measure of the characteristics. The first linear model has the following statistics indicated in the Table 3.

The hedonic modelling form depends on assumptions arbitrarily chosen by a researcher (Cassel and Mendelson 1985). Though there are many examples in the literature of linear hedonic models, there is a belief that hedonic model is nonlinear

Table 3 Results of the linear regression analysis runned on the 290 observations

	AR	LEV	DAT	LOC
	1126.559353	131.517	– 74.7756	9234.6833
R^2	0.819519449			
R^2 Adj	0.817935444			
F	536.31			
t	39.91239001	2.90955	– 0.34506	9234.6833
MAPE	0.055981574			
COD	5.70044303			
PRD	1,06,734,094			

(Ekeland et al. 2002; Kauko 2003; Kaklauskas et al. 2012). For this reason a semi-log or log-log form is often used as the nonlinear representative. In this case a log-log form (Cobb Douglas) form was selected whose formula is indicated below:

$$\ln(P) = \ln b_0 + b_1 \ln x_{i1} + b_2 \ln x_{i2} + \ldots + b_n \ln x_{in}$$

In the first formula ln (P) are the logarithm of the dependent variable or prices, while b are the marginal prices of the characteristics and ln x are the logarithm of the characteristics. The second log-log model (Cobb Douglas) showed the following statistics indicated in the Table 4.

The observations are spatially distributed as indicate in the Fig. 1.

It is well known in literature that AVMs without spatial effect may drive to a inaccurate results because of the different behaviour of the market across geographic space (Berry and Bednarz 1975). According to Anselin (1988), there are two major types of spatial effects: spatial autocorrelation and spatial heterogeneity. Spatial autocorrelation refers to a functional relationship between observations. Spatial heterogeneity, on the other hand, refers to the lack of uniformity arising from space, leading to spatial heteroscedasticity and spatially varying parameters. In our case spatial autocorrelation has been detected using Moran's I (Moran 1948; Moran 1950) test. This index measures autocorrelation between values of the x

Table 4 Results of the cobb douglas regression model runned on the 290 observations

	AR	LEV	DAT	LOC
	0.816573181	0.00861	–0.02192	2767.2159
R^2	0.838683113			
R^2 Adj	0.83698504			
F	493.9030062			
t	38.32024896	3.17611	–0.34353	240.42471
MAPE	0.060803507			
COD	5.987130797			
PRD	1.520257789			

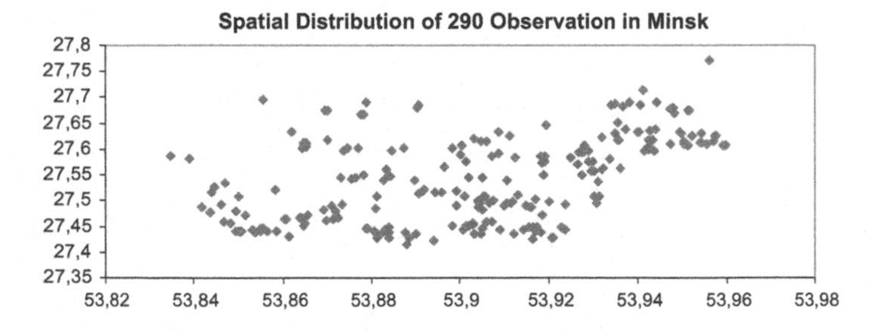

Fig. 1 Spatial distribution of 290 sample of properties transactions in Minsk

vector. It ranges from −1 to +1 and each observation is only compared with its relevant neighbourhood. Positive Moran's I indicates positive autocorrelation which means that high values for x should be located near other high values while values should be located near other low values. A significantly negative Moran's I implies spatial heterogeneity, or that high values are near low, or vice versa. Moran's test formula is indicated in the formula below

$$I = \frac{N\sum_{i=1}^{n}\sum_{i=1}^{n} w_{ij}(x_i - \bar{x})(x_j - \bar{x})}{\sum_{i=1}^{n}\sum_{i=1}^{n} w_{ij}(x_i - \bar{x})^2}$$

where: x is the variable and w_{ij} represents the set of neighbours j for observation i. In this case, as in previous examples in literature, inverse squared distance among the observations has been considered (Des Rosiers et al.,1991). The final result showed positive autocorrelation assuming a value of 0.8234. As a first approach to deal with spatial location a dummy variable has been introduced in the model (Kauko and d'Amato 2011; d'Amato 2010; d'Amato 2012; d'Amato 2004; d'Amato 2015). In particular the set of observations have been divided in two main areas using a dummy variable. Therefore the model varied as follows (Table 5).

In particular the area was divided in two part and in only dummy variable was added improving the quality pof the model (Table 6). The third log log model having a dummy variable shows the following statistics indicated in the Table 4.

Table 5 Dummy variable in the log model

Variables	
AR	Square meter of property
LEV	Level of floor
DAT	Date of sale
DU1	Dummy variable 1
LAT	Latitude
LON	Longitude

Table 6 Final result of the cobb douglas model using dummy variables

	DU1	AR	LEV	DAT	LOC
	−0.024376618	0.82465	0.006813	−0.015967	2688.01751
R^2	0.840154847				
R^2 Adj	0.837903506				
F	373.1798735				
t	−3.304955091	38.414	1.430231	−0.646347	81.5769583
MAPE	0.050803507				
COD	4.987130797				
PRD	1.020257789				

Table 7 Final result of mixed autoregressive models

	AR	LEV	DAT	LOC	WPRZ
	1128.79	135.75	−76.7698	−9121.3	0.0000267
R^2	0.848833				
z	40.1532	30.1506	−0.3551	1.3741	−0.000219117
MAPE	0.0497768				
COD	4.89696243				
PRD	1.01767129				

The last model is a spatial lag model. In particular a specific spatial lag model (Can 1990; Can 1992) a mixed regressive model including spatial variable in the traditional hedonic form has been selected. The final result of the model is indicated in the table (Table 7).

3 Conclusion

The paper showed a comparison among different Automated Valuation Modelling. The contribution that spatial approaches may give in presence of spatial autocorrelation clearly emerged. In particular the application of mixed autoregressive model to the context permitted a more satisfactory results for the most important indicators. The integration between the spatial dimension and the technical features of the property allowed a better understanding of the market (Renigier-Biłozor et al. 2014a, b). Further works can be required in order to include temporal variable inside the AM modelling in order to increase the predictive power of the model (Borst 2015).

References

Anselin, L. (1988). *Spatial Econometrics*. Methods and Model, Dordrecht: kluwer Academic Publisher

Anselin, L., & Rey, S. (1991). Properties of tests for spatial dependence in linear regression models. *Geographical Analysis, 23*(2), 112–131.

Berry, Brian J., & Bednarz, Robert S. (1975). A hedonic model of prices and assessments for single-family homes: Does the assessor follow the market or the market follow the assessor? *Land Economics, 51*(1), 21–40.

Borst, R., & McCluskey, W. (2007). *Comparative Evaluation of the Comparable Sales Method With Geostatistical Valuation Models Pacific Rim Property Research Journal, 13*(1), 106–129.

Borst, R. (2015). *Improving mass appraisal valuation models using spatio-temporal models*. International Property Tax Institute.

Can, A. (1990). The measurement of neighborhood dynamics in urban house prices. *Economic Geography, 66*, 254–272.

Can, A. (1992). Specification and estimation of hedonic housing price models. *Regional Science and Urban Economics, 22*, 453–474.

Cassel, E., & Mendelson, R. (1985). The choice of functional forms for hedonic price equations: Comment. *Journal of Urban Economics, 18*, 135–142.

Ciuna, M. (2010). L'Allocation Method per la stima delle aree edificabili. *AESTIMUM, 57*, 171–18.

Ciuna, M., Salvo, F., & Simonotti, M. (2015). Appraisal value and assessed value in italy. *International Journal of Economics and Statistics*, 24–31. ISSN: 2309-0685.

Ciuna, M., Salvo, F., & De Ruggiero, M. (2014). Property prices index numbers and derived indices. *Property Management, 32*(2), pp. 139–153. doi:10.1108/PM-03-2013-0021

Ciuna, M., Salvo, F., & Simonotti, M. (2014). Multilevel methodology approach for the construction of real estate monthly index numbers. *Journal of Real Estate Literature, 22*(2), pp. 281–302.

Conway, Delores, Christina Q. Li, Jennifer Wolch, Christopher Kahle, and Michael Jerrett. (2010) A Spatial Autocorrelation Approach for Examining the Effects of Urban Greenspace on Residential Property Values. *The Journal of Real Estate Finance and Economics, 41*(2) 150–169.

d'Amato. (2010). A location value response surface model for mass appraising: An "Iterative" location adjustment factor in Bari, Italy. *International Journal of Strategic Property Management, 14*, 231–244.

d'Amato, M. (2004). A comparison between rst and mra for mass appraisal purposes. A case in Bari. *International Journal of Strategic Property Management, 8*, 205–217.

d'Amato, M., & Kauko, T. (2012, March–April). Sustainability and risk premium estimation in property valuation and assessment of worth. *Building Research and Information, 40*(2), 174–185.

d'Amato, M. (2015). Income approach and property market cycle. *International Journal of Strategic Property Management, 29*(3), 207–219.

Des Rosiers, F., Lagana, A., Thériault, M., & Beaudoin, M. (1996). Shopping centres and house values: An empirical investigation. *Journal of Property Valuation and Investment, 14*(4), 41–62.

Dubin, R. A. (1988). Estimation of regression coefficients in the presence of spatially autocorrelated error terms. *Review of Economics and Statistics, 70*(3), 466–474.

Ekeland, I., Heckman, J. J., & Nesheim, L. (2002). Identifying hedonic models. *American Economic Review, 92*(2), May. 304–309.

Kaklauskas, A., Zavadskas, Edmundas Kazimieras, Kazokaitis, Paulius, Bivainis, Juozas, Galiniene, Birute, d'Amato, Maurizio, et al. (2012). Crisis management model and recommended system for construction and real estate. In Radosław Katarzyniak & Geun-Sik Jo (Eds.), *Ngoc Thanh Nguyen, Bogdan Trawi'nski* (pp. 333–343). Advanced methods for computational collective intelligence in studies in computational intelligence series edited by Janusz Kacprzyk: Springer.

Kauko, T. (2003). Residential property value and locational externalities: On the complementarity and substitutability of approaches. *Journal of Property Investment & Finance, 21*(3), 250–263.

Kauko, T. & d'Amato, M. (2011). *Neighbourhood effect*. Elsevier: International Encyclopedia of Housing and Home.

Krantz, D. P., Weaver, R. D., & Alter, T. R. (1982). Residential property tax capitalization: Consistent estimates using micro-level data. *Land Economics, 58*(4), 488–496.

Moran, P. A. P. (1948). The interpretation of statistical maps. *Journal of the Royal Statistical Society Series B-Statistical Methodology, 10*, 243–251.

Moran, P. A. P. (1950). Notes on continuous stochastic phenomena. *Biometrika, 37*(1–2), 17–23. doi:10.1093/biomet/37.1-2.17.

Quintos, Carmela. (2013). Spatial weight matrices and their use as baseline values and location-adjustment factors in property assessment models. *Cityscape, 15*(3), 295–306.

Renigier-Biłozor, M., Wiśniewski, R., Biłozor, A., & Kaklauskas, A. (2014a). Rating methodology for real estate markets—Poland case study. *International Journal of Strategic Property Management, 18*(2), 198–212. ISNN. 1648-715X.

Renigier-Biłozor, M., Dawidowicz, A., & Radzewicz, A. (2014b). An algorithm for the purposes of determining the real estate markets efficiency in land administration system. *Public Survey Review., 46*(336), 189–204.

Wilhelmsson, Mats. (2002). Spatial models in real estate economics. *Housing, Theory and Society, 19*(2), 92–101

Using Multi Level Modeling Techniques as an AVM Tool: Isolating the Effects of Earthquake Risk from Other Price Determinants

Richard Dunning, Berna Keskin and Craig Watkins

Abstract Automated valuation methods employ diverse methodologies. The appropriateness of the individual model is contingent upon the characteristics of the housing market in question. In circumstances where the housing market is spatially variegated and data is sparse, distinguishing impacts of an environmental event or externality on different spatial segments of the market is challenging. A multi-level model provides one possible approach. Here we apply the model to explore the impact of a natural disaster on the housing market in Istanbul, a city located in a region with relatively frequent seismic activity. The Chapter provides a relatively simple exemplar of the method and its utility. Two levels of influence emerge. We distinguish between the citywide and segmented neighbourhood impact of earthquakes on house prices. Appraisers working in segmented markets where the potential for natural disasters occur might consider the methodological advantages of using multi-level modeling to help isolate both the effects of the risk of damage and also to discern the spatial variations in such effects. Furthermore we contend that there is considerable scope to expand the modelling approach to take account of different levels.

Keywords Automated valuation methods · Multi level modeling · Earthquakes · Shocks · Hedonics · Regression

R. Dunning (✉)
University of Liverpool, Liverpool, UK
e-mail: r.j.dunning@liverpool.ac.uk

B. Keskin · C. Watkins
The University of Sheffield, Sheffield, UK
e-mail: b.keskin@sheffield.ac.uk

C. Watkins
e-mail: c.a.watkins@sheffield.ac.uk

M. d'Amato and T. Kauko (eds.), *Advances in Automated Valuation Modeling*,
Studies in Systems, Decision and Control 86, DOI 10.1007/978-3-319-49746-4_13

1 Introduction

> Strange and mysterious things, though, aren't they - earthquakes? We take it for granted that the earth beneath our feet is solid and stationary. We even talk about people being 'down to earth' or having their feet firmly planted on the ground. But suddenly one day we see that it isn't true. The earth, the boulders, that are supposed to be solid, all of a sudden turn as mushy as liquid
>
> —Haruki Murakami, *After the Quake,* from the short story "Thailand"

Automated valuation methods, housing markets and hedonic regression techniques can all appear to be strange and mysterious things, but when combined with the influence of natural disasters, such as earthquakes, their metaphorical ground may be less solid and stationary.

Haruki Murakami's fictional masterpiece *after the quake* explores the aftermath of the Kobe earthquake in Japan in 1995. The characters in the collection of six short stories are impacted by the earthquake, but not all directly affected by it. This sets the backdrop for Murakami's exploration of the enigma of social variation in self-reflection and changing expectations. Whilst the characters appraise their situation, they find themselves differentially affected by the earthquake, and differentially assessing their futures.

The shadows of natural disasters hang over all property markets. But the temporal regularity, scale and spatial impact of their occurrence is not homogenous. Like Murakami's characters, as predictors of future events, most mass appraisal methods suggests that homebuyers attempt to factor into their decision making the possibility of a natural occurrence and the scale of the impact on values. When the pricing behaviour of individual transactions is amalgamated in areas of natural disasters, appraisers hope that by analyzing the data they can disentangle the bundle of characteristics leading to particular price formations, and use this information to value properties.

Homebuyers, estate agents, lenders and policy makers are all concerned with accurately valuing properties in areas where natural disasters occur (whether infrequently or not). Not least because the locational advantages for urban areas may also correspond to the locus of natural disasters. In Istanbul for example the location of the urban settlement benefits from advantages bringing together two continents and the potential for crossing the Bosphorus which connects the Aegean Sea and the Black Sea. This natural locational advantage also carries a risk of earthquakes as the fault line, the North Atlantic Fault Zone, between the two continents causes regular seismic activity.

Whilst the resources of homebuyer, estate agents, policy makers, lenders and valuers all vary, they are balancing a trade off between accurate valuations of properties in areas of natural disasters and the cost of those valuations. Automated valuation methods offer one solution to this tension. AVM's methodologically neutrality is contingent upon one of the many forms of regression model utilized and the information entered into the model. The appropriate form of the regression model in each case depends on the characteristics of the housing market being conceptualized and the availability of data.

Automated valuation, using regression models, vary widely in complexity and in their conceptual categories, allowing for extensions of the model. Spatially, some AVMs consider house prices to be smoothed over Cartesian space, whilst others consider dummy variables to be adequate proxies for the distinctions between submarkets. Whilst both of these approaches may relate well to particularly structured stable housing markets, the impact of shocks on the housing market arguably requires a reconceptualization of the fundamental spatial characteristics of market behaviour.

This chapter considers the scope to apply new forms of regression model for an AVM within one set of specific circumstances. We consider the introduction of multi-level modeling as a tool to support automated valuations in markets where an environmental event or external market shock has taken place.

In the following sections we outline the complexity of modeling shocks (in particular natural disasters) on granular housing markets, considering spatially smooth distance-decay and discrete submarket effects. The chapter then outlines existing explanations of the impact on housing markets of a particular natural disaster, namely earthquakes. The context of the Istanbul housing market and its earthquake risk follows on, before an outline of the method used in the multi-level model. The results find that earthquake effects can be identified across the city region, and there is a varied spatial impact at the neighbourhood level too, suggesting that AVMs operating in areas of natural disasters could benefit from a multi-level approach. We conclude with a brief discussion of future extensions of AVM multi-level modeling.

2 Modeling the Impact of Shocks on Property Values

Risk in housing markets emerges for a wide variety of reasons, including as a result of the risk of environmental disasters. Natural disasters occur at different geographical scales, with different levels of predictability about their location and impact (e.g. compare a volcano and a cyclone). The type of disaster also has an unequal spatial impact upon the built environment and upon actor's perceptions of risk in the housing market. These variations mean that modeling the impact of event shocks on property values can be very spatially complex. In this chapter we focus on the extension of standard hedonic regression models as a basis for a novel, spatially richer modelling framework. This seeks to encapsulate the positive features of hedonic, whilst overcoming some of the weaknesses common in the applied literature.

Des Rosiers and Theriault (2008) outline three reasons for AVM's use of hedonic regression methods. First, multiple regression analysis uses probability theory to divide the impact of competing influences on house prices. Second, they argue that its calculative nature is objective (see Schulz et al. (2014) for some of the subjective trade offs in constructing a hedonic based AVM) and therefore more likely to produce the market value that fits a testable probability distribution. Third, hedonic approaches reveal the causal dimensions of house pricing, and when combined with GIS can discern the spatial dimensions too. Des Rosiers and Theriault (2008) highlight the conditions necessary for the data to accurately support this threefold rationale, acknowledging that they may not always be reproducible. Whilst we may wish to extend the critique of the detail of these reasons, the perception of hedonic methods to support these reasons does explain it is growing in use by both professionals and the general public.

As discussed extensively earlier in this book, the hedonic techniques undergirding many AVMs have been applied to human and natural phenomenon, as well as the interplay between the two (e.g. parks) and their positive and negative impacts on prices. Environmental variables have been a particularly frequent theme in the hedonic literature (see Ridker and Henning 1968 for an early example). This vast literature covers issues such as the impact of noise pollution, water contamination and the location of hazardous waste sites or powerlines (see Boyle and Kiel 2001). Hedonic models are also frequently utilized in understanding the impact of natural disasters, such as flooding (e.g. Macdonald et al. 1987; Bin et al. 2008), forest fires (e.g. Loomis 2004; Mueller et al. 2009; Stetler et al. 2010) and hurricanes (e.g. Simmons et al. 2002; Hallstrom and Smith 2005). Whilst the precise methodological form varies in many of these papers, the basic proposition encapsulated is that with a hedonic regression it should be possible to discern a discount in property prices that reflects the impact (or risk) of one of these negative external events.

The simple hedonic regression approach assumes that the impact on house prices of an event is constant over time. It views market actors as having the ability and stability to factor this information into their house price calculations consistently over time. In the case of natural disasters, this assumption is only valid if market actors use this information to inform bidding and selling strategies perfectly. It does not easily allow perceptions of the potential of a disaster to vary. Nor does it allow this to feed through into house price patterns.

In the context of this chapter, a key weakness of the most simplistic hedonic models is that they view space as a continuous plain, with the impact and perception of natural disasters equally distributed across the plain. These models have, of course, been extended in a variety of ways to respond to the challenge of spatial variation in attribute values. This includes attempts to model differences in prices that change evenly through Cartesian space by using augmented hedonic methods (e.g. Clapp 2003; Pavlov 2000). These approaches are appealing because they take away the need for the appraiser to possess, or obtain knowledge of spatial boundaries, and simply allow the spatial variance in price to emanate from the data, be smoothed and therefore be predicted quite readily.

The submarket literature offers some pointers for dealing with spatial complexity. Whilst some AVM's support data driven spatial segmentation, Bourussa et al. (2003) found that AVM's using existing spatial submarkets (including those defined by appraisers) produced more accurate price predictions than those based on principal component and cluster analysis approaches. Generally, the smaller the size of each group (whether submarkets, ZIP codes, neighbourhoods), the more accurate the house price predictions have been shown to be (Goodman and Thibodeau 2003). The long-standing argument appears to hold that neighbourhood or submarket boundaries should be used, where known, to improving the predictive power of AVMs (see Strazheim 1975; Schnare and Struyk 1976).

This observation has provided a platform for researchers to ask how, if we accept spatial submarkets as a given, should we seek to most effectively accommodate them in a model (Watkins 2012). Leishman et al. (2013) compare the outcomes from four different modelling strategies when applied to data from Perth, Western Australia. They apply a standard market-wide hedonic equation; a system of submarket specific hedonic models; and two different multi-level model specifications to predict house price patterns across space. Their results suggested that the most spatially granular multi-level specification generates the greatest explanatory power and reduces the instance of non-random spatial errors.

In this vein, we also argue that it seems reasonable to suggest that one way to overcome the problem of dealing with spatial differentiation in housing market models, whilst meeting the necessary conditions for AVMs, is to employ multi-level methods. Multi-level models are a variant on standard hedonic methods (Orford 1999; Leishman 2009). Use of multi-level methods is advised when the observations being analysed are clustered and correlated, the causal processes underlying the relationships operate simultaneously at multiple spatial scales and there is value in seeking to separate out the spatial and temporal effects of different attributes (Subramanian 2010). Their use has begun to expand within the quantitative geography field where the technique has been used to examine complex spatial impacts and interactions in a variety of arenas including in the measurement of social well-being and happiness (see Ballas and Tranmer 2008). We discuss the method more fully in Sect. 3 below.

The remainder of the chapter illustrates how this approach might be implemented. It is our view that this approach (at least partly) extends traditional models and is particularly useful in contexts where delineating the differences in the spatial impact of risk is especially desirable.

3 An Applied Case Study—The Impact of Earthquake Risk on Property Values in Istanbul

3.1 Case Study of Istanbul

The empirical analysis in this chapter focuses on Istanbul, the largest city in Turkey and home to almost fifteen per cent of the population. Its formal housing sector is dominated by market dwellings, much of which are located in high density, inner urban neighbourhoods where the stock often dates from the early twentieth century; or, in the case of newer stock, is found planned housing areas established since the start of the century. Many of the latter properties occupy the mid and higher end of the price scale and are promoted to potential sellers in a manner that draws on the growth in popularity of gated and semi-gated communities that benefit from very good links to transport infrastructure, employment centres and excellent public amenities (Alkay 2011). At the lower end of the market, there are significant numbers of unplanned dwellings, estimated by some to be over fifty per cent of the total, located within squatter settlements, known as 'Gecekondu' (see Gokmen et al. 2006). These informal parts of the sector are occupied by lower income groups and consist of dwellings in poor physical condition and with limited sales values.

As Keskin (2010) explains, although property values have been moving upwards across the market, there has been evidence of increasing divergence between the top and bottom of the market. Figure 1 illustrates the degree of spatial disaggregation

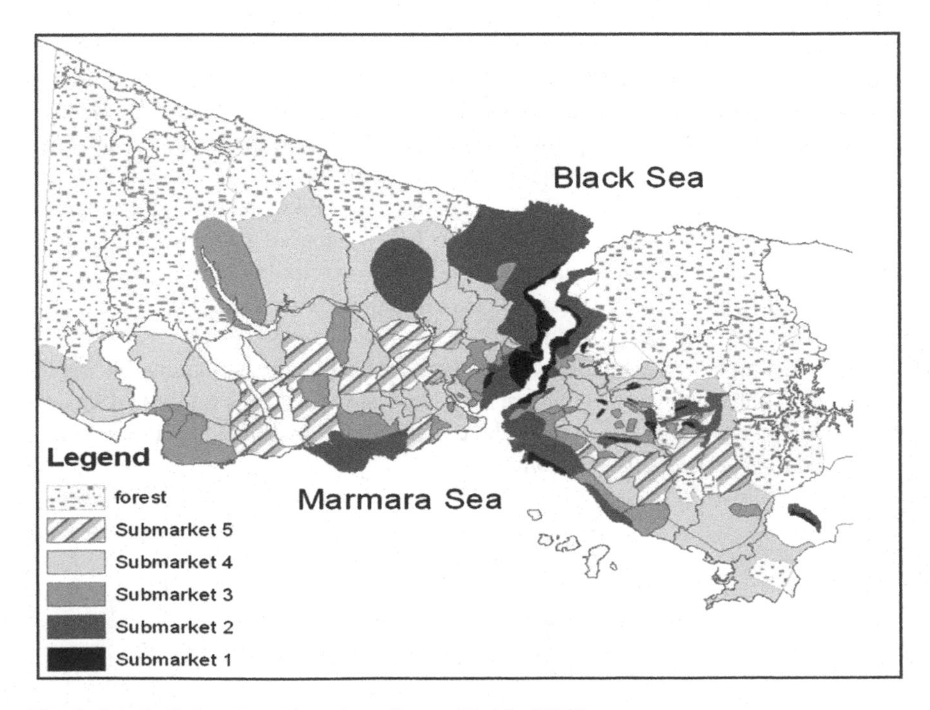

Fig. 1 Istanbul's housing submarkets. *Source* Keskin (2010)

within the market. The submarkets identified on the map comprise neighbourhood that act as close substitutes, even though they are not always spatially contiguous, and according to Keskin represent distinct market segments with their own unique price formation processes and price structure. The highest price neighbourhoods are in submarket 1 while the lowest price subareas are in submarket 5, which also happen to be the neighbourhoods traditionally perceived to be most likely to suffer earthquake damage.

3.2 *Data*

The house price and housing attributes data used in our applied modelling work are drawn from the internet listing services of two leading realtors, Turyap and Remax. This dataset includes details of 2175 housing transactions from 2007 and have been combined with socio-economic, neighborhood and locational attributes collected from a survey of households undertaken in 2006 by the Istanbul Greater Municipality (IGM). We have also added data on earthquake risk from the Japanese International Cooperation Agency (JICA) report (JICA 2002). Table 1 summarises the descriptive statistics for all of the variables used in the modelling process.

3.3 *Method of Estimation*

Some of the house price modelling literature that informs AVM approaches overlooks the hierarchical or clustered structure of the data. This can be a source of errors in these models. Multi-level models starts by recognizing the challenges in analysing hierarchical data structures or variables at different levels. The method models individual-level dependent variables by using combinations of individual-level and group-level independent variables. Multi-level models are also known in the literature as contextual models, hierarchical linear models, hierarchical linear regression, random coefficients models, hierarchical mixed linear models, or Bayesian linear models.

Usually in studies of social phenomena and social data, the hierarchical structure of data consists of lower and upper levels. The lower level consists of individuals or properties which are grouped in higher levels with respect to the context. Due to the fact that multi-level analysis involves individuals that are nested in a contextual level, this method often attempts to examine how the individual level (micro level) outcomes are affected by both the individual level and the group level (macro level or contextual level) variables (Blalock 1984). This statistical method helps to specify effects of contextual subjects on individual-level outcomes. Thus, it becomes possible to display the different relationships between the dependent and independent variables within different contextual groups. These kinds of relationships are referred to as contextual effects and these are the effects that a space has on

Table 1 Descriptive statistics for 2007

Variables	Description	N	Maximum	Mean	St. Dev
TRANSACTION PRICE ($)	Transaction price of the housing unit	2175	8,000,000	251,082.92	382,467.37
AREA (m^2)	Living area in the housing unit	2171	1920	170.08	123.063
AGE (year)	Age of the dwelling	1962	150	12.22	14.57
LOW (dummy)	Dummy that indicated a low-rise building (less than 5 storey)	2106	1	0.38	0.485
SITE (dummy)	Dummy that reflects the fact that housing unit is located within a gated or semi gated community	2132	1	0.17	0.38
GARDEN (dummy)	Dummy for presence of garden	2021	1	0.79	0.41
BALCONY (dummy)	Dummy for presence of balcony	2026	1	0.92	0.277
LIVPER (year)	Living period in the city	2175	73	29.51	9.48
INCOME ($)	Average income of the household	2113	6,000	1448.74	1095
HHSIZE (person)	Household size	2174	6.5	3.487	0.67
NEIGHSAT (1–7 likert scale)	The level of neighbour satisfaction revealed in the 2006 survey undertaken by the municipal authority	2175	7	5.79	0.79
SCHOOLSAT (1-7 likert scale)	Is the survey-based estimate of school satisfaction 2006 survey undertaken by the municipal authority	2175	7	4.35	1.29
HEALTHSAT (1-7 likert scale)	Is the survey-based estimate of health services satisfaction 2006 survey undertaken by the municipal authority	2175	7	4.103	1.375
TTW (minutes)	Is the travel time to local employment and education hub 2006 survey undertaken by the municipal authority	2034	95	28.67	15.19

(continued)

Table 1 (continued)

Variables	Description	N	Maximum	Mean	St. Dev
QUAKE (%)	Is the estimated risk of an earthquake and is computed as the % of buildings that will be highly damaged by an earthquake (based on JICA 2002)	1980	18.27	5.34	4.1
CONTINENT	Indicates whether the dwelling is in the European zone	2175	1	0.45	0.497

individuals. On the other hand, compositional effects are the effects that the characteristics of individuals in different geographical levels have.

Multi-level modelling is thus developed from hierarchical approaches that can include both fixed and random effects, and can be modelled at each and every level within the hierarchy. Fixed effects are the "permanent" or "unchanging/constant/fixed" elements of the model and, as such, one estimate is derived for the whole sample. Random effects are the part of the equation that is "allowed to vary" part and where there is potential for different results to occur within the sample (Jones and Bullen 1993).

Clearly, multi-level modelling can be considered a modified version of hedonic price modelling. Hedonic models contain only fixed effects- the intercept and coefficients describe the sample as a whole. Arguably the spatial pattern of house price is not very effectively represented by fixed effects/regression models, given that they assume that the same intercept and slopes characterizes all neighbourhoods or submarkets. An alternative approach to deal with the tendency towards uneven spatial distribution of housing prices is to allow each of the segments to have their own random intercept. In doing this, multi-level models allow us to decompose the residuals and to expose the random intercepts and the hedonic slope parameters unique to each separate geographic unit (Leishman et al. 2013).

In general, a multi-level equation is formulated as:

$$Y_{ij} = \alpha_j + \Sigma\beta_i X_{ij} + (\varepsilon_{ij} + \mu_j \alpha + \mu_j \beta X_{ij})$$

Here, Y_{ij} represents the price of the house i in area j; α, β and μ are the parameters to be estimated, ε is the error term and X_{ij} is a set of explanatory variables which include housing attributes, socio-economic data and earthquake risk of the house i in area j.

4 Model Results and Research Findings

Our multi-level model includes standard property and neighbourhood attributes and a measure of earthquake risk. The results, shown in Table 2, highlight the fixed effects as well as model fit statistics. The coefficients shown in Table 2 are analogous to hedonic coefficients from a regression model. As with standard hedonics, this aspect of the model allows us to isolate the influence of earthquake risk from other price determinants. It also allows the differentiation between market-wide effects (−0.19 % discounts) and random neighbourhoods effects (which can be + where risk is below the market average or − where risk is above the average).

As might be expected the results show that many of the standard variables, including living area of the housing unit, being located at a semi gated, the age of the building, or gated community, income of the household and earthquake risk have a significant impact on prices. The Wald chi-squared test suggests strong explanatory power. The model shows that, on average, we should expect a 0.164 % discount in house price for every 1 % increase in the likelihood that a dwelling might be damaged.

Table 3 provides additional information about the random effects. The likelihood ratio test shows that the random intercepts model offers significant improvement

Table 2 Multi level model: fixed effects and model fit statistics

Variable	Coefficient (2007)
Constant	2.196529
Area	1.045023*
Age	−0.0024381
Residence	–
Garden	0.0347619*
Low	0.0096652
Site	0.113051*
Income	0.195355*
Hhsize	−0.486838*
Schoolsat	0.0931491
Neigsat	0.081995
Culturesat	–
Livper	0.1556817
Ttwork	–
Quake	−0.1642557*
Continent	0.1085348*
Wald chi(2)	2690.26*
Log restricted likelihood	632.96615*
Groups	270
N	1825

Notes *indicates significant at the 5 per cent level
–Indicates that the variable is excluded due to multicollinearity

Table 3 Multi-level model random effects

	2007	
	Estimate	Std. Error
Constant	0.1254122	0.0260374
Neighsat	0.1405585	0.0508316
Schoolsat	7.11E-08	0.0001425
Quake	6.62E-11	7.82E-11
Lr Test	chi2(4) = 680.93	
	prob > chi2 = 0000	

Table 4 Impact of earthquake risk on high price 5 neighbourhoods

Neighbourhood	R effect (quake) 2007	Average transaction price $ 2007	Earthquake risk (% highly damaged buildings)
Kanlica	0.018	1,649,500	1.7
Cubuklu	0.076	1,289,320	1.7
Beylerbeyi	0.11	1,196,079	2.31
Alkent (Etiler)	0.061	1,068,550	4.06
Bebek	0.198	957,942	4.1

over a standard linear regression model, of course, which includes only fixed effects (Table 3).

This basis multi-level formulation can be used to determine the spatial variations in the impact of key variable. Table 4 summarises the impact of earthquake risk on the five neighbourhoods that had the highest house price levels. It shows that these neighbourhoods with have positive R effect values and that the risk of earthquake is relatively low.

Table 5 shows the impact of earthquake risk on the five lowest price neighbourhoods. This shows that the cheapest neighbourhoods have mainly negative R effect values, implying significant price discounts.

Thus, it is earthquake risk impacts on price in the most pronounced manner in the neighbourhoods where the lowest price properties are found.

Table 5 Impact of earthquake risk (The top 5 Neighbourhoods with lowest transaction price-2007 Period)

Neighbourhood	R effect (quake) 2007	Average transaction price $ 2007	Earthquake risk (% highly damaged buildings)
Molla Serif	0.0008	47,619	1.7
Havaalani	−0.143	53,512	1.7
Nenehatun	−0.089	57,823	2.31
Birlik	−0.11	62,625	4.06
Gumuspala	0.054	69,727	4.1

5 Conclusion

Modelling house prices across spatially segmented housing markets is a major challenge. It is, however, an important task where the market being analysed is highly spatially differentiated and/or exposure to negative environmental externalities is uneven across the market. The ability to model the likely impacts of house price determinants in a granular way is limited in AVMs that rely on standard hedonic regression methods. Using Istanbul as a case study, we seek to illustrate how a multi-level model can be used for AVM purposes. We seek to illustrate the general robustness of the approach and the way in which it allows us to detect the market wide and neighbourhood specific effects of particular price determinants, in this case the perceived risk of earthquake damage. This approach has been shown in these circumstances to be able to distinguish between the different level impacts and to have high predictive power.

Like other forms of AVM regression, changes in house price caused by perceived risk of earthquake damage is assumed to be internal to the home purchasing decision-making process. This standard assumption, common to the hedonic framework, means that in practice earthquake effects can be observed to have an impact at both the citywide and the neighbourhood level. The effects at neighbourhood level vary widely with clear evidence of discrete spatial impacts on house prices. This finding is significant for all AVM's in earthquake zones. Importantly it implies that the use of a single variable for earthquake risk, with a constant parameter across the market, will adversely affect the predictive power of the model as a whole and will misrepresent the likely scale of the impact within specific neighbourhoods. Whilst it is foolhardy to suggest that the ground beneath any AVM is unshakeable, this research supports the extension of AVM models by using multi-level methods to separate neighbourhood effects from higher level (urban or regional) effects. Furthermore, multi-level modeling also benefits from being able to cope with fewer numbers of individual observations as long as the number of groups is high enough to analyse the variation between groups at a particular level.

References

Alkay, E. (2011). The residential mobility pattern in the Istanbul Metropolitan Area. *Housing Studies, 26*(4), 521–539.

Ballas, D., & Tranmer, M. (2008). *Happy places or happy people? A multi-level modelling approach to the analysis of happiness and well-being*. Mimeo: University of Sheffield.

Blalock, H. M. (1984). Contextual-effects models: Theoretical and methodological issues. *Annual Review of Sociology, 10,* 353–372.

Bin, O., Brown Kruse, J., & Landry, C. E. (2008). Flood hazards, insurance rates, and amenities: Evidence from the coastal housing market. *The Journal of Risk and Insurance, 75*(1), 63–82.

Boyle, M., & Kiel, K. (2001). A survey of hedonic studies of the impact of environmental externalities. *Journal of Real Estate Literature, 9,* 117–144.

Bourassa, S. C., Hoesli, M., & Peng, V. C. (2003). Do housing submarkets really matter? *Journal of Housing Economics, 12,* 1, 12–28.

Case, B., Colwell, P., Leishman, C., & Watkins, C. (2006). The impact of environmental contamination on condo prices: A hybrid repeat-sale/hedonic approach. *Real Estate Economics, 34*(1), 77–107.

Clapp, J. M. (2003). A semiparametric method for valuing residential locations: Application to automated valuation. *Journal of Real Estate Finance and Economics, 27*(3), 303–320.

Des Rosiers, F., & Theriault, M. (2008). Mass appraisal, hedonic price modelling and urban externalities: Understanding property value shaping processes. In T. Kauko, & M. d'Amato (Eds.), *Mass appraisal methods, an international perspective for property valuers*. UK: RICS Research, Wiley-Blackwell.

Goodman, A., & Thibodeau, T. (2003). Housing market segmentation and hedonic prediction accuracy. *Journal of Housing Economics, 12*(3), 181–201.

Grigsby, W. (1963). *Housing markets and public policy*. Uni of Penn Press.

Hallstrom, D. G., & Smith, K. (2005). Market response to hurricanes. *Journal of Environmental Economics and Management, 50*(3), 541–561.

JICA. (2002). *A disaster prevention/mitigation plan for Istanbul*.Report to Istanbul Municipal Authotiy, Istanbul. Available at: http://ibb.gov.tr/tr-TR/SubSites/DepremSite/PublishingImages/JICA_ENG.pdf (accessed 20th July 2014)

Jones, K., & Bullen, N. (1993). A multilevel analysis of the variations in domestic property prices—Southern England, 1980–87. *Urban Studies, 30*(8), 1409–1426.

Keskin, B. (2010). *Alternative approaches to modelling housing submarkets: Evidence from Istanbul,* Unpublished Ph.D. thesis, Sheffield: University of Sheffield.

Keskin, B., & Watkins, C. (2014). *The impact of earthquake risk on property values*. London: Royal Institution of Chartered Surveyors.

Lamond, J., Proverbs, D., & Hammond, F. (2010). The impact of flooding on the price of residential property: A transactional analysis of the UK market. *Housing Studies, 25*(3), 335–356.

Leishman, C. (2009). Spatial change and the structure of urban housing sub-markets. *Housing Studies, 24*(5), 563–585.

Leishman, C., Costello, G., Rowley, S., & Watkins, C. (2013). The predictive performance of multi-level models of housing submarkets: A comparative analysis. *Urban Studies, 50*(6), 1201–1230.

Loomis, J. (2004). Do nearby forest fires cause a reduction in residential property values? *Journal of Forest Economics, 10*(3), 149–197.

MacDonald, D. N., Murdoch, J. C., & White, H. L. (1987). Uncertain Hazards. *Insurance, and consumer choice: evidence from housing markets, land economics, 63*(4), 361–371.

Malpezzi, S. (2003). Hedonic pricing models: A selective and applied review. In T. O'Sullivan, & K. Gibb (Eds.), *Housing Economics and Public Policy*. Blackwells.

Mueller, J., Loomis, J., & Gonzalez-Caban, A. (2009). Do repeated wildfires change homebuyers' demand for homes in high-risk areas? A hedonic analysis of the short and long-term effects of repeated wildfires on house prices in Southern California. *Journal of Real Estate Finance and Economics, 38*(2), 155–172.

Orford, S. (1999). *Valuing the built environment: GIS and house price analysis*. Aldershot: Ashgate.

Pavlov, A. D. (2000). Space-varying regression coefficients: A semi-parametric approach applied to real estate markets. *Real Estate Economics, 28*(2), 249–283.

Pryce, G. (2013). Housing submarkets and the lattice of substitution. *Urban Studies, 50*(13), 2682–2699.
Ridker, R. G., & Henning, J. A. (1968). The determination of residential property value with special reference to air pollution. *Review of Economics and Statistics, 49*, 246–257.
Rosen, S. (1974). Hedonic prices and implicit markets: Product differentiation in pure competition. *The Journal of Political Economy, 82*(1), 34–55.
Schnare, A., & Struyk, R. (1976). Segmentation in urban housing markets. *Journal of Urban Economics, 3*, 146–166.
Schulz, R., Wersing, M., & Werwatz, A. (2014). Automated valuation modeling: A specification exercise. *Journal of Property Research, 31*(2), 131–153.
Simmons, K. M., Kruse, J. B., & Smith, D. A. (2002). Valuing mitigation: Real estate response to hurricane loss reduction measures. *Southern Economic Journal, 68*(3), 660–671.
Straszheim, M. (1975). *An econometric analysis of the urban housing market*. New York: National Bureau of Economic Research.
Subramanian, S. V. (2010). Multilevel modelling. In M. M. Fischer & A. Getis (Eds.), *Handbook of spatial analysis: Software, tools, methods and applications*. Berlin: Springer.
Stetler, K. M., Venn, T. J., & Calkin, D. E. (2010). The effects of wildfire and environmental amenities on property values in northwest Montana, USA. *Ecological Economics, 69*(11), 2233–2243.
Tu, Y. (2003). Segmentation, adjustment and disequilibrium. In T. O'Sullivan, K. Gibb (Eds.), *Housing Economics and Public Policy*. Blackwells.
Watkins, C. A. (2001). The definition and identification of housing submarkets. *Environment and Planning A, 33*(12), 2235–2253.
Watkins, C. (2012). Housing submarkets. In S. Smith et al. (Eds.), *International encyclopedia of housing and home*. North-Holland: Elsevier.
Willis, K. G., & Asgary, A. (1997). The impact of earthquake risk on housing markets: Evidence from tehran real estate agents. *Journal of Housing Research, 8*(1), 125–136.

Author Biographies

Richard Dunning is a Lecturer in Geography and Planning at the University of Liverpool. His principal research interest is applying behavioural analysis approaches to the study of housing and real estate markets. Prior to working as a researcher, Richard worked in the real estate industry. Richard has undertaken research for the EU, RICS, the Department for Communities and Local Government, the French Government, the Joseph Rowntree Foundation, Sheffield City Council and Rotherham City Council.

Berna Keskin is a University Teacher in the Department of Town and Regional Planning at the University of Sheffield. Her research interests focus on understanding the structure of urban housing market and specifically exploring the relative merits of different approaches to capturing neighbourhood segmentation. Following the completion of her PhD in Housing Economics at the University of Sheffield, she has worked on funded projects for the INTERREG NWE Programme, Royal Institution of Chartered Surveyors, Department of Communities and Local Government, Department for Environment, Food & Rural Affairs and Investment Property Forum.

Craig Watkins is Director of Research and Innovation for the Faculty of Social Sciences, and Professor of Planning and Housing at the University of Sheffield. Craig has written extensively on the economic structure and operation of housing and commercial property markets, urban development and on the relationship between public policy and markets. He has produced more than 130 research outputs including 3 books, 12 book chapters and more than 40 peer reviewed journal articles and has undertaken around 50 funded projects (half as Principal Investigator) including research for ESRC, Technology Strategy Board, EU, Joseph Rowntree Foundation, various Central and Local Government departments in the UK, RICS, Investment Property Forum and the Royal Town Planning Institute.

The Multilevel Model in the Computer-Generated Appraisal: A Case in Palermo

Marina Ciuna, Francesca Salvo and Marco Simonotti

Abstract The construction of a mass appraisal model requires the preliminary study of the real estate market, the sampling of sold properties, the development of a forecasting model and the verification of the appraisal results. They are generally computerised methods, that work with geo-referenced data. This experimental work has proceeded to build a mass appraisal model, collecting a data sample of sales of apartments in the city of Palermo, in the five years 2008–2012, using a multivariate statistical model (multilevel), testing the results and providing the operating applications in a scheme of online real estate valuations.

Keywords Computer-generated appraisal · Multilevel models · AVM

1 Introduction

The mass appraisal deals with the valuation of the market value of a plurality of properties, by collecting the real estate market data and applying statistical appraisal methodologies. These valuations are used mainly by tax agencies for the valuation of real estate income for tax purposes, by financial institutions for the appraisal of the real estate portfolios of mutual funds, by banks in periodic revaluations of assets for security of loan, and from companies specialized in on line valuations of real estate. Among the uses of mass appraisal are included the valuations for the acquisition of

M. Ciuna (✉) · M. Simonotti
University of Palermo, Palermo, Italy
e-mail: marina.ciuna@unipa.it

M. Simonotti
e-mail: marco.simonotti@unipa.it

F. Salvo
University of Calabria, Rende, Italy

M. d'Amato and T. Kauko (eds.), *Advances in Automated Valuation Modeling*,
Studies in Systems, Decision and Control 86, DOI 10.1007/978-3-319-49746-4_14

real estate in urban renewal, the valuations of benefits in the acquisition of expropriated properties and the valuations of compensation in the imposition of servitudes, generally for the construction of a network. There is a gap between the real estate valuations in Italy (Ciuna et al. 2014a; Simonotti et al. 2015) and those undertaken by international organizations and institutions: the Office of Revenue does not apply the mass appraisal in the valuation of cadastral income; the investment funds are not very important in the field of investment and there are experts that formulate periodic valuations; web sites use typically real estate listings of directories already available, rather than appraisal statistical models (Ciuna et al. 2013a). The construction of a mass appraisal model requires the preliminary study of the real estate market, sampling of properties sold, the development of a forecasting model and verification of the appraisal results. There are generally of computerised methods that work with georeferenced data (Orford 2000, 2002). The present experimental study proceeded to build a model of mass appraisal, collecting a data sample of sales of apartments in the city of Palermo, in the five years 2008–2012, using a multivariate statistical model (multilevel), testing the results and providing the operating applications in a scheme of online real estate valuations (Ciuna 2007; Ciuna et al. 2014b). The statistical model was built to provide the economic framework of the segments of the real estate market and to provide potentially the valuation of every single property hanging in each segment. For both of these purposes, the model is based on the cross section analysis of data, aiming to provide the tool for automatic valuation of individual buildings (d'Amato 2008, 2010; d'Amato and Kauko 2012).

2 Automated Valuation Modelling

AVM refers to the process of valuating a universe of real estate, using common real estate data and standard appraisal methodologies. The units are appraised individually. The real estate data are collected regularly and systematically for samples. The use of the appraisal methodology is required and the choice of the appraisal models is related to the goal of the valuation, the means and the resources available and the housing market. These conditions allow the application of standard of survey, routine of statistical and appraisal analysis and control of the valuations results (Salvo et al. in print). AVM allows to appraise the income and market values of the properties in the manner as faithful as possible to the rents and market prices, mainly due to the fact that these are the actual data and the reference point for any decision and valuation of savings and investment (Renigier-Biłozor et al. 2014a, b).

The purposes of the AVM are multiple and related: directly, to the valuations of properties in the case of the valuations for land taxation, the administrative valuations, the valuations for renegotiation of mortgage loans, the valuations for periodic revaluations of assets, the valuations for urban renewal; and indirectly, to the development of economic indicators of the real estate market taking into account that investment properties are part of the financial markets and are an important component of the insurance companies, pension funds and private funds that invest in real estate.

The mass appraisal methods are designed to provide reliable real estate valuations, which are readily available and cheap; in the tax valuations are intended to offset the tax limitations in budget, in time and in personnel employed in the analysis of real estate data. It is generally computerized methods that use geo-referenced data in a GIS environment (Renigier-Biłozor and Biłozor 2016a, b). The basic methodological requirements of mass appraisal are: the fixing of the appraisal criterion or criteria based on the purpose of valuation, to determine the market value of properties in the current destination; the collection of market prices and multiple real estate characteristics with a standard of survey, containing the requirements of technical and methodological practices for the detection and collection of the parameters of the market segments, of the real estate data and other punctual market information; the choice of valuation models able to reflect and simulate the market in accordance with recognized valuation standards and procedures; the verification of the results of the valuation using the same valuation standards; the immediate understanding by operators, professional technicians, customers.

3 Valuation Standards

The increasing use of mass appraisal worldwide in different sectors, from the private to the public, has pushed international associations of professionals and the technical valuators to define specific standards that are implementing procedures articulated in quantitative analysis and quality control in order to prevent errors and unforeseen complications. The main international valuation standards on the mass appraisal are (IVSC 2007; USPAP 2006; IAAO 2012): International Valuation Standards (IVS), Valuation Guidance note n. 13 Mass Appraisal for Property Taxation, International Valuation Standards Committee (IVSC 2007); Standard 6: Mass Appraisal, Development And Reporting, Uniform Standard Of Professional Appraisal Practice (USPAP 2006); Standard On Mass Appraisal of Real Property, International Association Of Assessing Officers (IAAO 2012).

The italian valuation standards on the mass appraisal is (Tecnoborsa 2011): Stime su Larga Scala (Mass Appraisal), Codice delle valutazioni immobiliari IV, Capitolo 17, (Tecnoborsa 2011). The mass appraisal process include (IVSC 2007): identifying properties to be appraised; defining market area in terms of consistent behaviour on the part of property owners and would-be purchasers; identifying characteristics of supply and demand that affect the creation of value in that market area; developing a model structure that reflects the relationship among the characteristics affecting value in the market area; calibrating the model structure to determine the contribution of the individual characteristics affecting value; applying the conclusions reflected in the model to the characteristics of the property(ies) being appraised; validating the adopted mass appraisal process, model, measurements or other reading including the performance measures, on an ongoing basis and/or at discrete stages throughout the process; the reviewing and reconciling the results. Around the second half of the 60s, the spread of computer programs and electronic computers led some operators in the real estate industry in the United States, to exploit the potential of information systems, to implement procedures to calculate the market value of the property: it was assisted calculations rather than computer-generated, which represent the first applications of so-called CAMA (Computer Assisted Mass Appraisal). This generic term is used today to indicate the software used by tax agencies to make valuations aimed at the calculation of property taxes. In the United States, around the late 90s they developed the first Automated valuation methods (Downie and Robson 2007) initially used by lenders to determine the risk connected with the granting of mortgage loans. Although these tools designed to calculate the value of the property, the automated methods differ significantly from CAMA for many aspects: the CAMA are used for tax and the assessment purposes and the are mainly used by financial institutions and private clients; the CAMA models are essentially multiple regression models, while those of the AVM are more versatile and to multiple regression models they add different techniques, such as multivariate calculations, simulations, etc.; the CAMA relate the valuations to a specific date, which is the same for all properties that fall within the area of interest, the AVM fix the valuation date for each specific need, trying to simulate the dynamics of the recent real estate market. In general it can be said that while the scope of the AVM is essentially the private sector, that of CAMA is exclusively public. Alongside the various applications of economic, financial and fiscal, the technology of the AVM is used to provide a public service of online consultation and real estate advice (International Association of Assessing Officers 2003). There are numerous websites that allow citizens to know in a few seconds, some free and others upon payment of a small fee others, the value of a property on a certain date by simply entering in a screen of input characteristics and the location of real estate. The same websites, in addition to the valuation, offer statistics and economic indicators designed to support investment decisions and real estate management, with a clear effect on the transparency of the market property of the concerned area (Bahjat-Abbas et al. 2005).

4 On Line International Valuations

In international websites are proposed online appraisals of the real estate value. The value expresses a computer-generated valuation carried out on market data (of contracts). The valuation is based on the detection of individual data of comparable properties with the property we're considering. The properties of comparisons are simply called comparable (or comps). The comps and the property to be appraised are usually georeferenced and we can know the characteristics (and often the floor plans and photos). The preparation of the valuation is performed by comparing the prices and characteristics of the comps with the characteristics of the property to be appraised. The comparison is done with the data collected in the lists of properties recently sold in the same market segment of the property being appraised (sold listings), assuming market conditions remain unchanged. These sales are the same as those included in the valuation of market value performed by an accredited assessor, whereas the latter valuation is carried out with an inspection and in the specific market segment of the property to be appraised, sometimes with little or no comparable property. In summary, the on line value is an appraised value from the site, using a proprietary formula, applied to market data and to the data of the property, entered by the user. The valuations are available on line for single family homes and condominiums (coverage). The accuracy of these valuations is measured by comparing the sales prices achieved in recent months with the appraisals. According to the information of the websites, the median error made by these valuations swings between 5 and 20 % of the final selling prices.

5 Real Estate Data Collection

The real estate market is a complex market divided into markets or segments that have different characteristics according to the area, the building type, the destination, the characteristics of demand and supply and market arrangements. In the same urban building, for instance, there are market segments of the garages, shops, lofts and apartments; the apartments themselves are divided into segments for their size (small, medium and large) and sometimes to the floor (the first floor and others). The buildings are by their nature non-homogeneous goods traded generally in partially competitive markets. The market segment is the set not further divisible in the appraisal and economic analysis of the property market (Goodman and Thibodeau 2003, 2007). A property is part of a market segment if it has the same parameters of the properties in the same segment. In addition, properties with the same parameters can be grouped into market segments. So the analysis of the real estate market is to identify the market segments that compose it and to survey sample data consistent in the real estate market prices and the technical and economic real estate characteristics (Ciuna et al. 2015a, b, c).

6 Data Sample

The detection center real estate data covered the purchase prices of apartments in buildings of different districts of the city of Palermo. The data sample is composed of n. 426 full data detected in 18 districts of the city in the period between 2008 and

2012 (Ciuna et al. 2014a, b). The data sample is the data collected by the students of the Faculty of engineering appraisals of the University of Palermo in recent academic years. For the purposes of data analysis overview have been supplemented by ad hoc surveys carried out for this study of n. 21 real estate data in 2011 and no 29 real estate data in 2012. The data collection covered the apartments of 18 municipal districts: Acqua dei Corsari; Borgo Nuovo; Brancaccio-Ciaculli; Cruillas-C.E.P.; Cuba-Calatafimi; Libertà; Malaspina-Palagonia; Mezzo Monreale-Villatasca; Montepellegrino; Noce; Oreto-stazione; Palazzo Reale-Monte di Pietà; Politeama; Resuttana-San Lorenzo; Santa Rosalia Montegrappa; Settecannoli; Tribunali-Castellammare; Zisa. The delimitation of the market segments in the districts was carried out with reference to the qualitative and quantitative parameters represented by: the location: city centre, inner suburbs, suburbs of the city of Palermo; the type of contract: sale negotiations with private; residential; the type of property: in condominiums; the type of building: multi-storey buildings and small block of flats; the interval of time between 2008 and 2012; the form of the market: monopolistic competition; the size of the property between 50 and 200 sqm; the characters of supply and demand: the motivation to sell (transfer or liquidation) and the motivation to buy (before transfer or home); the level of the average unit price. The collection of real estate data has used the appropriate survey form, which includes: the locational, typological, economic and positional real estate characteristics; the nomenclators for the qualitative real estate characteristics; the graphical presentation with floor plans; the photographic representation of the collected properties.

The sources of the data collected are official acts, those directly involved, namely buyers and sellers in particular, and market practitioners (solicitors, accountants, tax consultants, bank officials working in the real estate credit, technical aspects of construction companies, engineers professionals, administrators of condominiums and real estate agents). The data were geo-referenced in a separate map (Ciuna et al. 2013b). The statistical analysis was applied in accordance with the 'Linee Guida per la rilevazione dei dati del mercato immobiliare' (Ciuna and Simonotti 2011a, b). For the analysis we have considered the following characteristics of real estate: Date (DAT), the date of the contract measured in months; Type of building (TYP), the apartments in small block of flats are measured at 0 and apartments in multi-storey buildings with 1; Main area (SUP), the main surface of the area of major importance in the apartment is measured in square meters; Surface balconies (BAL), the surface of the balconies is measured in square meters; Surface terraces (TER), the surface of the terraces is measured in square meters; Commercial area (COM), the commercial area includes the main surface and secondary areas of the apartment, multiplied for their commercial ratios; the commercial area is measured in square meters; Toilets (TOI), the toilets are numbered numerically; Floor level (LEV), the level of plan is measured as the number of floors above street level; Lift (LIF), the lift is measured in the presence or absence of a dichotomous scale (0–1); State of maintenance (STM), the maintenance status is representing the degree of physical deterioration of a

Table 1 Nomenclator: state of maintenance

Real estate characteristic	Class	Nomenclator	Point
State of external maintenance (MAE)	Very good	The walls, the roof, the windows and the plants do not require any maintenance work or ordinary or extraordinary	3
	Good	The same characteristics are in fairly good condition	2
	Sufficient	The same characteristics show signs of deterioration that require specific and limited maintenance	1
	Poor	The same characteristics are degrading situations	0
State of internal maintenance (MAI)	Very good	The floors, ceilings, walls, fixtures and equipment shall be such as not requiring any maintenance work or ordinary or extraordinary	3
	Good	The same characteristics are in fairly good condition	2
	Sufficient	The same characteristics show signs of deterioration that require specific and limited maintenance	1
	Poor	The same characteristics present situations of degradation applicants to remake it again to restore the ordinary and original features of the home	0

building; the maintenance status is divided into internal and external; for the status of the internal maintenance (MAI) and the external (MAE) were defined four levels of the ordinal scale: 0 poor state of repair, 1 sufficient maintained, 2 good condition of maintenance, 3 very good state of repair (Table 1).

Views (VIW), the views are intended to represent the number of fronts with windows both in terms of air and brightness; Car parking (PAR), the parking space is measured according to its presence or absence on a dichotomous scale (0–1). The market price (PRZ) is expressed in euro. For the calculation of the commercial area of the apartments have been detected commercial ratios of surfaces, distinct for market segments of districts. Should be noted that for the purposes of the analysis the subdivision in districts is purely nominal, because the spatial delimitation have to be understood as a set of segments that fall in an area, however defined (area, district, block, etc.).

7 Sample Statistics

The statistic sample considers the characteristics detected and the nominal grouping in districts. For the characteristic date (DAT) the time sequence is the following: in 2008 were found n. 105 apartments; in 2009 were found n. 95 apartments; in 2010 were found n. 79 apartments; in 2011 were found n. 49 apartments; in 2012 were found n. 98 apartments (d'Amato 2015). For the characteristic typology (TYP) there

Table 2 Frequency characteristics of real estate properties on an ordinal scale and a dichotomous scale

Characteristics	Absolute frequency														
	0	1	2	3	4	5	6	7	8	9	10	11	12	13	14
TOI (1, 2, 3)	–	268	188	1											
LEV	12	82	90	92	69	41	34	14	12	4	–	4	2	1	–
GRO	–	–	17	42	66	40	78	73	60	37	17	9	9	8	1
LIF (0, 1)	113	344													
MAI (0, 1, 2, 3)	20	122	272	43											
MAE (0, 1, 2, 3)	24	95	305	33											
VIW (1, 2, 3, 4)	–	46	264	117	30										
PAR (0, 1)	372	85													

are n. 96 apartments in small block of flats and n. 330 apartments in multi-storey buildings. For the main surface (SUP), the range of variation of the sample is between a minimum of 50.00 sqm and a maximum of 195.00 sqm, the average is equal to 109.88 sqm and the standard deviation is equal to 26.70 sqm. For balconies (BAL), the range of variation of the sample is between a minimum of 0 sqm, or property without balcony, and a maximum of 55.00 sqm, the average is 11.59 sqm and the standard deviation is equal to 7.66 sqm. For the surface of the terraces (TER), the range of variation of the sample is between a minimum of 0 sqm, or apartments without terraces, and a maximum of 300.00 sqm, the average is 2.01 sqm and the standard deviation is equal to 16.98. The commercial area (COM) is calculated with the commercial ratios of the balconies and terraces defined by district (Ciuna and Simonotti 2014). For the commercial area the range of variation of the sample is between a minimum of 51.20 sqm and a maximum of 234.43 sqm, the average is equal to 114.28 sqm and the standard deviation is equal to 28.31 sqm. For the characteristics: toilet (TOI) floor level (LEV), number of floors above ground (GRO), lift (LIF), internal state of maintenance (MAI), exterior state of maintenance (MAE), views (VIW) and parking (PAR) shows the absolute frequencies (Table 2).

The total market price of the apartments varies between a minimum of 45,000.00 and a maximum of 550,000.00 €. The unit price of the apartments varies from a minimum of 726.04 €/sqm to a maximum of 3793.34 €/m (Table 3).

The correlation between the real estate characteristics, which define the data into consideration and are the variables in the model, is shown in the multiple correlation matrix, considering the commercial area of the apartment computed with the commercial ratios of surfaces, defined by district (Table 4).

Table 3 Market price and total price per district

	District	Total price (€)		Unit price (€/sqm)	
		Mean	Dev. st.	Mean	Dev. st.
1	ACQUA DEI CORSARI	187,500.00	15,545.63	1,450.61	185.99
2	BORGO NUOVO	164,627.20	67,150.96	1,531.89	508.89
3	BRANCACCIO-CIACULLI	160,250.00	31,159.54	1,426.83	170.03
4	CRUILLAS-CEP	272,942.22	41,236.20	2,172.08	212.82
5	CUBA-CALATAFIMI	212,108.70	76,900.00	1,819.58	541.91
6	LIBERTA'	368,933.33	54,818.49	2,554.97	444.31
7	MALASPINA PALAGONIA	292,272.93	87,135.63	2,237.79	301.34
8	MEZZOMONREALE-VILLA TASCA	210,000.00	50,990.20	1,804.49	206.24
9	MONTEPELLEGRINO	427,000.00	54,781.38	2,981.62	40.69
10	NOCE	97,166.67	66,303.59	1,709.55	297.32
11	ORETO-STAZIONE	165,235.32	46,621.80	1,522.31	267.30
12	PALAZZO REALE-MONTE DI PIETA'	95,100.00	31,959.87	1,117.62	169.81
13	POLITEAMA	382,393.33	69,881.48	2,641.26	366.14
14	RESUTTANA-SAN LORENZO	325,250.00	60,098.53	2,692.71	310.55
15	SANTA ROSALIA MONTEGRAPPA	188,552.55	53,957.70	1,748.51	330.77
16	SETTECANNOLI	209,250.00	41,177.46	1,841.91	236.03
17	TRIBUNALI-CASTELLAMMARE	119,962.50	51,894.78	1,074.01	406.32
18	ZISA	219,640.00	63,780.14	1,866.36	389.79

Table 4 Correlation matrix

	Property characteristics and price								
	COM	TOI	LEV	PIA	MAI	MAE	VIW	PAR	PRZ
COM	1	0.4759	0.2380	0.3491	0.0836	0.2248	0.2507	0.1743	0.7138
TOI		1	0.2174	0.3488	0.1059	0.2507	0.1533	0.1997	0.4923
LEV			1	0.5727	0.0882	0.1234	0.0590	0.1019	0.2824
PIA				1	0.1804	0.2705	0.1005	0.2664	0.4730
MAI					1	0.4735	−0.0257	0.1819	0.1422
MAE						1	0.0433	0.2434	0.2782
VIW							1	0.0663	0.2739
PAR								1	0.2095
PRZ									1

8 Multilevel Model

The multilevel analysis is a statistical methodology for the analysis of data that have a complex structure of variability and a hierarchical configuration. The model is a generalization of the multilevel regression methods, and as such can be used for various purposes, including prediction, data reduction, and the causal inference (Kreft and De Leeuw 1998; Snijders and Bosker 1999; Raudenbush and Bryk 2002; Hox 2002). Compared to the regression classical, multilevel analysis is usually an improvement, but to different degrees: for predicting the multilevel modeling may be essential, for data reduction and for the causal inference can be useful (Gelman 2006). In this study, multilevel analysis is applied to the real estate market. This statistical method is within the market-oriented models as bases its forecasts on market data. The multilevel model has the advantage of separating the effects of the characteristics (or variables) to different levels for groups present in the data structure. When it is necessary to study the relationships between the variables in the presence of a group structure (hierarchical structure of the data), such as the segmentation of the market, we need a statistical model that considers these relationships and establishes a link between the phenomena under study. The model allows to define the effect of macro units (top-level units) on the lower level and vice versa, because the data arrays that define it have a hierarchical structure characterized by relationships between variables at different levels. The multilevel model takes into account the relationships between the different levels, and the presence of relationships between variables belonging to each level. There are several multilevel regression models, the most popular are: the random coefficient models; the variance component models; the hierarchical linear models. These models are based on a dependent variable explained and on one or more independent variables, one measured at a lower level and the other at each level of the structure. The random coefficient model, is a probabilistic model that allows to represent the variability within and between groups; in fact, it is considered as a random variable, both that is not explained by variability between groups and that within the groups. For example, if we analyze the properties belonging to a market segment, the model considers as random variables both the variability unexplained between the apartments and that between market segments. The random coefficient models can be a variable intercepts (random intercept models) or random coefficients (random scope models) or intercepts and random coefficients (random parameters models). By way of example we consider a model with two levels, with the possibility to extend the concepts to models with more levels. We study the relationship between the market price P of the property and their commercial area COM. We can consider a linear regression that does not take into account the effect of belonging to different single market segment or define the hierarchical structure of the data considering housing units at the first level, and market segments to the second level, and studying the residual variability for each level. In the model of simple linear regression analysis is carried out micro, thus not considering the grouping into districts. In the random intercept model we consider the grouping into districts and we examine the random intercept. In the random parameter model we consider the intercept and the regression coefficient random. We can study the functional relationships of the linear regression model, the random intercept models and the random parameter model with respect to changes in the price of real estate compared to the commercial area in the event of a hierarchical structure of data at two levels: the first level is represented by the given property, the second level by market segment reported to the district. The linear regression model considers only the fixed effects of the explanatory variables as follows:

$$P_i = \beta_{0i} + \beta_{1i} \cdot COM_i + e_i$$

where:

P_i is the market price; i is the generic property (con i = 1, 2, …, n); β_{0i} is the constant coefficient;

β_{1i} is the coefficient relating to the commercial area; e_i is the residual random.

This equation is not valid for a multilevel data structure, because it does not consider the grouping of data that can vary the coefficients of the regression from group to group. The random intercept model examines the group effect of the predictor by means of changes in the intercept: then estimate a model in which the regression coefficient is constant in the groups but the intercepts are different varying from group to group, per district. Whereas, for example, only the variable commercial area at level 1, we can write the model as follows:

$$P_{ij} = \beta_{0j} + \beta_{1i} \cdot COM_{ij} + e_i,$$

in which the intercept β_{0j}, is variable at the group level and can be decomposed into:

$$\beta_{0j} = \beta_0 + u_{0j};$$

where:

β_0 is the mean intercept between all districts;

u_{0j} is the measure of its deviation based on the average between the districts.

So substituting the previous expression:

$$P_{ij} = \beta_0 + \beta_{1i} \cdot COM_{ij} + u_{0j} + e_i.$$

The parameter u_{0j} is the variance, and in particular it is possible to obtain: a model where u_{0j} are N fixed parameters; a model where u_{0j} are random variables independent and identically distributed.

The first case occurs when the groups (districts) have interpreted separately; is possible to consider $\sum u_{oj} = 0$ and then the parameters of the model due to the groups are $N - 1$. In the second case u_{0j} are the effects of group, this occurs if the effects of the groups are interchangeable. The random parameter model considers a two-level model that describes a separate relationship for each of the districts: the n apartments are units of level 1, divided into J market areas that define the unit of level 2. The following relationship allows to simultaneously describe the relationship between different properties, for J groups, for j district. Then the response variable P_{ij} is given by the sum of a fixed part and a random part:

$$P_{ij} = \beta_{0j} + \beta_{1i} \cdot COM_{ij} + e_{ij};$$

where: P_{ij} is the response variable of the apartment i of the district j; COM_{ij} is the explanatory variable of the apartment i of the district j; e_{ij} is the error. We get the random parameter model when: the data are hierarchically structured on several levels; the parameters β_{0j} and β_{1j} may vary in the different groups; imposing the following equalities:

$$\beta_{0j} = \beta_0 + u_{0j};$$

$$\beta_{1j} = \beta_1 + u_{1j};$$

where: u_{0j} and u_{1j} are random variables with mean zero; β_0 and β_1 are fixed coefficients, since they do not vary between the units belonging to the macro level. Both the intercept and the coefficient of the regression are variables at the group level and both can be divided into two parts: a constant, or average coefficient, and a part that varies at the group level and measure the distance from the average coefficient. Based on these considerations it is possible to rewrite the equation:

$$P_{ij} = \beta_0 + \beta_1 \cdot COM_{ij} + \left(u_{0j} + u_{1j} \cdot COM_{ij} + e_{0ij}\right).$$

Given a model with random parameters, the relationship between the dependent variable and the explanatory can vary between the groups, in fact we may have a heterogeneity of regressions between the different groups. The coefficients β_0 and β_1 are random variables, and each group has a different intercept β_{0j} and a different regression coefficient β_{1j}.

The random variables are the residuals of the model. In particular in the case of a model with a single level, the residual of first level e_{0ij} is the residual term of a linear model.

Each J group has its own regression function and all the functions are connected by the parameters from the common hyper-distributions, so all the groups present in the data are a random sample of a hypothetical population of groups. Ultimately in the case considered it may have different β_{0j} intercepts and different β_{1j} regression coefficients. Obviously, the following relations hold:

If the coefficients β_{0j} and β_{1j} are constant, then we have the OLS regression (Ordinary Least Squares regression), the hierarchical structure has no effect; if the coefficients β_{0j} and β_{1j} depend on j, then the OLS regression can not be used; if only the β_{0j} coefficient varies with the variations of j, then we have the random intercept model; if the coefficients β_{0j} and β_{1j} vary when j varies, then we have the random coefficient model. We can say that the model implies not only that the apartments within the same group have values of P correlated, but also that this correlation (such as the variance of P) is dependent on the value of COM, in fact the term u_{1j} is connected with COM_{ij}. Then the total error is different according to the values of COM_{ij}. Using this equation it is possible to value, as a function of the coefficients, the independent effects of the variables of the second level, of the first level and their interaction. Also we can calculate: the variability within the group defined as a function of variance σ^2; the variability between groups defined in terms of the variances of the random effects τ_{02} and τ_{12}. For the terms u_{0j}, u_{1j} and e_{0ij} are valid the standard assumptions: the errors in the first level e_{ij} are normally distributed with mean zero and constant variance σ^2_{e0};

$$e \approx N(0, \sigma^2_{e0});$$

The random coefficients of second level u_{0j}, u_{1j} are correlated between groups but may also be related within each group; is attributed to a multivariate normal distribution with mean zero and variance-covariance matrix constant and equal to:

$$\Omega = \begin{bmatrix} \sigma^2_{u0} & \sigma_{u01} \\ \sigma_{u01} & \sigma^2_{u1} \end{bmatrix};$$

then we can write:

$$\begin{bmatrix} u_{0j} \\ u_{1j} \end{bmatrix} \approx N\left(\begin{bmatrix} 0 \\ 0 \end{bmatrix}, \begin{bmatrix} \sigma^2_{u0} & \sigma_{u01} \\ \sigma_{u01} & \sigma^2_{u1} \end{bmatrix} \right);$$

the residual vector $\begin{bmatrix} u_{0j} \\ u_{1j} \end{bmatrix}$ is uncorrelated with the errors e_{ij} i.e. the random parameters of the first and second level are stochastically independent:

$$e \perp \begin{bmatrix} u_{0j} \\ u_{1j} \end{bmatrix}.$$

The covariance, for example, between two properties i and i' of two groups (districts) j and j' is given by:

$$\operatorname{cov}(V_{ij}, V_{i'j'}) = \sigma_{u01};$$

while the covariance between the two properties i and i' of the same group j is equal to:

$$\operatorname{cov}(V_{ij}, V_{i'j'}) = \sigma^2_{u0};$$

The total variability when the units are grouped together (such as buildings within districts), can be decomposed into: variability due to the group; variability between groups. Ultimately the total variance is given by the sum of a quadratic function of the independent variable linked to the random component and the variance of the first level.

$$\text{var}(V_{ij}) = \sigma_{e0}^2 + (\sigma_{u0}^2 + 2\sigma_{u01} COM_{ij} + \sigma_{u1}^2 COM_{ij}^2).$$

To test hypotheses about the parameters of the fixed part of the model we take into account the Wald test (Goldstein 2003):

$$T(\beta_h) = \frac{\hat{\beta}_h}{s.e.(\hat{\beta}_h)};$$

under the null hypothesis, the test has approximately a t-distribution with (d.f.) degrees of freedom, which depend on the multilevel analysis structure. The criteria for testing hypotheses vary depending on whether we want to verify a hypothesis relating to a coefficient of an explanatory variable of the level 1 and in this case we have: M = total number of observations at level 1; r = total of the explanatory variables on the level 1; d.f. = $M - r - 1$; or if we want to test a hypothesis of a coefficient of an explanatory variable at level 2, and in this case we have:

N total units level 2;
q total explanatory variables to the level 2;
d.f. N – q − 1.

The method used for the valuation of the parameters is the maximum likelihood method. To test a hypothesis about the random part of the model we are using the test on the deviance D (Deviance test). Once appraised the model and determined the likelihood, i.e. $-2 \ln L$, is measured the goodness of fit of the model to the data. D is usually interpreted in differential terms calculating the difference between the deviances of alternative models. The test considered, if we denote by M_0 the model with parameters m_0 and deviance D_0 and with M_1 the model with parameters m_1 and deviance D_1, is:

$$D_0 - D_1 = -2 \ln L_0 + 2 \ln L_1$$

Under the hypothesis H_0 the difference between the deviations is distributed as a χ^2 with $m_1 - m_0$ degrees of freedom.

$$D_0 - D_1 \approx \chi^2(m_1 - m_0)$$

The test D can be applied to the fixed part and to that random of the model.

9 The Model Construction

Multilevel analysis was used to value an appraisal model to be applied in mass appraisal and in particular in the field of online valuations. The model plays a computer-generated estimate of the real estate and, because it is based on market prices, predicting that offers can be traced back to market value. Collected data and real estate features defined to apply multilevel model was identified the hierarchical data structure consists of: from apartments, constituting the first level units; the type of separate small block of flats and multi-storey buildings, which comprise the second level units districts as proxies for market segments that make up the third-level units. The multilevel model applied to the hierarchical structure of data allows the estimation of regression functions according to the level considered. These functions are defined based on regression coefficients given by the sum of a fixed part and a variable part depending on the hierarchical level under consideration. The model is applied multilevel random parameter where the random covariate is espressa from the surface. The model used is that which occurs with more degrees of freedom and is the smallest value of −2 log likelihood. From an intercept only model without explanatory variables, these were included in the model based on the correlation with the variable reply. Explanatory variables bring significant information, which are inserted while those who do not make improvements to the template are deleted from the analysis. In particular, when the value of the log-likelihood improves, fit new predictors, while if the value decreases the model is built with explanatory variables but without angular coefficients randomized. This procedure is performed iteratively until the introduction of all the explanatory variables. In the hierarchical structure of data: are individual apartments (with $i = 1, \ldots, 426$) and sicknesses are no top-level units; (j) are the second-level unit (with $j = 1, \ldots, 31$) representing group membership by type and construction of real estate properties, distinguished for apartments in the small block of flats and real estate in multi-storey building; k the districts seen as proxies of nominal market segments (with $k = 1, \ldots, 18$) which are the third-level units. Object variable of interest represented by the market value of the property (PRZ_{ijk}), observed the elementary units, investigates about the link with the explanatory variables. It is supposed that this bond is linear and can vary from group to group, regarding the action of the explanatory variables that intervene in the second and third level. The multilevel model aims to connect, with a single statistical formulation, regression models can be specified separately for different groups. Explanatory variables are represented by real estate characteristics, in particular the estimated model is considered: the commercial surface (COM), the piano (LEV), the number of toilets (TOI), the status of internal maintenance (ever), the number of views defined as number of fronts with openings (VIW) and uncovered parking (PAR). Among the explanatory variables of the multilevel not the variable data (DAT) that has been misplaced; for this variable is the array of variances and covariance matrix has supplied null values and not allowed to calculate its equation coefficients. The generic multilevel function with respect to the variables considered can be written as follows:

$$PRZ_{ijk} = \beta_{0k} \cdot CON + \beta_{1ik} \cdot COM_{ijk} + \beta_{2j} \cdot LEV_{ijk} + \beta_{3j} \cdot TOI_{ijk} \\ + \beta_4 \cdot MAI_{ijk} + \beta_5 \cdot VIW_{ijk} + \beta_6 \cdot PAR_{ijk};$$

where:

$\beta_{0k} = \beta_0 + v_{0k}$ is the coefficient of the constant;
$\beta_{1ik} = \beta_1 + v_{1k} + e_{1ijk}$ is the coefficient of the surface;
$\beta_{2j} = \beta_2 + u_{2jk}$ is the coefficient of the floor level;
$\beta_{3j} = \beta_3 + u_{3jk}$ is the coefficient of the number of toilets;
β_4 is the coefficient of the state of maintenance;
β_5 is the coefficient of the number of views;
β_6 is the coefficient of the parking space.

Function coefficients represent the marginal prices of real estate properties and express the variation of the total price of an apartment at different characteristic considered. The regression coefficients are given by the sum of a fixed part and a variable that varies with the level considered to be units. The fixed part consists of the constant components, while the random components of 1°, 2° and 3° level are called respectively $e_{..ijk}$, $u_{..jk}$ e $\upsilon_{..k}$. In particular, In the equation the coefficients for internal maintenance status (MAI), the views of the (VIW) and parking (PAR) are the only fixed part was not significant individual variation coefficient for the real estate and construction and typology for the districts, while the coefficients for the other features included are composed from the fixed and random parameters that vary at different level unit in question; in reference to the equation, these are: e_{1ijk}, u_{2jk}, u_{3jk}, υ_{0k} e υ_{1k}. The component υ_{0k} is the random component of the constant and measures the distance of the individual groups and the media. The random component then expresses the locational characteristics, for which two samples of real estate data, extracted from two different districts while presenting the same characteristics show two different price levels. The other coefficients are positional, typological characteristics and economic. The component υ_{1k} is the random component refers to the marginal price of commercial surface (COM), in relation to districts considered (third level) and represents the deviations of the individual groups from the marginal price of commercial area. The random component e_{1ijk} is always related to the marginal price of commercial area, but it represents the random per-apartment compound detected (first level). The random components u_{2jk} and u_{3jk}, instead, refer to the marginal price of floor level characteristics (LEV) and toilet (TOI) in relation to the typology (second level).

10 Model's Results

For the multilevel analysis was used the software package MLwiN version 2.27 produced by the Centre for Multilevel Modelling, University of Bristol. The general equation of the multilevel model, once explicit the values of the estimated fixed parameters, is the following:

$$\begin{aligned} PRZ_{ijk} = & (-350.61 + v_{0k}) \cdot CON + (1,318.01 + v_{1k} + e_{1ijk}) \cdot COM_{ijk} \\ & + (916.05 + u_{2jk}) \cdot LEV_{ijk} + (18,860.52 + u_{3jk}) \cdot TOI_{ijk} \\ & + 7,879.44 \cdot MAI_{ijk} + 6,654.60 \cdot VIW_{ijk} + 9,468.19 \cdot PAR_{ijk}; \end{aligned}$$

where:

$$\begin{aligned} & \beta_{0k} = \beta_0 + v_{0k} = -350.61 + v_{0k}; \\ & \beta_{1ik} = \beta_1 + v_{1k} + e_{1ijk} = 1,318.01 + v_{1k} + e_{1ijk}; \\ & \beta_{2j} = \beta_2 + u_{2jk} = 916.05 + u_{2jk}; \\ & \beta_{3j} = \beta_3 + u_{3jk} = 18,860.52 + u_{3jk}; \\ & \beta_4 = 7,879.44; \\ & \beta_5 = 6,654.60; \\ & \beta_6 = 9,468.19. \end{aligned}$$

It is possible to explain all the regression functions obtained from the analysis also appraising the random components depending on the considered level. The equations are built according to the districts considered for apartments in multi-store buildings (Table 5) and property in the small block of flats (Table 6).

Ultimately the multilevel analysis allowed to determine a number of regression functions equal to the districts considered, in this case 18 for apartments in multistorey buildings for which data are available for district, and 12 for the apartments in the small block of flats for which lacked data in 6 of the 18 districts analyzed. The coefficients in these equations were calculated by summing to the fixed part the residues of the second or third level, where these results are statistically significant. The value of the fixed part, is appraised for all of the real estate characteristics in the model (Table 7).

The calibration of the multilevel model aims to determine the contribution of individual property characteristics. The marginal prices of the floor level and the number of toilets vary according to the typology within each districts considered. The residuals of the second level and the coefficients (marginal prices) of the characteristics floor level (LEV) and toilet (TOI) are calculated for the typology related to apartments in the small block of flats with no lift (Table 8).

To the floor level the marginal price is negative and between −593.12 €/lev in Palazzo reale-Monte di pietà district and −4,472.55 €/lev in Cuba-Calatafimi, because typically in buildings in the absence of lift the increase in floor level leads to a decrease

Table 5 Equations of multilevel model for apartments in multi-store buildings per district

District	Multilevel function (€)
ACQUA DEI CORSARI	$-12,068.39 + 1,208.95 \cdot COM + 2,050.85 \cdot LEV + 23,585.20 \cdot TOI + 7,879.44 \cdot MAI + 6,654.60 \cdot VIW + 9,468.19 \cdot PAR$
BORGO NUOVO	$-31,862.39 + 1,183.57 \cdot COM + 2,893.35 \cdot LEV + 26,058.22 \cdot TOI + 7,879.44 \cdot MAI + 6,654.60 \cdot VIW + 9,468.19 \cdot PAR$
BRANCACCIO-CIACULLI	$-19,922.39 + 1,118.57 \cdot COM + 869.24 \cdot LEV + 19,069.73 \cdot TOI + 7,879.44 \cdot MAI + 6,654.60 \cdot VIW + 9,468.19 \cdot PAR$
CRUILLAS-CEP	$-4,363.09 + 1,409.29 \cdot COM + 1,717.91 \cdot LEV + 26,018.92 \cdot TOI + 7,879.44 \cdot MAI + 6,654.60 \cdot VIW + 9,468.19 \cdot PAR$
CUBA-CALATAFIMI	$-6,062.49 + 997.76 \cdot COM + 6,529.25 \cdot LEV + 44,481.52 \cdot TOI + 7,879.44 \cdot MAI + 6,654.60 \cdot VIW + 9,468.19 \cdot PAR$
LIBERTA’	$64,038.61 + 1,524.02 \cdot COM + 1,647.19 \cdot LEV + 24,850.32 \cdot TOI + 7,879.44 \cdot MAI + 6,654.60 \cdot VIW + 9,468.19 \cdot PAR$
MALASPINA-PALAGONIA	$-240.55 + 1,457.37 \cdot COM + 4,541.55 \cdot LEV + 37,753.52 \cdot TOI + 7,879.44 \cdot MAI + 6,654.60 \cdot VIW + 9,468.19 \cdot PAR$
MEZZOMONREALE-VILLA TASCA	$-10,300.39 + 1,313.04 \cdot COM + 1,900.46 \cdot LEV + 23,703.32 \cdot TOI + 7,879.44 \cdot MAI + 6,654.60 \cdot VIW + 9,468.19 \cdot PAR$
MONTEPELLEGRINO	$42,870.61 + 1,725.73 \cdot COM + 3,458.65 \cdot LEV + 32,086.52 \cdot TOI + 7,879.44 \cdot MAI + 6,654.60 \cdot VIW + 9,468.19 \cdot PAR$
NOCE	$-14,626.39 + 1,281.91 \cdot COM + 1,293.95 \cdot LEV + 19,296.07 \cdot TOI + 7,879.44 \cdot MAI + 6,654.60 \cdot VIW + 9,468.19 \cdot PAR$
ORETO-STAZIONE	$-29,274.39 + 1,189.37 \cdot COM + 2.055.35 \cdot LEV + 23,835.22 \cdot TOI + 7,879.44 \cdot MAI + 6,654.60 \cdot VIW + 9,468.19 \cdot PAR$

(continued)

Table 5 (continued)

District	Multilevel function (€)
PALAZZO REALE-MONTE DI PIETA'	− 33,715.39 + 1,085.89 · COM + 656.40 · LEV + 18,784.49 · TOI + 7,879.44 · MAI + 6,654.60 · VIW + 9,468.19 · PAR
POLITEAMA	58,913.61 + 1,740.27 · COM + 2,412.35 · LEV + 18,715.46 · TOI + 7,879.44 · MAI + 6,654.60 · VIW + 9,468.19 · PAR
RESUTTANA-SAN LORENZO	39,803.61 + 1,708.38 · COM + 3,569.35 · LEV + 18,842.15 · TOI + 7,879.44 · MAI + 6,654.60 · VIW + 9,468.19 · PAR
SANTA ROSALIA MONTEGRAPPA	− 15,915.61 + 1,083.13 · COM - 724.25 · LEV + 30,738.52 · TOI + 7,879.44 · MAI + 6,654.60 · VIW + 9,468.19 · PAR
SETTECANNOLI	− 5,039.79 + 1,268.85 · COM + 1,462.48 · LEV + 19,340.12 · TOI + 7,879.44 · MAI + 6,654.60 · VIW + 9,468.19 · PAR
TRIBUNALI-CASTELLAMMARE	− 34,626.39 + 984.29 · COM + 2,266.75 · LEV + 27,139.82 · TOI + 7,879.44 · MAI + 6,654.60 · VIW + 9,468.19 · PAR
ZISA	− 13,126.39 + 1,443.80 · COM + 1,415.36 · LEV + 28,769.42 · TOI + 7,879.44 · MAI + 6,654.60 · VIW + 9.468.19 · PAR

Table 6 Equations of the multilevel model for apartments in the small block of flats per district

District	Multilevel function (€)
ACQUA DEI CORSARI	$-12{,}068.39 + 1{,}208.95 \cdot COM + 144.58 \cdot LEV + 14{,}131.62 \cdot TOI$ $+7{,}879.44 \cdot MAI + 6{,}654.60 \cdot VIW + 9{,}468.19 \cdot PAR$
BRANCACCIO-CIACULLI	$-19{,}922.39 + 1{,}118.57 \cdot COM + 469.72 \cdot LEV + 15{,}834.62 \cdot TOI$ $+7{,}879.44 \cdot MAI + 6{,}654.60 \cdot VIW + 9{,}468.19 \cdot PAR$
CRUILLAS-CEP	$-4{,}363.09 + 1{,}409.29 \cdot COM - 1{,}145.95 \cdot LEV + 6{,}220.52 \cdot TOI$ $+7{,}879.44 \cdot MAI + 6{,}654.60 \cdot VIW + 9{,}468.19 \cdot PAR$
CUBA-CALATAFIMI	$-6{,}062.49 + 997.76 \cdot COM - 4{,}472.55 \cdot LEV - 9{,}975.48 \cdot TOI$ $+7{,}879.44 \cdot MAI + 6{,}654.60 \cdot VIW + 9{,}468.19 \cdot PAR$
MALASPINA-PALAGONIA	$-240.55 + 1{,}457.37 \cdot COM - 1{,}261.05 \cdot LEV + 7{,}099.52 \cdot TOI$ $+7{,}879.44 \cdot MAI + 6{,}654.60 \cdot VIW + 9{,}468.19 \cdot PAR$
MEZZOMONREALE-VILLA TASCA	$-10{,}300.39 + 1{,}313.04 \cdot COM + 102.40 \cdot LEV + 14{,}204.22 \cdot TOI$ $+7{,}879.44 \cdot MAI + 6{,}654.60 \cdot VIW + 9{,}468.19 \cdot PAR$
NOCE	$-14{,}626.39 + 1{,}281.91 \cdot COM - 1{,}286.85 \cdot LEV + 5{,}357.52 \cdot TOI$ $+7{,}879.44 \cdot MAI + 6{,}654.60 \cdot VIW + 9{,}468.19 \cdot PAR$
ORETO-STAZIONE	$-29{,}274.39 + 1{,}189.37 \cdot COM + 1{,}092.96 \cdot LEV + 16{,}936.52 \cdot TOI$ $+7{,}879.44 \cdot MAI + 6{,}654.60 \cdot VIW + 9{,}468.19 \cdot PAR$
PALAZZO REALE-MONTE DI PIETA’	$-33{,}715.39 + 1{,}085.89 \cdot COM - 593.15 \cdot LEV + 9{,}372.82 \cdot TOI$ $+7{,}879.44 \cdot MAI + 6{,}654.60 \cdot VIW + 9{,}468.19 \cdot PAR$

(continued)

Table 6 (continued)

District	Multilevel function (€)
SANTA ROSALIA MONTEGRAPPA	$-15{,}915.61 + 1{,}083.13 \cdot COM - 3{,}796.65 \cdot LEV + 7{,}489.52 \cdot TOI + 7{,}879.44 \cdot MAI + 6{,}654.60 \cdot VIW + 9{,}468.19 \cdot PAR$
TRIBUNALI-CASTELLAMMARE	$-34{,}626.39 + 984.29 \cdot COM - 1{,}263.55 \cdot LEV + 6{,}057.52 \cdot TOI + 7{,}879.44 \cdot MAI + 6{,}654.60 \cdot VIW + 9{,}468.19 \cdot PAR$
ZISA	$-13{,}126.39 + 1{,}443.80 \cdot COM - 551.65 \cdot LEV + 10{,}017.32 \cdot TOI + 7{,}879.44 \cdot MAI + 6{,}654.60 \cdot VIW + 9{,}468.19 \cdot PAR$

Table 7 Multilevel analysis: fixed part

Fixed part	Unit	Amount
Constant (CON)	€	350.61
Commercial area (COM)	€/sqm	1,318.01
Floor level (LEV)	€/lev	916.05
Toilet (TOI)	€/n	18,860.52
State of internal maintenance (MAI)	€/n	7,879.44
Views (VIW)	€/n	6,654.60
Parking (PAR)	€/n	9,468.19

Table 8 Residuals of the second level

No.	District	Floor level (LEV)		Toilet (TOI)	
		Residual (€/lev)	Coefficient (€/lev)	Residual (€/n)	Coefficient (€/n)
1	ACQUA DEI CORSARI	−771.47	144.58	−4,728.90	14,131.62
2	BRANCACCIO-CIACULLI	−446.33	469.72	−3,025.90	15,834.62
3	CRUILLAS-CEP	−2,062.00	−1,145.95	−12,640.00	6,220.52
4	CUBA-CALATAFIMI	−5,388.60	−4,472.55	−28,836.00	-9,975.48
5	MALASPINA-PALAGONIA	−2,177.10	−1,261.05	−11,761.00	7,099.52
6	MEZZOMONREALE-VILLA TASCA	−813.65	102.40	−4,656.30	14,204.22
7	NOCE	−2,202.90	−1,286.85	−13,503.00	5,357.52
8	ORETO-STAZIONE	176.91	1,092.96	−1,924.00	16,936.52
9	PALAZZO REALE-MONTE DI PIETÀ	−1,509.20	−593.15	−9,487.70	9,372.82
10	SANTA ROSALIA MONTEGRAPPA	−4,685.70	−3,769.65	−11,371.00	7,489.52
11	TRIBUNALI-CASTELLAMMARE	−2,179.60	−1,263.55	−12,803.00	6,057.52
12	ZISA	−1,467.70	−551.65	−8,843.20	10,017.32

Typology: small block of flats

in the market price. For districts Acqua dei Corsari, Brancaccio-Ciaculli and Mezzomonreale-Villa Tasca the positive marginal price the poorly affects and is between 102.40 and 469.72 €/lev, in the district Oreto-Stazione reaches 0.7 % of total price average. The marginal prices of the toilets in the buildings of the districts are considered positive and ranged from 5,357.52 €/n for the Noce district to a maximum of 16,936.52 €/n for the district Oreto-Stazione. In the district Cuba-Calatafimi the marginal price is negative and equal to −9,975.48 €/n. Typically the presence of a second toilet increases the value of the apartment, and can induce a shift of the apartment in a higher market segment, except in some circumstances in which for example the second toilet subtracts surface in smaller apartments or is in precarious conditions that require expensive interventions (Tables 9 and 10).

The marginal price of floor level for apartments in multi-storey buildings have a positive sign. The maximum marginal price of floor level is found in the

Table 9 Residuals of the second level

No.	District	Floor level (LEV)		Toilet (TOI)	
		Residual (€/lev)	Coefficient (€/lev)	Residual (€/n)	Coefficient (€/n)
1	ACQUA DEI CORSARI	1,134.80	2,050.85	4,724.50	23,585.02
2	BORGO NUOVO	1,977.30	2,893.35	7,197.70	26,058.22
3	BRANCACCIO-CIACULLI	−46.81	869.24	209.21	19,069.73
4	CRUILLAS-CEP	801.86	1,717.91	7,158.40	26,018.92
5	CUBA-CALATAFIMI	5,613.20	6,529.25	25,621.00	44,481.52
6	LIBERTA'	731.14	1,647.19	5,989.80	24,850.32
7	MALASPINA-PALAGONIA	3,625.50	4,541.55	18,893.00	37,753.52
8	MEZZOMONREALE-VILLA TASCA	984.41	1,900.46	4,842.80	23,703.32
9	MONTEPELLIGRINO	2,542.60	3,458.65	13,226.00	32,086.52
10	NOCE	377.90	1,293.95	435.55	19,296.07
11	ORETO-STAZIONE	1,139.30	2,055.35	4,974.70	23,835.22
12	PALAZZO REALE-MONTE DI PIETÀ	−259.65	656.40	−76.03	18,784.49
13	POLITEAMA	1,496.30	2,412.35	−145.06	18,715.46
14	RESUTTANA-SAN LORENZO	2,653.30	3,569.35	−18.37	18,842.15
15	SANTA ROSALIA MONTEGRAPPA	−1,640.30	−724.25	11,878.00	30,738.52
16	SETTECANNOLI	546.43	1,462.48	479.60	19,340.12
17	TRIBUNALI-CASTELLAMMARE	1,350.70	2,266.75	8,279.30	27,139.82
18	ZISA	499.31	1,415.36	9,908.90	28,769.42

Typology: multistorey building

Cuba-Calatafimi with 6529.25 €/lev, while the lowest value occurs in the district Palazzo Reale-Monte di pietà with 656.40 €/lev. In the district Santa Rosalia-Montegrappa the marginal price is negative and equal to −724.25 €/lev corresponding to 0.8 % of the total price average. In more remote areas, or greater prestige, we note that the positive values are comparatively high. The marginal prices of toilets in the apartments of multi-storey buildings, with reference to the districts concerned, all assume a positive sign. The highest marginal price of toilets amounted to 44,481.52 €/n found within Calatafimi-Cuba, while the lowest value occurs within Politeama with 18,715.46 €/n.

With regard to the marginal prices of the constant, referring to the district, they assume both positive and negative values. Analyzing the results of the multilevel should be noted that the marginal price of the constant variable is in accordance to the district considered.

The highest values are found in the most prestigious areas, such as districts Libertà, Politeama, Monte pellegrino, and also the districts Resuttana-San Lorenzo, that have values respectively of 64,038.61 €, 58,913.61 €, 42,870.61 €,

Table 10 Residuals of the third level: district

No.	District	Constant (CON)		Commercial area (COM)	
		Residual (€)	Coefficient (€)	Residual (€/sqm)	Coefficient (€/sqm)
1	ACQUA DEI CORSARI	−12,419.00	−12,068.39	−109.06	1,208.95
2	BORGO NUOVO	−32,213.00	−31,862.39	−134.44	1,183.57
3	BRANCACCIO-CIACULLI	−20,273.00	−19,922.39	−199.44	1,118.57
4	CRUILLAS-CEP	−4,713.70	−4,363.09	91.28	1,409.29
5	CUBA-CALATAFIMI	−6,413.10	−6,062.49	−320.25	997.76
6	LIBERTA’	63,688.00	64,038.61	206.01	1,524.02
7	MALASPINA-PALAGONIA	−591.16	−240.55	139.36	1,457.37
8	MEZZOMONREALE-VILLA TASCA	−10,651.00	−10,300.39	−4.97	1,313.04
9	MONTEPELLIGRINO	42,520.00	42,870.61	407.72	1,725.73
10	NOCE	−14,977.00	−14,626.39	−36.10	1,281.91
11	ORETO-STAZIONE	−29,625.00	−29,274.39	−128.64	1,189.37
12	PALAZZO REALE-MONTE DI PIETÀ	−34,066.00	−33,715.39	−232.12	1,085.89
13	POLITEAMA	58,563.00	58,913.61	422.26	1,740.27
14	RESUTTANA-SAN LORENZO	39,453.00	39,803.61	390.37	1,708.38
15	SANTA ROSALIA MONTEGRAPPA	15,565.00	15,915.61	−234.88	1,083.13
16	SETTECANNOLI	−5,390.40	−5,039.79	−49.16	1,268.85
17	TRIBUNALI-CASTELLAMMARE	−34,977.00	−34,626.39	−333.72	984.29
18	ZISA	−13,477.00	−13,126.39	125.79	1,443.80

39,803.61 €. The lowest values were in districts Borgo Nuovo, Palazzo Reale-Monte di Pietà and Tribunali-Castellammare that have values respectively of −31,862.39 €, −33,715.39 € and −34,626.39 €. The marginal prices of commercial are, referring to the districts, take on positive values. The marginal price of commercial are assuming the highest values in the districts of greater prestige: Libertà 1,524.02 €/sqm, Montepellegrino 1,725.73 €/sqm, Politeama 1,740.27 €/sqm and Resuttana-San Lorenzo 1,708.38 €/mq; while the districts Cuba-Calatafimi and Tribunali-Castellammare have a marginal price of the commercial area below than 1,000.00 €/sqm and equal respectively to 997.76 and 984.29 €/sqm. For all districts the marginal price of commercial area was less than the average price according to the position ratio in a manner consistent with the expectations (Table 11).

So for the districts considered the membership in the group has a significantly different effect on the average price of the property, net of other characteristics. To understand the significance of the random component of the constant (v_{0k}) is necessary to refer to the average coefficient β_{0k}. According to the model the coefficient of the variable k for the group is the following:

Table 11 Position ratio of commercial area per district

No.	District	Average unit price (€/sqm) (1)	Marginal Price (€/sqm) (2)	Position Ratio (1)/(2)
1	ACQUA DEI CORSARI	1,450.61	1,208.95	0.83
2	BORGO NUOVO	1,531.89	1,183.57	0.77
3	BRANCACCIO-CIACULLI	1,426.83	1,118.57	0.78
4	CRUILLAS-CEP	2,172.08	1,409.29	0.65
5	CUBA-CALATAFIMI	1,819.58	997.76	0.55
6	LIBERTA'	2,554.97	1,524.02	0.60
7	MALASPINA PALAGONIA	2,237.79	1,457.37	0.65
8	MEZZOMONREALE-VILLA TASCA	1,804.49	1,313.04	0.73
9	MONTEPELLEGRINO	2,981.62	1,725.73	0.58
10	NOCE	1,709.55	1,281.91	0.75
11	ORETO-STAZIONE	1,522.31	1,189.37	0.78
12	PALAZZO REALE-MONTE DI PIETA'	1,117.62	1,085.89	0.97
13	POLITEAMA	2,641.26	1,740.27	0.66
14	RESUTTANA-SAN LORENZO	2,692.71	1,708.38	0.63
15	SANTA ROSALIA MONTEGRAPPA	1,748.51	1,083.13	0.62
16	SETTECANNOLI	1,841.91	1,268.85	0.69
17	TRIBUNALI-CASTELLAMMARE	1,074.01	984.29	0.92
18	ZISA	1,866.36	1,443.80	0.77

$$\beta_{0k} = \beta_0 + v_{0k};$$

it is a random variable with mean β_{0k} and standard deviation equal to var $(v_{1k})^{1/2} = (\sigma^2_{u1})^{1/2}$. Approximately 95 % of the market areas has a regression coefficient between $\beta_1 \pm 2\sigma_{u1}$.

The same considerations apply for the other explanatory variables considered.

The confidential intervals at 95 % for the random components on the intercept (CON) are plotted on a graph indicating in the horizontal axis units of the third level (districts) by its rank and in the y-axis the value of the standardized random parameter (Fig. 1).

The intervals of the random component of the constant that do not overlap, i.e., that differ significantly concern the districts: Tribunali-Castellammare, Palazzo Reale-Monte di Pietà, Borgo Nuovo, Oreto-Stazione, Santa Rosalia Montegrappa,

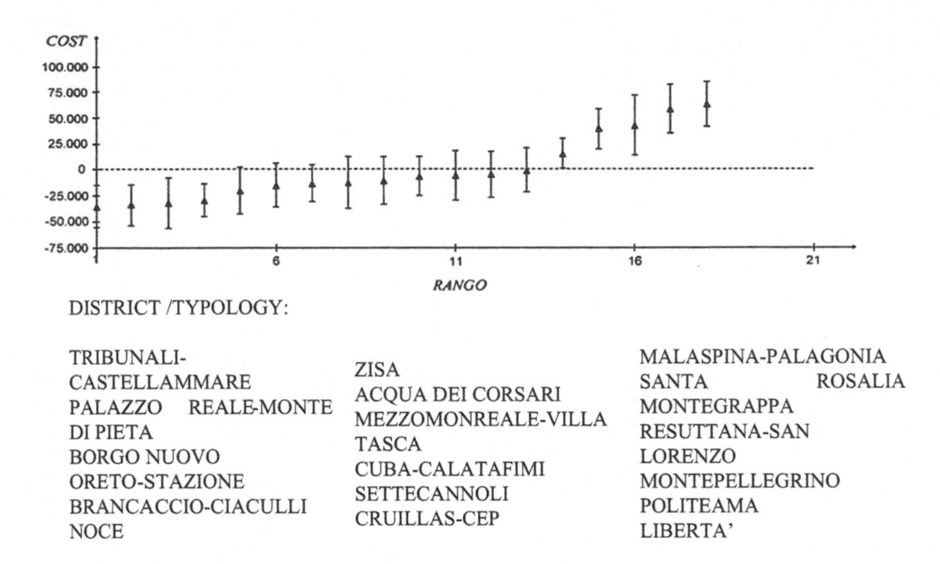

Fig. 1 Confidential intervals for residues of the third level-constant (CON)

Resuttana-San Lorenzo, Montepellegrino, Politeama and Libertà. For these districts the membership to the group has a significantly different effect on the average starting price of the apartment. Also the other districts induce a differentiation of the average prices for the districts. The confidential intervals at 95 % for the random components of the commercial area (v_{1k}) are plotted on a graph indicating the horizontal axis units of the third level (districts) by its rank and the y-axis the value of the standardized random parameter (Fig. 2).

The districts in which the confidential intervals do not superimpose, or that differ significantly are: Tribunali-Castellamare, Cuba-Calatafimi, Santa Rosalia Montegrappa, Palazzo Reale-Monte di Pietà, Brancaccio-Ciaculli, Libertà, Resuttana-San Lorenzo, Montepellegrino and Politeama. The districts Politeama, Montepellegrino, Resuttana-San Lorenzo, Libertà have a greater positive effect of the district varying the commercial area; in the districts Tribunali-Castellammare, Cuba-Calatafimi, Santa Rosalia Montegrappa there is a minor effect of commercial area on the market price of the property. The confidential intervals at 95 % for the random components related to the level floor (u_{2jk}) are plotted on a graph indicating in the horizontal axis the units of the second level (districts differentiated by the type of buildings, P buildings, multi-storey M) by its rank and the y-axis the value of the standardized random parameter (Fig. 3).

The districts in which the confidential intervals do not superimpose, or that differ significantly are: Cuba-Calatafimi_P, Santa Rosalia Montegrappa_P, Santa Rosalia Montegrappa_M, Resuttana-San Lorenzo_M, Malaspina-Palagonia_M and Cuba-Calatafimi_M. The confidential intervals at 95 % for the random components of the toilets (u_{3jk}) are plotted on a graph indicating in the horizontal axis the units

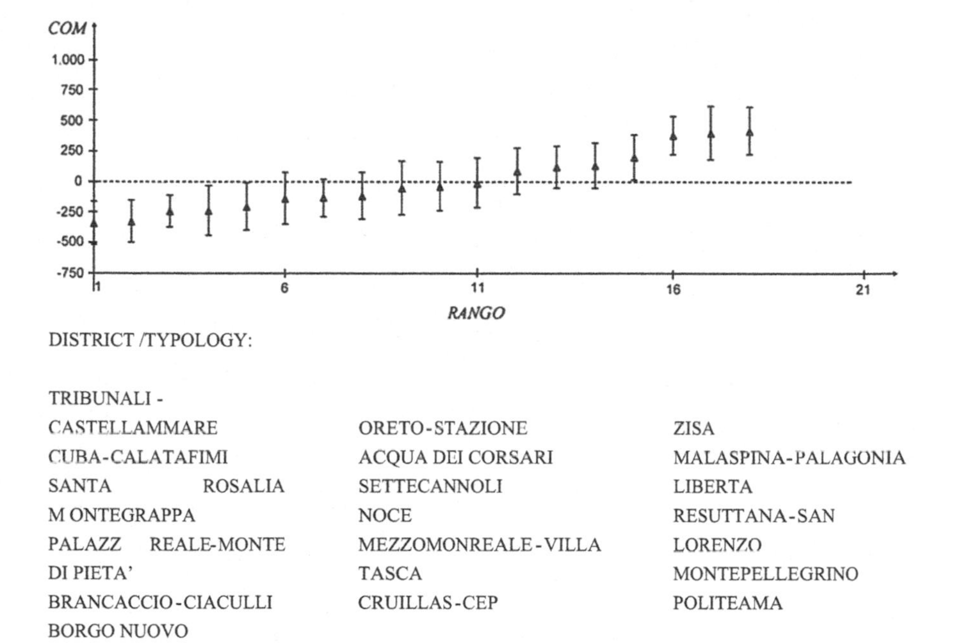

Fig. 2 Confidential intervals for residues of the third level-commercial area (COM)

of the second level (districts differentiated by type of buildings P buildings, M multi-storey) by its rank and the y-axis the value of the standardized random parameter (Fig. 4).

The districts in which the confidential intervals do not superimpose, or that differ significantly are: Cuba-Calatafimi_P, Santa Rosalia Montegrappa_P, Zisa_M, Santa Rosalia Montegrappa_M, Montepellegrino, Malaspina-Palagonia_M and Cuba-Calatafimi_M.

11 Multilevel Model Validation

The multilevel model built is a purely statistical model, based on data collection and processing them with a statistical tool. This model has addressed the statistical tests to check for all variables, except for the time variable (DAT) that was excluded from the preliminary model as irrelevant. The multilevel statistical model to be applied in the real estate valuation, must be validated to ensure that it has achieved the required standard for its use. This is done through the study of the ratio (ratio study), in which the values appraised by the model are compared to the prices observed in the market. In mass appraisal for tax purposes, the test is carried out on

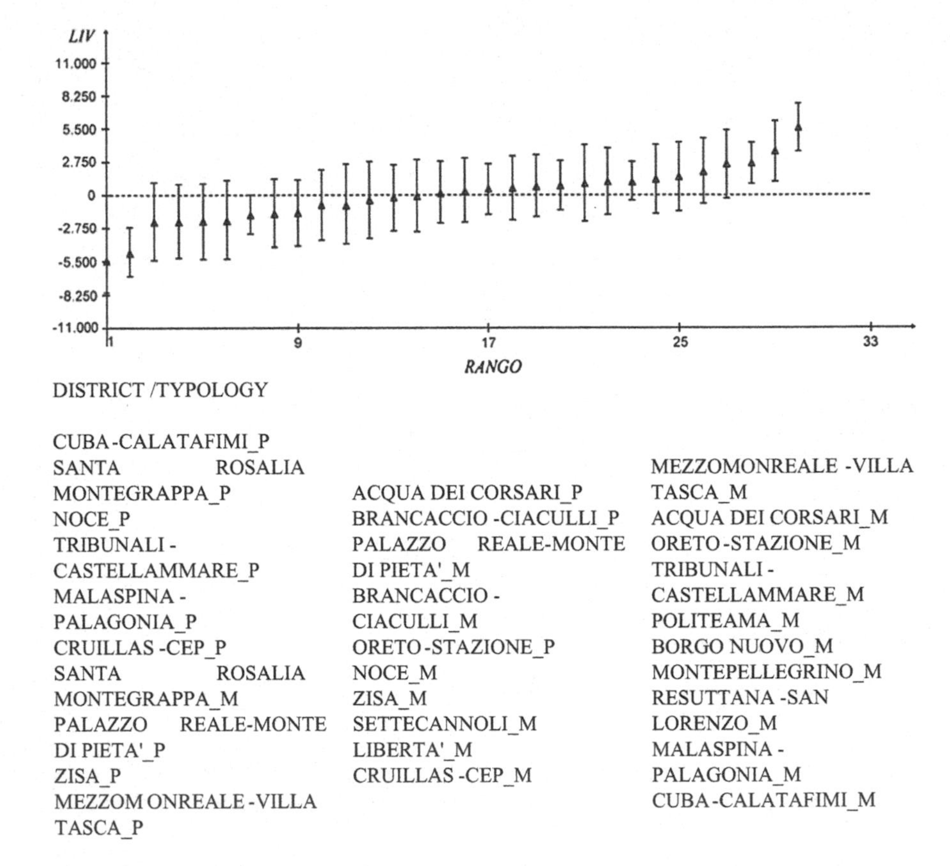

Fig. 3 Confidential intervals for residues of the second level-level floor (LEV)

a separate sample, represented by a group of properties that have not been used in analyzing and have market prices known; in the on line valuations with AVM is not necessary to use this additional data collection, as the original algorithm is reviewed periodically (2–3 years) with completely new versions and we can also update several times a week. On rare occasions the service delivery of the on line valuations is interrupted by operations connected with the changes of the algorithms or with the distribution of new analytical capabilities. The study compares the appraised values of the ratios A and the market price S defining the elementary ratio for each property. The ratio for the generic apartment i, with $i = 1, 2, \ldots, k$, with an appraised value A_i and the market price S_i is equal to:

$$A_i/S_i.$$

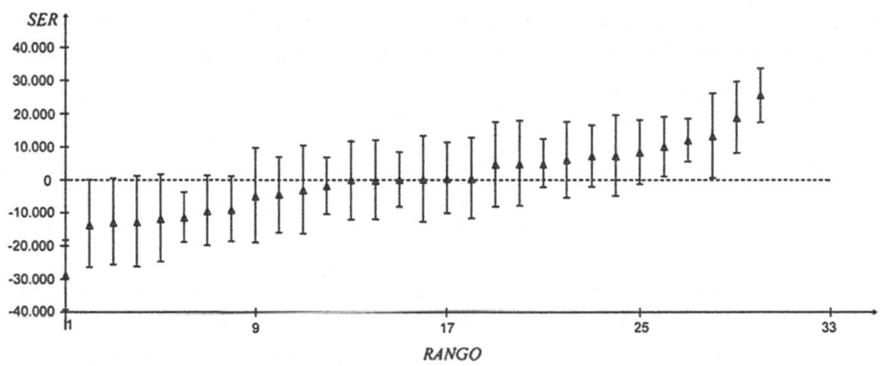

DISTRICT /TYPOLOGY:

CUBA-CALATAFIMI_P
NOCE_P
TRIBUNALI-
CASTELLAMMARE_P
CRUILLAS-CEP_P
MALASPINA-
PALAGONIA_P
SANTA ROSALIA
MONTEGRAPPA_P
PALAZZO REALE-MONTE
DI PIETA' P
ZISA_P
ACQUA DEI CORSARI_P
MEZZOMONREALE-VILLA
TASCA_P

BRANCACCIO-CIACULLI_P
ORETO-STAZIONE_P
POLITEAMA_M
PALAZZO REALE -MONTE
DI PIETA'_M
RESUTTANA-SAN
LORENZO_M
BRANCACCIO-
CIACULLI_M
NOCE_M
SETTECANNOLI_M
ACQUA DEI CORSARI_M
MEZZOMONREALE-VILLA
TASCA_M

ORETO-STAZIONE_M
LIBERTA'_M
CRUILLAS-CEP_M
BORGO NUOVO_M
TRIBUNALI-
CASTELLAMMARE_M
ZISA_M
SANTA ROSALIA
MONTEGRAPPA_M
MONTEPELLEGRINO_M
MALASPINA-
PALAGONIA_M
CUBA-CALATAFIMI_M

Fig. 4 Confidential intervals for residues of the second level-toilets (TOI)

The performance is measured with respect to: the level of appraisal represented by the appraisal error:

$$\mathrm{A_i} - \mathrm{S_i};$$

the uniformity of the valuation represented by the variability of the appraisal error; the percentage average difference absolute:

$$e\% = \frac{\sum_{i=1}^{k} |A_i - S_i|}{k} \cdot 100.$$

Statistically, the central measures provide an indication of the overall level of valuations for a group of properties. The measures of the level of valuation are: the median $A\tilde{/}S$, the average $A\bar{/}S$ and the weighted average $\bar{A}/\bar{S}$, which expresses the average ratio of the group of properties weighted by market prices:

$$\bar{A}/\bar{S} = \frac{\sum_{i=1}^{k} A_i}{\sum_{i=1}^{k} S_i}.$$

The uniformity between the groups can be analyzed in terms of horizontal equity and vertical equity: horizontal equity concerns the comparison of the ratios between groups of properties; the vertical equity concerns the range of variation in property prices. The measures of the uniformity of the valuation are: the range of variation calculated from the difference between the maximum value and the minimum ratio of the group of properties; the dispersion coefficient *COD* measures the average percentage deviation of the ratios from the median ratio and is calculated as follows:

$$COD = \frac{100}{A\tilde{/}S} \cdot \frac{\sum_{i=1}^{k} \left| A_i/S_i - A\tilde{/}S \right|}{k};$$

the coefficient of variation *COV* calculated based on deviations from the average of the ratios:

$$COV = \frac{100}{A\bar{/}S} \cdot \left[\frac{\sum_{i=1}^{k} \left(A_i/S_i - A\bar{/}S \right)^2}{k-1} \right]^{\frac{1}{2}};$$

and the price-related differential *PRD* calculated as the ratio between the average and the weighted average:

$$PRD = \frac{A\bar{/}S}{\bar{A}/\bar{S}} = \frac{\sum_{i=1}^{k} A_i/S_k}{k} \cdot \frac{\sum_{i=1}^{k} S_i}{\sum_{i=1}^{k} A_i}.$$

Measurements of the *PRD* significantly greater than 1 indicate regressivity of the valuation; measures less than 1 suggest progressivity of the valuation. Generally the valuations of the properties with higher value have lower ratios compared to the properties with lower value, in this case the valuations are considered regressive, while in the opposite case, the valuations are considered progressive if the properties of higher value are overestimated compared to properties with lower value.

12 Appraisal Test

According to the Standard on Ratio Studies (IAAO 2013) the appraisal tests were conducted on the sample of market data in the districts for which we have a larger number of data (Table 12).

The percentage average difference for the residential property should be less than 10 ÷ 15 % (Ciuna 2011). For the sample of the collected data the percentage difference calculated respect to median is equal to 0.099; the standard deviation calculated from the average percentage is equal to 0.084. For the measurements of the appraisal level (columns 1, 2 and 3 of Table 5), the test provides an indication of the overall level of valuations for the apartments and districts considered and for the apartments of all districts. The test result is generally favorable. For measurements of the uniformity, the ranges, for apartments and districts considered and for the apartments of all districts, show considerable variability, however, offset by measure of the appraisal level. The coefficient of dispersion *COD*, calculated based on deviations from the mean of the ratios, for residential property should be below 15 % in the older areas and heterogeneous and below 10 % of residences in areas newest and quite similar. The test is passed favorably for the apartments and considered districts and for apartments in all districts. The coefficient of variation *COV*, calculated based on deviations of the ratios by the average, is exceeded for the full sample of data. The districts Cuba-Calatafimi and Oreto-Stazione do not pass the test. The price-related differential *PRD* calculated as the ratio between the average and the weighted average ratio should be close to 1, in particular between 0.98 and 1.03. The standard for the *PRD* are not absolute when the samples are small or when there are wide variations in the prices. The test is passed for the apartments and districts considered for apartments in all districts. The standard suggests that the level of the ratios of a group of properties must be within 5 % of the total value of all the groups considered. The level of overall ratio should be within 10 % less or more than the level of 100 % (0.90 ÷ 1.10). It is not necessary that the analyst knows the details or be able to explain the algorithm of the statistical model or the complexity of its mathematical and statistical formulas. We

Table 12 Results of the ratio study per district

District	$A\tilde{/}S$ (1)	$A\bar{/}S$ (2)	$\bar{A}/\bar{S}$ (3)	Range	COD	COV	PRD
CUBA-CALATAFIMI	1.033	1.035	1.010	0.983	10.946	16.374	1.025
ORETO-STAZIONE	1.025	1.024	0.999	0.693	13.351	16.000	1.025
RESUTTANA-SAN LORENZO	1.012	1.005	0.999	0.371	7.058	8.829	1.006
SANTA ROSALIA MONTEGRAPPA	1.015	1.018	1.004	0.019	9.881	12.408	1.014
ZISA	1.001	1.017	0.992	0.556	10.885	14.617	1.025
CAMPIONE DEI DATI	1.014	1.019	1.001	1.001	10.053	12.978	1.020

must instead that the analyst is able to describe the process of development and to ensure that its results are constant and faithfully reflect the behavior of the market for the property to be appraised (Appraisal Foundation 2003; Borst et al. 2008).

13 Multilevel Estimate

The multilevel model applied to the valuation of the apartments of the districts of the city of Palermo, creates a computer-generated valuation, designated as multilevel estimate (MLe), based on the collection of market data and the development of multivariate statistical model. The MLe is looming, so as the result of a proprietary formula provided online by a specialist website. Its accuracy depends on the number of available data, their variability and the ability acquired by the model to estimate similar properties. The MLe is not a valuation performed by a professional appraiser who performs the inspection in the property to be assessed nor, the opposite extreme, the expertise of a professional formula that summarizes the valuation on subjective bases without the recognition of market data. The MLe if it can not be used in place of a professional valuation, it can be shared with professionals and operators of real estate.

14 Pattern of the AVM

The multilevel model and the MLe applied to the valuation of the apartments can be configured with a pattern of automated valuation model (AVM), which is a calculation software with bases statistical valuation, to produce appraisals of the value of the buildings. The construction of the scheme of AVM is focused on the creation of a database, continuously monitored and updated, and the application of an algorithm, which considers the characteristics of real estate as variables that affect price formation. So that, the technology of the AVM can be used to provide a public service of consultation and advice online real estate. The schema of AVM for the MLe's calculation is based on the database constituted by data collected from the sample and on the algorithm developed with the functions appraised with the multilevel analysis. The 426 data of sales used for statistical-estimative analysis formed the database of real estate data. The single report (profile of the property) considers the characteristics of real estate and the market price of the collected property. Defined the AVM was made a Microsoft Excel interface, compatible to HTML, in which the MLe is calculated once we have entered the characteristics of the property being appraised and according to the data contained in the database. In particular, to determine the MLe of an apartment preliminarily is necessary to: select the district where falls the apartment to be appraised, based on the districts defined above (Fig. 5); select the type of the apartment (Fig. 6): apartments in small block of flats; apartments in multi-storey building.

ML estimate

Typology of apartment: Apartment in multi-storey building
Market area: BRANCACCIO-CIACULLI

ACQUA DEI CORSARI
BORGO NUOVO
✓ BRANCACCIO-CIACULLI
CRUILLAS-CEP
CUBA-CALATAFIMI
LIBERTA'
MALASPINA-PALAGONIA
MEZZOMONREALE-VILLA TASCA
MONTEPELLEGRINO
NOCE
ORETO-STAZIONE
PALAZZO REALE-MONTE DI PIETA'
POLITEAMA
RESUTTANA-SAN LORENZO
SANTA ROSALIA MONTEGRAPPA
SETTECANNOLI
TRIBUNALI-CASTELLAMMARE
ZISA

Characteristics	…inal price	Appraised value
		- 19.922,39
Main surface SUM	1.118,57	55.928,50
Balcony BAL	369,13	5.536,92
Terrace surface TER	391,50	7.829,99
Commercial surface COM		
Level floor LEV	869,24	2.607,72
Toilets TOI	19.069,73	38.139,47
Internal state of maintenance MAI	7.879,44	15.758,88
Views VIW	6.654,60	19.963,80
Parking PAR	9.468,19	9.468,19
	Value	135.311,08

Fig. 5 Input ML estimate for the district

Then we insert the data of the real estate characteristics of the property being appraised. The surfaces of the apartment are divided into the main surface (SUP), the surface of the balconies (BAL) and the surface of the terraces (TER). The software returns the commercial area (COM), which is used by the multilevel model, considering the commercial ratios recognized in the market segments of the districts. The other charcteristics of the property to be appraised are related at the level floor (LEV), toilets (TOI), the state of the internal maintenance (MAI), the views (VIW) and parking (PAR). Place the characteristics of property to be appraised, the scheme of AVM calculates the MLe Then enter all the characteristics of the property being appraised, the user gets the MLe. In the on line valuation of the apartments the MLe presents the trend of the prices of the apartments in the district in recent months.

ML estimate

Typology of apartment: Apartment in multi-storey building
Market area: BRANCACCIO-CIACULLI

Characteristics	Amount	Unit of measurement	Marginal price	Appraised value
				- 19.922,39
Main surface SUM	50,0	sqm	1.118,57	55.928,50
Balcony BAL	15,0	sqm	369,13	5.536,92
Terrace surface TER	20,0	sqm	391,50	7.829,99
Commercial surface COM	62,0	sqm		
Level floor LEV	3	n.	869,24	2.607,72
Toilets TOI	2	n.	19.069,73	38.139,47
Internal state of maintenance MAI	2	from 1 to 3	7.879,44	15.758,88
Views VIW	3	n.	6.654,60	19.963,80
Parking PAR	1	n.	9.468,19	9.468,19
			Value	135.311,08

Fig. 6 Input ML estimate for typology

15 Conclusions

This experimental study has applied a mass appraisal model for the valuation of real estate apartments located in 18 districts of the city of Palermo: (Acqua dei Corsari; Borgo Nuovo; Brancaccio-Ciaculli; Cruillas-C.E.P.; Cuba-Calatafimi; Libertà; Malaspina-Palagonia; Mezzo Monreale-Villatasca; Montepellegrino; Noce; Oreto-stazione; Palazzo Reale-Monte di Pietà; Politeama; Resuttana-San Lorenzo; Santa Rosalia Montegrappa; Settecannoli; Tribunali-Castellammare; Zisa). The sample of market data revealed the prices and characteristics of the real estate apartments in the five years between 2008 and 2012. The mass appraisal models are able to value simultaneously a plurality of properties and for this reason are used in cadastral valuations (Kauko and d'Amato 2008a, b). In the private sector the mass appraisal models are a widely used in the real estate online valuations proposed from specialized web sites for families and investors and in general for the real estate operators. The main purpose of the study concerns the creation of a model (or algorithm) that allows to calculate the computer-generated value of a property with a method of mass appraisal. These computer models are called automated valuation models (AVM) (d'Amato and Kauko 2008; d'Amato and Siniak 2008). The adopted AVM is the multilevel model which is a generalization of regression methods when the data have a complex structure of variability and a hierarchical configuration (Ciuna 2007). The hierarchical structure of the data is made up of: the apartments collected; the type of property distinct in small block of flats and in multi-storey buildings; and districts as proxies of the market segments. The model also considered the real estate characteristics represented: from the main surface, from balconies and terraces (in the commercial area), the level of floor, toilettes, internal state of maintenance, the views and the parking. The model is presented as a set of linear equations relating to each district in which the explained variable is the market price and the explanatory variables are characteristics of the apartment. For each variable, or characteristic, the equation shows the coefficients calculated from the model and for the price (marginal) of the corresponding characteristic; the equation also contains a constant which takes into account the locational characteristics of the apartment (Salvo and De Ruggiero 2011, 2013). The valuation of the apartments is carried out by replacing the variables in the equation with the characteristics (interpolation). The multilevel model provides a computer-generated value called, as in the current language, Multilevel estimate (MLe). The functions of the multilevel model are readily usable. The MLe can be considered to all effects a market value because it is based on the detection of market prices and the pure statistical model without the imposition of any extra-statistical assumptions (Salvo et al. 2013a, b). The sample survey was carried out by applying the guidelines for the collection of data in the housing market and the multilevel model has passed the statistic and estimative test provided by the valuation standards. Meet these conditions, it was possible to propose a scheme of AVM with an automated software process to be implemented in the on line valuations. The statistical model has the advantage of offering an independent indication of the real estate market situation in the districts, in particular on the phase of the market, the trends and variation rates. The multilevel model has shown, since the beginning, that in the five years, the market prices of the apartments of the districts reported remained stable: in fact ruled out the date of sale as irrelevant variable in the model (Salvo et al. 2014).

References

Appraisal Foundation. (2003). *Uniform standards of professional appraisal practice (USPAP)*. Washington, D.C.

Bahjat-Abbas, N., Carron, A., & Johnstone, V. (2005). *Guidelines for the use of automated valuation models for U.K. RMBS transactions, standard and poor's*.

Borst, R. A., Des Rosiers, F., Renigier, M., Kauko, T., & d'Amato, M. (2008) Technical comparison of the methods including formal testing accuracy and other modelling performance using own data sets and multiple regression analysis. In T. Kaukom & M. d'Amato (Eds.), *Mass appraisal an international perspective for property valuers*. Wiley Blackwell.

Ciuna, M. (2007). La stima immobiliare su larga scala: L'analisi multilevel. *Estimo e Territorio, 1*, 9–19.

Ciuna, M. (2011). The valuation error in the compound values. *AESTIMUM [S.l.]*, 569–583, Aug. 2013. ISSN:1724-2118.

Ciuna, M., Salvo, F., & D'amato, M. (2013a.) *Lo smoothing estimativo nelle quotazioni immobiliari*. In *MediMond, International Proceedings "Dynamics of land values and agricultural policies"* (pp. 63–71).

Ciuna, M., Salvo, F., & D'amato, M. (2013b). Repeat values model per la stima dei numeri indici dei prezzi delle aree edificabili nel comune di Paternò (CT). In *MediMond, International Proceedings "Dynamics of land values and agricultural policies"* (pp. 73–83).

Ciuna M., Salvo F., & De Ruggiero, M. (2014). Property prices index numbers and derived indices. *Property Management, 32*(2), 139–153.

Ciuna, M., Salvo, F., & Simonotti, M. (2014a). The expertise in the real estate appraisal in Italy. Recent Advances in civil engineering and mechanics. In *Proceedings of the 5th European Conference of Civil Engineering (ECCIE '14)* (pp. 120–129). Florence, Italy. November 22–24, 2014. North Atlantic University Union. Series: Mathematics and Computers in Science and Engineering Series, 36. ISBN:978-960-474-403-9. ISSN:2227-4.

Ciuna, M., Salvo, F., & Simonotti, M. (2014b). Multilevel methodology approach for the construction of real estate monthly index numbers. *Journal of Real Estate Literature: 2014, 22* (2), 281–302. ISSN:0927-7544.

Ciuna, M., Salvo, F., & Simonotti, M. (2015a). Appraisal value and assessed value in Italy. *International Journal of Economics and Statistics*, 24–31. ISSN:2309-0685.

Ciuna, M., Salvo, F., & Simonotti, M. (2015b). Compensation appraisal processes for the realization of hydraulic works in an agricultural area. In *Proceedings of XLIV INCONTRO DI STUDI Ce.S.E.T. Il danno. Elementi giuridici, urbanistici e economico-estimativi* (pp. 69–82). Bologna, Italy. November 27–28, 2014. ISBN:978-88-99459-21-5.

Ciuna M., Salvo F., & Simonotti, M. (2015c). Parametric measurement of partial damage in building. In *Proceedings of XLIV INCONTRO DI STUDI Ce.S.E.T. Il danno. Elementi giuridici, urbanistici e economico-estimativi*. Bologna, Italy. November 27–28, 2014. pp. 171–188. ISBN:978-88-99459-21-5.

Ciuna M., & Simonotti, M. (2011a). *Linee Guida per la rilevazione dei dati del mercato immobiliare*. Seconda parte. Geocentro, vol. 2011–16; pp. 88–97.

Ciuna, M., & Simonotti, M. (2011b.) *Linee Guida per la rilevazione dei dati del mercato immobiliare*. Prima parte. Geocentro, vol. 2011–15; pp. 86–93,

Ciuna, M., & Simonotti, M. (2014). Real estate surfaces appraisal. *AESTIMUM, 64*, Giugno 2014, 1–13.

d'Amato, M. (2010). A location value response surface model for mass appraising: An "iterative" location adjustment factor in Bari, Italy. *International Journal of Strategic Property Management, 14*, 231–244.

d'Amato, M. (2015). Income approach and property market cycle. *International Journal of Strategic Property Management, 29*(3), 207–219.

d'Amato, M., & Kauko, T. (2012). Sustainability and risk premium estimation in property valuation and assessment of worth. *Building Research and Information, 40*(2), 174–185 (March–April 2012).

d'Amato, M. (2008). *Rough set theory as property valuation methodology: The whole story*. In T. Kauko & M. d'Amato (Eds.), *Mass appraisal an international perspective for property valuers* (Chap. 11, pp. 220–258). Wiley Blackwell.

d'Amato, M., & Kauko, T. (2008). Property market classification and mass appraisal methodology. In T. Kauko & M. d'Amato (Eds.), *Mass appraisal an international perspective for property valuers* (pp. 280–303). Wiley Blackwell.

d'Amato, M., & Siniak, N. (2008). Using fuzzy numbers in mass appraisal: The case of belorussian property market. In T. Kauko, M. d'Amato (Eds.), *Mass appraisal an international perspective for property valuers* (Chap. 5, pp. 91–107). Wiley Blackwell.

Downie, M. L., & Robson, G. (2007). *Automated valuation models: An international perspective*. London: Council of Mortgage Lenders.

Gelman, A. (2006). Multilevel (hierarchical) modeling: What it can and cannot do. *Technometrics, 48*(3).

Goldstein, H. (2003). *Multilevel statistical models* (3rd ed.). London: Arnold.

Goodman, A. C. & Thibodeau, T. (2003). Housing market segmentation and hedonic prediction accuracy. *Journal of Housing Economics,* 12.

Goodman, A. C., & Thibodeau, T. (2007). The spatial proximity of metropolitan housing submarkets. *Real Estate Economics, 35*.

Hox, J. J. (2002). *Multilevel analysis. Techniques and applications*. Mahwah, NJ: Lawrence Erlbaum Associates

International Association of Assessing Officers. (2003). *Standard on automated valuation models (AVMs)*. Chicago: IAAO.

International Association of Assessing Officers. (2012). *Standard on mass appraisal of real property*. Kansas City: IAAO.

International Association of Assessing Officers. (2013). *Standard on ratio studies*. Kansas City: IAAO.

International Valuation Standards Committee. (2007). *International valuation standards IVS*. London: IVSC.

Kauko, T., & M. d'Amato. (2008a). Introduction: Suitability issues in mass appraisal methodology. In T. Kauko & M. d'Amato (Eds.), *Mass appraisal an international perspective for property valuers* (pp. 1–24). Wiley Blackwell.

Kauko, T., & M. d'Amato. (2008b). Preface. In T. Kauko & M. d'Amato (Eds.), *Mass appraisal an international perspective for property valuers* (p. 1). Wiley Blackwell.

Kreft, I., & de Leeuw, J. (1998). *Introducing multilevel modeling*. London: Sage Pubns Ltd.

Orford, S. (2000). *Modelling spatial structures in local housing market dynamics: A multilevel perspective*, Urban Studies (Vol. 37).

Orford, S. (2002). *Valuing locational externalities: A GIS and multilevel modelling approach* (p. 29). Environment and Planning B: Planning and Design, vol.

Renigier-Biłozor, M., Wiśniewski, R., Biłozor, A., & Kaklauskas, A. (2014a). Rating methodology for real estate markets—Poland case study. *Pub. International Journal of Strategic Property Management, 18*(2), 198–212. ISNN:1648-715X.

Renigier-Biłozor, M., Dawidowicz, A., & Radzewicz, A. (2014b). An algorithm for the purposes of determining the real estate markets efficiency in Land Administration System. *Pub. Survey Review, 46*(336), 189–204.

Renigier-Biłozor, M., & Biłozor, A. (2016a). Proximity and propinquity of residential market area-Polish and Italian case study. In *16th International Multidisciplinary Scientific GeoConferences SGEM*. Bułgaria. (web of science).

Renigier-Biłozor, M., & Biłozor, A. (2016b). The use of geoinformation in the process of shaping a safe space. In *16th International Multidisciplinary Scientific GeoConferences SGEM Bułgaria*.

Raudenbush, S. W., & Bryk, A. S. (2002). *Hierarchical linear models* (2nd ed.). Thousand Oaks: Sage Publications.

Salvo, F., Ciuna, M., & d'Amato, M. (2013a). Appraising building area's index numbers using repeat values model. A case study in Paternò (CT). In *Dynamics of land values and agricultural policies* (pp. 63–71), Bologna: Medimond International Proceedings, Editografica. ISBN:978-88-7587-690-6, Palermo, 22–23/11/2012.

Salvo, F., Ciuna, M., & d'Amato, M. (2013b). The appraisal smoothing in the real estate idices. In: (a cura di): *Maria Crescimanno, Leonardo Casini and Antonino Galati, Dynamisc of land values agricultural policies* (pp. 99–111), Bologna: Medimond Monduzzi Editore International Proceeding Division. ISBN:978-88-7587-690-6, Palermo, 22–23/11/2012.

Salvo, F., & De Ruggiero, M. (2011). Misure di similarità negli Adjustment Grid Methods. *AESTIMUM, 58*, 47–58. ISSN:1592-6117.

Salvo, F., & De Ruggiero, M. (2013). Market comparison approach between tradition and innovation. A simplifying approach. *AESTIMUM, 62*, 585–594. ISSN:1592-6117.

Salvo, F., De Ruggiero, M., & Ciuna, M. (2014). Property prices index numbers and derived indices. *Property Management, 32*, pp. 139–153, ISSN:0263-7472, doi:10.1108/PM-03-2013-0021.

Salvo, F., Simonotti, M., Ciuna, M., & De Ruggiero, M. (in print). Measurements of rationality for a scientific approach to the Market Oriented Methods. *Journal of Real Estate Literature*. ISSN:0927–7544.

Simonotti, M., Salvo, F., & Ciuna, M. (2015). Appraisal value and assessed value in Italy. *International Journal of Economics and Statistics, 3*, 24–31. ISSN:2309-0685.

Snijders, Tom A. B., & Bosker, Roel J. (1999). *Multilevel analysis: An introduction to basic and advanced multilevel modeling*. London etc.: Sage Publishers.

Tecnoborsa. (2011). Codice delle valutazioni immobiliari IV, Stime su Larga Scala (Mass Appraisal) Roma, Tecnoborsa.

Uniform Standard of Professional Appraisal Practice (USPAP). (2006). Standard 6: Mass Appraisal, Development and Reporting.

Part IV
AVM Methodological Challenges: Non Deterministic Modelling

Automated Valuation Methods in Real Estate Market—a Two-Level Fuzzy System

Marco Aurélio Stumpf González

Abstract This chapter presents a scheme to Automated Valuation Modelling that joined a two level-fuzzy system with multiple regression hedonic price models in Automated Valuation Modelling. This scheme is based in two main characteristics of real estate market: location and segmentation in sub-markets. Location is recognized as a very important element of real estate market and price of each unity is severely affected by its urban position. By another side, market is divided in several segments (based on diversity of real properties), although there is not a clear division point. Actually, there is a trespass among market segments, which is not abrupt, but continuous. Fuzzy logic permits to take in account continuity of variables. Then, fuzzy logic may contribute to improve Automated Valuation Modelling systems.

Keywords AVM · Hedonic price model · Fuzzy system

1 Difficulties in Using Conventional Hedonic Models to Mass Appraisal

The main challenge in Automated Valuation Modelling (AVM) is to build models with a reasonable precision in order to estimate values for a single case or for a large number of properties (a property portfolio, for instance). AVM may contribute to save time and money in real estate appraisal, but it need be based in mass appraisal techniques, using tested models and a sound database.

In general, individual real estate appraisal (sometimes named as "commercial appraisal") uses hedonic price models, calculated by multiple regression analysis and based in a small sample of market cases. The result is an equation, used then to estimate values. Precision of estimated values is linked to a small level of errors. Automated Valuation Modelling may follow the same scheme, but using a large

M.A.S. González (✉)
UNISINOS, São Leopoldo, Brazil
e-mail: mgonzalez@unisinos.br

M. d'Amato and T. Kauko (eds.), *Advances in Automated Valuation Modeling*,
Studies in Systems, Decision and Control 86, DOI 10.1007/978-3-319-49746-4_15

quantity of market data and more general statistical models. However, build a single model to evaluate a large quantity of properties is not a trivial task, because real properties may be very different among them; there are several difficulties on measuring location characteristics, and so on. Furthermore, the coefficients of regression models are calculated as base on average values of each variable included in the model. Then, properties with characteristics too different of average values will have a major error level. One alternative is to build a set of models—one model to each sub-market, for instance—but in this case the problem is the segmentation criterion. The analyst needs to divide the set of properties in some clear and well-defined classes. He needs to define the limits or frontiers of each model. Nevertheless, different models will potentially presents different estimates to similar properties in borderline. For instance, neighbours regions A and B may present differences to estimate values to two similar properties P_A and P_B, when calculated by Model A and Model B, respectively, creating a gap or step on estimated values (Fig. 1).

These differences may occur because frontiers in sub-markets commonly are not abrupt, but continuous. Values to similar properties probably are similar too. There are a continuous range of values among neighbour sub-markets. Fuzzy logic offers means to make soft these frontiers. So, a possible solution to mass appraisal is to divide a market in some sub-markets (and their correspondents' models) but in a fuzzy way. In the sequence, we introduce elements of theory, propose and present an application of a fuzzy system designed to real estate appraisal in AVM.

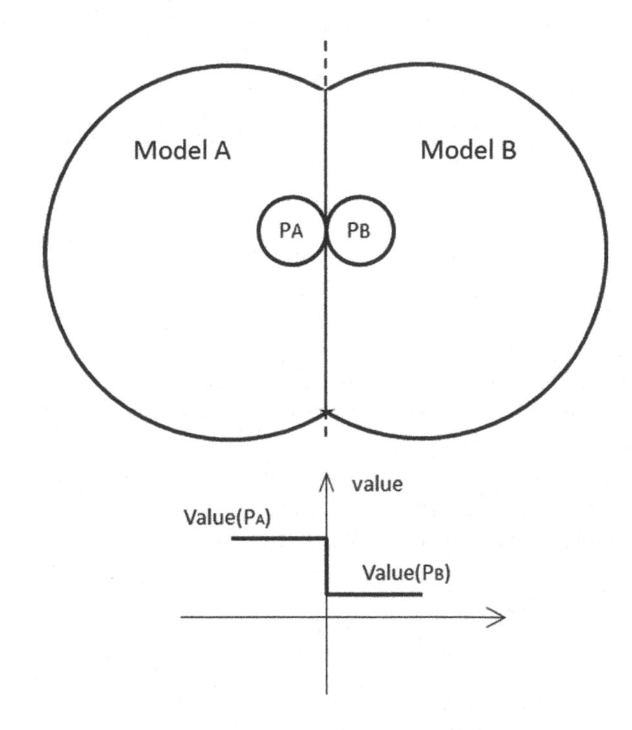

Fig. 1 Neighbour models with different valuations

2 Hedonic Price Models

The representation of a housing market may be developed with econometric models, which include measures representing the most important attributes. The hedonic price models are widespread in the urban economy, based on the theory founded by Court on 30s (Court 1939) with important contributions of Griliches (1971) and Rosen (1974). These models basically present a relationship among observed prices and property characteristics (Robinson 1979). The analyst should establish hypotheses of relationships between these characteristics (explanatory variables) and sale price (explained variable), proposing a format to the model. Transaction data (evidences of the market behaviour) should be collected to develop the models, which should be tested to verify if the models are able to represent the market segment in question. Statistical tests allow evaluate the model itself and the individual importance of the variables included, within a certain degree of accuracy, indicating the overall quality of the formulated model. A conventional model assumes a format such as (Eq. 1):

$$Y = \alpha_0 + \alpha_1 X_1 + \alpha_2 X_2 + \cdots + \alpha_k X_k + \varepsilon = \hat{Y} + \varepsilon \quad (1)$$

This format is known as 'classical linear model', in which Y is the explained variable (usually sale price), X_1, …, X_k are the explanatory variables (the characteristics of the property, location and sale conditions), α_0 is the intercept, α_1, …, α_k are partial regression coefficients (also known as implicit hedonic prices), $\hat{Y}$ is the estimate for the explained variable, and ε is the error term ($\varepsilon = Y - \hat{Y}$). The coefficients α_i are often fitted using Ordinary Least Squares (OLS). Of course, there are several assumptions to obtain unbiasedness and efficiency of OLS, such as linearity, constant variance and normally distributed errors (Gujarati 2009; Kutner et al. 2010).

3 Classic and Fuzzy Logics

A conventional hedonic model is based in a sample set that follow the classic set theory. In classic sets, membership is simple: a case belongs or not belongs to a particular set. A classic set follows a 'true or false' scheme (in mathematical terms, membership may be indicated as {0, 1}, as based on Aristotelic or bivalent logic). Following this reasoning, the limits of the sample set define the limits of model validity (Fig. 1) (Cordón et al. 2001). Properties that may be evaluated by a hedonic model need be in the same range of sample set. For example, a division in three sub-markets by size provokes an association of each case to a one single set. Using this scheme, a property could be small (or) medium (or) large, for instance, and its value is calculated by a single model, following a format as Eq. (1). Therefore, the value of a 'medium' property P is estimated by Eq. (2):

Fig. 2 Example of a group of classical sets divide by property size with a case P

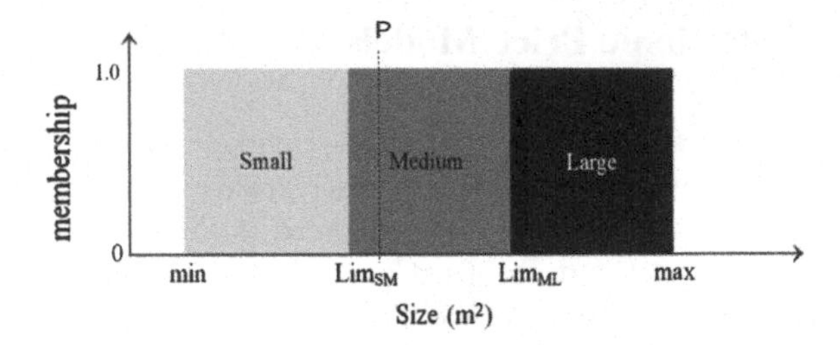

$$\text{Value(P)} = \text{Model\_M(P)} \tag{2}$$

where Value(P) is the estimation to the particular characteristics of property P and Model_M is a model estimated using a sample of market cases classified as 'medium'. It could have also independent models to estimate small (Model_S) and large properties (Model_L), shared by size-limits, such as Lim_{SM} and Lim_{ML}. In graphical terms, a group of classic sets looks as Fig. 2. Properties with size between [Lim_{SM}; Lim_{ML}]—like 'P' in Fig. 2—are named 'medium' properties, and so on.

Fuzzy sets have a different scheme. There are some works about application of fuzzy logic in property valuation, such as Byrne (1995), Smith and Bagnoli (1997), d'Amato and Siniak (2003, 2008). Membership in fuzzy sets may assume any value in a real interval [0,1]. There is a degree of truth in each fuzzy set. A fuzzy system —basically an organized group of fuzzy sets—has not a simple membership, but a multiple membership. Each case belongs simultaneously to two or more fuzzy sets, with different membership degrees. The general membership of a case to the fuzzy system is the sum of partial memberships (Cordón et al. 2001; Zadeh 1965). A fuzzy system applied to mass appraisal with some fuzzy sets follows a weighted sum, such as (Eq. 3):

$$\text{Value(P)} = \sum^{i} [\mu_i(P)^* \text{Model}_i(P)] \tag{3}$$

where Value(P) is the valuation using the particular characteristics of property P, $Model_i(P)$ is the value calculated to P using $Model_i$, $\mu_i(P)$ is the membership of case P to fuzzy set i, i = {1, …, n}, n is the number of fuzzy sets and $\sum^i \mu_i = 1$. For instance, in a fuzzy system with three fuzzy sets based on property size (such as in Fig. 3), a property P could have a $\mu_S = 0.30$ membership degree in the 'small' set (S), a $\mu_M = 0.70$ membership degree in the 'medium' set (M) and zero membership ($\mu_L = 0$) in the 'large' fuzzy set (L). Other property, with a little bit larger floor area would have $\mu_S = 0.29$, $\mu_M = 0.71$, and $\mu_L = 0$ memberships, and so on. Transitions among two classes became soft. Value (P) is a weighted average of three valuations and therefore permits a continuous value range.

Definitions about a fuzzy system begin by chose the number of fuzzy sets. In the sequence, some questions are to define the shape and limits to each fuzzy set. It's very common to use triangular or trapezoidal functions to fuzzy sets. Other useful

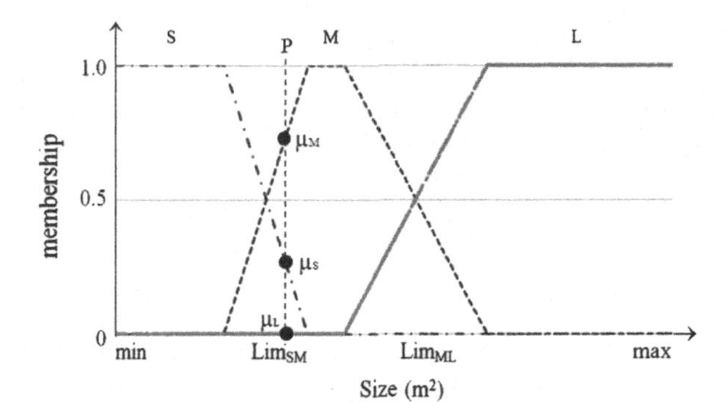

Fig. 3 Example of a fuzzy system divide by property size with a case P

formats are Sigmoid and Gaussian (Cordón et al. 2001). A group of trapezoidal fuzzy sets has a form like in Fig. 3, and limits could be Lim_{SM} and Lim_{ML}. If trapezoidal shape is choose, the next step is to define limit values for which two neighbour fuzzy sets have equal membership (a specific value to reach $\mu_S = \mu_M = 0.5$ and other value to $\mu_M = \mu_L = 0.5$), and the second is to define transition rules (the slope of each transition). The slope of each lateral of trapezium may be determined using equal angles or equal rate of value changing. Based on these values, the analyst can define the ranges for total membership in each set (values to $\mu_S = 1.0$, $\mu_M = 1.0$, and $\mu_L = 1.0$).

4 Fuzzy Sets Designed to a Two-Level Fuzzy-Hedonic Mass Appraisal System

The proposed system to AVM is a combination of conventional hedonic price functions and two fuzzy levels. We choose property size and location as criteria to create these fuzzy systems. In this particular case, both are known be important on property values, and they are significant to define sub-markets. The first has variation in one direction (or axis) and the second is spatial. These characteristics make a difference on fuzzy functions, calculations of memberships, and other elements.

4.1 Fuzzy Sets Based on Size

To consider the influence of size, the range of floor area may be divided in some groups. In this case, it was divided in three groups, but can divide in 4, 5, or more classes. Following this, Eq. (3) may be written as on Eq. (4):

$$\text{Value (P)} = \mu_S{}^*\text{Model\_S(P)} + \mu_M{}^*\text{Model\_M(P)} + \mu_L{}^*\text{Model\_L(P)} \quad (4)$$

where Model_S(P), Model_M(P), Model_L(P) are the estimates of models S, M and L applied to property P, and μ_S, μ_M, μ_L are membership degrees of property P to each class.

We chose a slope of 20 % of size-limits in each crossing (0.8 and 1.2). In this case, membership functions use a scheme such as the presented in Fig. 3 and detailed in Exhibit 1:

Exhibit 1—Membership functions to trapezoidal fuzzy sets
$\mu_S = 1$, $\mu_M = 0$ and $\mu_L = 0$, if size < Lim_{SM}*0.8
μ_M = (size-Lim_{SM}*0.8)/(Lim_{SM}*0.4), $\mu_L = 0$ and $\mu_S = 1$-μ_M, if size $\in$ [Lim_{SM}*0.8; Lim_{SM}*1.2]
$\mu_M = 1$, $\mu_S = 0$ and $\mu_L = 0$, if size $\in$ (Lim_{SM}*1.2; Lim_{ML}*0.8)
μ_M = (Lim_{ML}*1.2-size)/(Lim_{ML}*0.4), $\mu_S = 0$ and $\mu_L = 1$-μ_M, if size $\in$ [Lim_{ML}*0.8; Lim_{ML}*1.2]
$\mu_L = 1$, $\mu_S = 0$ and $\mu_M = 0$ if size > Lim_{ML}*1.2

Actually, using the size of each property, only one or two memberships function has value different from zero.

4.2 Fuzzy Sets Based on Distance

In the case of location, relationship of properties with the neighbour properties occur in all directions (360°). General influence is the weighted sum of influences in all directions. The participation of neighbour cases in the final values is dependent of the weighting scheme defined to fuzzy sets. In this format, participation is more significant to next units. A format based on inverse of distance to weighting cases is an interesting option. Figure 4 present two inverse-distance fuzzy sets, using $1/d^k$, with k = 1 and k = 2 (1/d and $1/d^2$). Increase k will reinforce membership values to neighbours points (weighting more strongly neighbour cases). Therefore, the importance of neighbours cases on final value increases.

A fuzzy system to distance composed by m fuzzy sets may be described like in Eq. (5):

$$\text{Value(P)} = \sum^{j}[\mu_j(P)^*\text{Value}_j(P)] \quad (5)$$

where Value(P) is the valuation applied to property P; $\mu_j(P)$ is the membership of P to each fuzzy set j; $Value_j(P)$ is the value calculated to P using $Model_j$; $\mu_j(P) = distance(P, j)^{-k}/w$, with $w = \sum^j[distance(P, j)^{-k}]$, and w is calculated to reach $\sum\mu_j(P) = 1$; distance(P, j) is the linear distance from case P to the reference centre of the $Model_j$, which have coordinates (x_j, y_j); k is the exponent that give a weight to the influence of distance; and j = {1, …, m}.

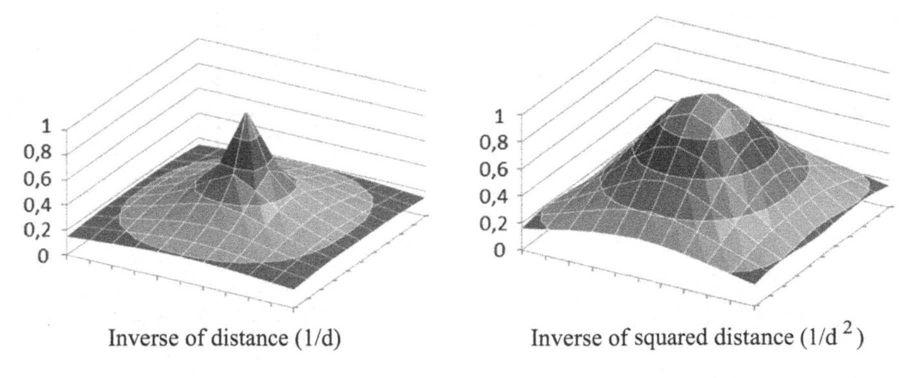

Fig. 4 Two conical fuzzy sets with membership determined by distance to centre point

Fuzzy systems consist on a weighted sum of partial estimates to size, weighted then by distance. The double-fuzzy value is calculated combining Eqs. (4) and (5) —generating Eq. (6):

$$\begin{aligned} \text{Double} - \text{fuzzy(P)} &= \sum^{j}\{\mu_j(\text{P})^*\sum^{i}[\mu_i(\text{P})^*\text{Model}_i(\text{P})]\} \\ &= \sum^{j}\{\mu_j(\text{P})^*[\mu_S{}^*\text{Model\_S(P)} + \mu_M{}^*\text{Model\_M(P)} + \mu_L{}^*\text{Model\_L(P)}]\} \end{aligned} \tag{6}$$

5 Developing the Fuzzy-AVM System

In the sequence, we present an example of this system. This process may be viewed as the first step to build an AVM, and because that may seem a bit complex and time consuming. However, after this step, it's only need to include new market cases to feed the database and to adapt the models to market changes, with minor effort.

The proposed fuzzy system is compared with conventional hedonic and surface models. We use a modular system, considering a lattice of squared modules of 1 × 1 km (such as Fig. 5). They are based on distance of the case in valuation to each module centre of the lattice and use a weighting based on the inverse of squared distance.

The centre of each module is the reference to model location. In each module were calculated 3 models, designed to measure values in 3 different classes of property size (small, medium and large properties). The limits to each size-class were determined to each module in an independent way.

Data collected to develop the example. The system uses a database with more than 160 thousand cases of apartment sales in Porto Alegre, a southern Brazilian city. We use a cross section/time series sample of urban area, considering all the

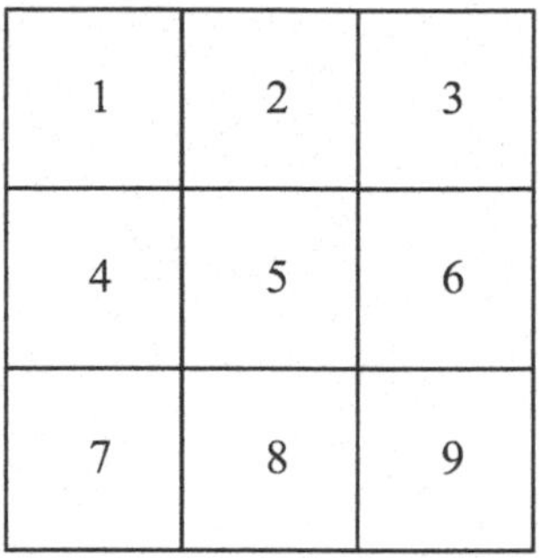

Fig. 5 Spatial configuration to proposed system

1	2	3
4	5	6
7	8	9

urban space in the city. The sales occur between 1998 and 2014. Data available include physical, location and sales characteristics. Table 1 presents the variables and basic statistical data and Fig. 6 illustrate location of properties and positioning of the lattice of modules.

Sale price was originally in Brazilian Real (R$) and was converted to Euros, using exchange rate of July, 2014. Size, quality and age of the properties are like in municipal cadastre. District quality is a subjective attribute, which is determined as base on author' experience. Coordinates are measured using as reference point the Central Business District (CBD, the historical and commercial city centre, which is located as 0.0 in Fig. 6) Distance from CBD is the linear distance, in km, from to each property to CBD. Sale date was converted to a continuous scale of months, in which the month of first sale represents Time = 0. Characteristics like number of rooms and parking spaces were not available in source data. As based on Fig. 5, we use 9 fuzzy sets to consider location in the proposed systems. The central module (5) is the target to evaluate the system. Demonstration reaches a part of the city (almost 6 % of urban area).

A sample was extracted from database to the region in analysis (in white in Fig. 6). Nine individual samples of equal size (6,000 cases) were obtained of each

Table 1 Variables and statistical properties for data collected

Variable	Unit	Minimum	Maximum	Average	Standard deviation
Sale price	€	5,795.55	1,810,021.10	63,957.25	74,130.71
Property size	m^2	20.000	2,306.000	94.722	70.155
Property quality	–	1	7	2.334	0.959
Age	years	0	90	20.079	12.294
District quality	–	10	90	30.229	17.095
Distance from CBD	km	0.000	16.283	5.237	3.578
N-S Coordinate (X)	km	−2.950	13.634	3.739	3.748
E-W coordinate (Y)	km	−14.000	6.585	−1.196	3.257
Sale date	–	07 August 98	23 June 14	–	–
Time	Month	1	202	89.817	38.849

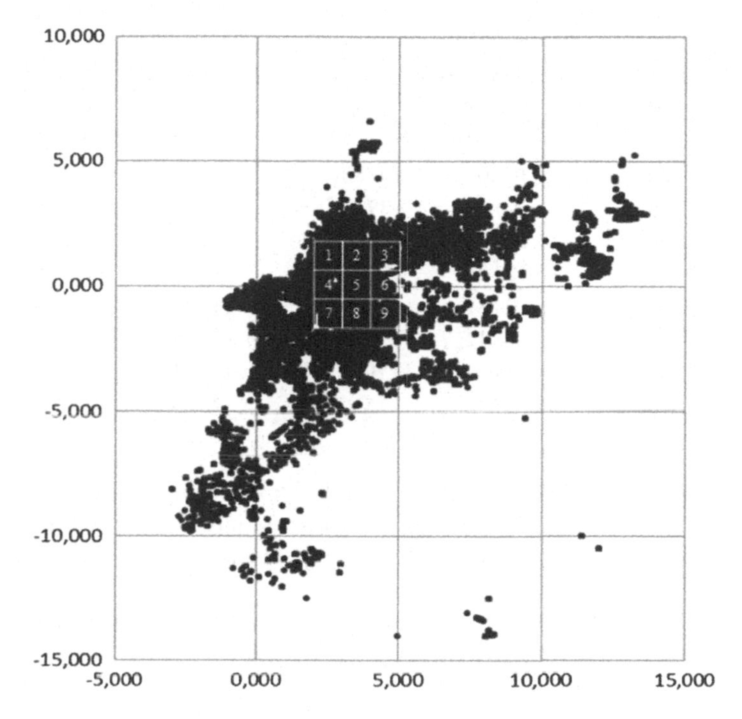

Fig. 6 Positioning of sample cases and the 9-module lattice (The point 0.0 represents the CBD; scale: approximated 1:10,000)

module centre (the 6,000 nearest cases using distance to the module' centres as neighbour measurement). Cases were classified by size in three classes, resulting in three sub-samples of equal size (2,000 in each size-class). The limits by size are designed as Lim_{SM} (division between properties with small and medium sizes) and Lim_{ML} (between medium and large size)—(see Fig. 3 and Table 2). Then 80 % of the cases in each sub-sample were randomly selected to calculate the models and the other 20 % were separated to verify statistical quality of these models. Limits are different in each module because the differences on size in each module probably represent real market differences among properties and make matter in appraisals.

6 Models to be Estimated

Modelling starts with an exploration of available data. After study about some alternatives in model and variable formats, a set of variables were defined. Initial estimations using a Box-Cox procedure indicates better statistical results using an exponent of $\theta = 1.25$ (Eq. 7—it's equivalent to use $\text{Value}^{0.8}$). The same general format was used for all equations, to make the system a little bit simpler. Distance

Table 2 Property size in each module

Module	Minimum (m^2)	Lim_{SM} (m^2)	Average (m^2)	Lim_{ML} (m^2)	Maximum (m^2)
1	21.00	82.91	145.56	142.55	1471.00
2	23.00	91.00	140.87	156.00	940.00
3	24.00	104.00	148.12	164.00	872.73
4	23.00	103.00	172.75	179.60	1471.00
5	23.00	118.28	198.57	214.00	1028.00
6	24.00	129.34	200.68	226.00	984.00
7	23.00	90.87	132.09	135.80	808.00
8	24.00	113.12	182.36	193.71	1028.00
9	24.00	124.29	191.25	212.23	984.00
Averages	21.00	87.16	153.31	76.66	1471.00

from Central Business District was the main location characteristic. Distance from other points, such as shopping centres and leisure points, were tested, but had not better results. Hedonic Price Models follow a quite simple format (Eq. 6).

$$\text{Value} = (a_0 + a_1.\text{Property\_Size} + a_2.\text{Property\_Quality} + a_3.\text{Age} + a_4.\text{Time} + a_5.\text{Distance\_CBD} + a_6.\text{District\_Quality})^{\theta} \quad (7)$$

Surface model is very similar (Eq. 8). Location variables (Distance_CBD and District_Quality) are substituted by terms combining coordinates (X, Y), to the degree 2. Surfaces using degrees 3 and 4 also were estimated, but they were no better than degree 2 surface.

$$\text{Value} = (b_0 + b_1.\text{Property\_Size} + b_2.\text{Property\_Quality} + b_3.\text{Age} + b_4.\text{Time} + b_5.X + b_6.Y + b_7.X^2 + b_8.XY + b_9.Y^2)^{\theta} \quad (8)$$

In each module were estimated 5 models (Table 3), resulting in 45 models, which were mounted using the same routine. Furthermore, two general models (G.C and G.S) were estimated using all cases in the sample (31,618 cases).

Where M = {1, … 9}. All models were examined at light of assumptions of regression, and test hypothesis to model and regressors, at $\alpha = 0.05$. In most cases, models and regressors reaches $\alpha = 0.01$ level.

Table 3 Scheme for HPM estimated in each module

Model	Training data	Test data
M.C—Conventional HPM	4800	1200
M.S—Surface model	4800	1200
M.1a—Conventional HPM—small properties	1200	400
M.1b—Conventional HPM—medium properties	1200	400
M.1c—Conventional HPM—large properties	1200	400

7 Results Obtained to the Example-System

Initially we present a detailed example of calculations to individual properties. A case located in module 5 was randomly selected to demonstrate the procedure to obtain fuzzy values. The case has 185.00 m^2 of floor area. Its sale price is €158,401.12. The first stage uses hedonic models. Initial estimation using general conventional model (G.C) indicates a value of €160,758.32. The valuation using general surface model (G.S) reaches €162,922.06. On a similar way, conventional (5.C) and surface (5.S) models of module 5 have estimative of €160,758.38 and €162,922.08, respectively. In this case, P is a medium property, then the valuation by 5.1 is equivalent to using only the equation to medium properties (5.1b) and reach €159,254.28 (see Table 4). The second stage consists on fuzzy system. Values calculated by fuzzy schemes are presented in Tables 5, 6 and 7. The estimative using fuzzy model based on size (from module 5) is €159,375.89 (Table 4). Value calculated using fuzzy model based on location, which includes weighting the values of the nine modules by the inverse of squared distance among property and reference centre of the modules (without effects of size' fuzzy sets) is €159,245.14 (Table 6). After all, the value using double-fuzzy model (combining size and distance fuzzy sets) is €159,378.86 (Table 7). Table 4 illustrates the estimative calculated by the three equations (models 5.1a, 5.1b, and 5.1c), weighted by fuzzy sets based on property size. In module 5, the limits among small/medium and medium/large sizes are defined by 118.28 and 214.00 m^2, respectively (Table 2). The membership values are defined by trapezoidal fuzzy numbers (such as in Fig. 3) and value is calculated by Eq. (4).

The second fuzzy system is based only on distance fuzzy sets. The distance among property and each module is calculated using your coordinates of reference. The estimative are calculated by nine equations (models 1.1 to 9.1, which have 3 models: S, M, L), weighted by fuzzy memberships based on inverse of squared distance from property to each module-centre, corrected to reach $\sum(\mu_j) = 1$. For instance, to the fuzzy set 5 in Table 5, $\sum(1/d_5^2) = 346.03504$, then $\mu_5 = 0.47598/346.03504 = 0.001376$, and the $\text{Value}_5(P)$ is €159,254.24, then partial value to module 5 is 0.001376* 159,254.24 = €156,486.08. Each $\text{Value}_j(P)$ is calculated by Eq. (4) and the final value is calculated by Eq. (5).

Table 6 present the estimative to double-fuzzy system, which is calculated by twenty seven equations (models 1.1a to 9.1c—see Table 3). These values are weighted in a first stage by fuzzy sets based on property size. In a second stage

Table 4 Fuzzy estimate by property size

Sub-set i	$\mu_i(P)$	$\text{Model}_i(P)$ (€)	Partial estimates $\mu_i(P)*\text{Model}_i(P)$ (€)
Small	0.000000	152,001.77	0.00
Medium	0.838785	159,254.28	133,580.11
Large	0.161215	160,008.60	25,795.78
Sum	1.000000	–	159,375.89

Table 5 Fuzzy estimate by distance

Sub-set j	Distance	$1/d_j^2$	$\mu_j(P)$	$Value_j(P)$ (€)	Partial estimates $\mu_j(P)*Value_j(P)$ (€)
1	1.449462	0.47598	0.001376	159,503.42	219.48
2	0.996464	1.00711	0.002910	149,888.60	436.18
3	1.372932	0.53052	0.001533	170,561.48	261.47
4	1.054012	0.90014	0.002601	172,627.88	449.01
5	0.054231	340.02024	0.982618	159,254.24	156,486.08
6	0.946013	1.11739	0.003229	158,432.75	511.58
7	1.456345	0.47149	0.001363	150,003.27	204.45
8	1.006450	0.98722	0.002853	154,024.33	439.43
9	1.380196	0.52495	0.001517	156,536.75	237.47
Sum	–	346.03504	1.000000	–	159,245.14

Table 6 Double fuzzy estimate—by size and distance

Sub-set j	Distance	μ_j	Fuzzy-size estimates (€)	Partial double fuzzy estimates (€)
1	1.449462	0.001376	136,840.75	188.29
2	0.996464	0.002910	156,660.33	455.88
3	1.372932	0.001533	169,174.17	259.34
4	1.054012	0.002601	168,006.61	436.99
5	0.054231	0.982618	159,375.84	156,605.57
6	0.946013	0.003229	159,836.40	516.11
7	1.456345	0.001363	163,907.78	223.41
8	1.006450	0.002853	155,275.78	443.00
9	1.380196	0.001517	164,977.72	250.27
Sum	–	1.000000	–	159,378.86

values are weighted by fuzzy sets based on distance from property (P) to each module-centre, using also the inverse of squared distance. The values based on fuzzy-size sets are not presented in Table 6. However, for instance, the estimative to module 5 (€159,375.84) is the same presented in Table 4. Values are calculated by Eq. (6).

In the sequence we present a view of the system to mass appraisal purposes, like to valuate a property portfolio (represented in this case by test sample of module 5). RMSE of test sample are similar to training figures, confirming statistical quality of these models. Values for model 5.1 are determined using the equation adequate to each property size—small or medium or large models (Table 7).

Table 7 Results to general sample and models from module 5

Model	RMSE	
	Training. data (€)	Test data (€)
G.C—Conventional HPM—general—Eq. (7)	–	30,515.10
G.S—Surface model-general—Eq. (8)	–	30,880.24
5.C—Conventional HPM—Eq. (7) to module 5	28,144.73	29,504.55
5.S—Surface model—Eq. (8) to module 5	27,835.82	28,145.40
5.1—Conventional system—by size (one of three estimates) 5.1a—Hedonic model—Eq. (7) to small properties in module 5 5.1b—Hedonic model—Eq. (7) to medium properties in module 5 5.1c—Hedonic model—Eq. (7) to large properties in module 5	25,821.91	27,502.03
5.2—Fuzzy system by size (three estimates)—Eq. (4)	25,633.55	27,314.92
5.3—Fuzzy system by location (nine estimates)—Eq. (5)	–	27,283.33
5.4—Double level fuzzy system by size and location (twenty seven estimates)—Eq. (6)	–	27,151.69

8 Final Comments

The proposed system has a reasonable statistical behaviour, regarding the size of sample. Of course, a detailed treatment of outliers and estimate of models with different formats maybe conduce to better results. In case, we intent to maintain a general format in order to permit comparisons among error figures and value estimates. One advantage of fuzzy logic in valuations is to add flexibility to conventional variables and introduce variables based on common language in regression analysis. Some alternatives to the presented mass appraisal system:

Regarding sample size Sample size proportional to variability: number of cases to each module is k * standard deviation in the module; Size semi-proportional to variability: a fixed number (minimum) plus k * standard deviation; In both cases, using round (k * standard deviation), in thousand cases; Size proportional to units in municipal cadastre: sample = k * number of existing units in each module; Regarding number of models in each module Number of models semi-proportional to variability: number of models is 2 + p * standard deviation.

In this cases, using round (p * standard deviation), in number of models; Regarding fuzzy levels In case of detected importance to other variables, such as property type or property quality, one would be to include a third level, using 3 models to consider quality.

References

Byrne, P. (1995). Fuzzy analysis: A vague way of dealing with uncertainty in real estate analysis? *Journal of Valuation and Investment, 13*(3), 737–750.

Cordòn, O., Herrera, F., Hoffman, F., Magdalena. (2001). *L. Genetic fuzzy systems: Evolutionary tuning and learning of fuzzy knowledge bases*. Singapore: World Scientific.

Court, T. A. (1939). Hedonic price indexes with automotive examples. *The dynamics of automobile demand* (pp. 98–119). New York: General Motors.

d'Amato M., Siniak, N. (2003). An application of fuzzy number for property valuation and investment. *International Journal of Strategic Property Management, 7*(3), 129–143.

d'Amato, M., Siniak, N. (2008). Using fuzzy numbers in mass appraisal: The case of the Belorussian property market. In M. D'Amato, & T. Kauko (Eds.), *Mass appraisal methods. an international perspective for property valuers, RICS Real Estate Issue* (pp. 91–107). Wiley Blackwell Publishers.

d'Amato. (2010). A location value response surface model for mass appraising: An "Iterative" location adjustment factor in Bari, Italy. *International Journal of Strategic Property Management, 14*, 231–244.

d'Amato, M. (2004). A comparison between RST and MRA for mass appraisal purposes. A Case in Bari. *International Journal of Strategic Property Management, 8*, pp. 205–217.

d'Amato, M., & Kauko, T. (2012, March, April). Sustainability and risk premium estimation in property valuation and assessment of worth. *Building Research and Information, 40*(2), 174–185.

Griliches, Z. (1971). *Price indexes and quality change*. Cambridge: Harvard University Press.

Gujarati, D. N. (2009). *Basic econometrics*. Noida, India: Tata McGraw-Hill Education.

Kutner, M. H., Nachtsheim, C.J., Neter, J., LI, W. (2010). *Applied linear statistical models*. Burr Ridge, IL: McGraw-Hill Irwin.

Renigier-Biłozor, M., Wiśniewski, R., Biłozor, A., & Kaklauskas, A. (2014). Rating methodology for real estate markets—Poland case study. *Publication International Journal of Strategic Property Management, 18*(2), 198–212. ISNN. 1648-715X.

Renigier-Biłozor, M., Dawidowicz, A., & Radzewicz, A. (2014b). An algorithm for the purposes of determining the real estate markets efficiency in Land Administration System. *Publication Survey Review, 46*(336), 189–204.

Renigier-Biłozor, M., & Biłozor, A. (2016). Proximity and propinquity of residential market area —Polish and Italian case study. In *16th International Multidisciplinary Scientific GeoConferences SGEM*. Bułgaria.

Renigier-Biłozor, M., & Biłozor, A. (2016). The use of geoinformation in the process of shaping a safe space. In *16th International Multidisciplinary Scientific GeoConferences SGEM*. Bułgaria.

Robinson, R. (1979). *Housing economics and public policy*. London: McMillan.

Rosen, S. (1974). Hedonic prices and implicit markets: Product differentiation in pure competition. *Jornal of Political Economy, 82*, 34–55.

Smith, H. C., & Bagnoli, C. (1997). The theory of fuzzy logic and its application to real estate valuation. *Real Estate Issues, 22*(2), 35–41.

Zadeh, L. A. (1965). Fuzzy sets. *Information and Control, 8*, 338–352.

An Application of RST as Automated Valuation Methodology to Commercial Properties. A Case in Bari

Maurizio d'Amato and Malgorzata Renigier-Biłozor

Abstract The lack or unavailability of data poses one of the greatest obstacles hindering the exploration of real estate market information. Due to the small number of observations (cases), there are limited possibilities of using statistical methods, which are generally based on the assumption of a larger number of cases compared to the data describing them. This work provides an application of Rough Set Theory to a small sample of commercial properties as an Automatic Valuation Method. RST has been proposed as an automatic procedure for small sample (d'Amato 2002). It may happen in the application of RST that it is possible to deal with small samples. The rough set theory was applied taking into account the specific nature of information referring to the real property market. This theory is dedicated to examine imprecision, generality and unavailability in the process of data analysis that often occurs on the real estate market. It has been highlighted that one of the problem in AVM application may be the scarcity of data (Downie and Robertson 2007). The model allows the opportunity to reach the results of a single point estimate using if then rules providing a causal non deterministic relationship between price and property characteristics. The model has been applied to a small sample of commercial properties in the city of Bari.

Keywords Automated valuation methodologies · Rough sets · Valued tolerance relation · Market of real estate/analysis of real estate market

The paper has been written in strict cooperation between the two authors. Therefore the credit of the article should be equally divided between them.

M. d'Amato (✉)
DICATECh Technical University Politecnico Di, Bari, Italy
e-mail: madamato@fastwebnet.it

M. Renigier-Biłozor
Department of Real Estate Management and Regional Development,
University of Warmia and Mazury in Olsztyn, Poland, Faculty of Geodesy,
Geospatial and Civil Engineering, Olsztyn, Poland
e-mail: malgorzata.renigier@uwm.edu.pl

M. d'Amato and T. Kauko (eds.), *Advances in Automated Valuation Modeling*,
Studies in Systems, Decision and Control 86, DOI 10.1007/978-3-319-49746-4_16

1 Introduction

In the 21st century, a real estate market cannot be evaluated without the involvement of effective systems for gathering and processing information. The popularity of computerized systems for collecting and processing of real estate market data has soared in recent years. Despite the above, comprehensive and effective systems that facilitate analyses of real estate market data, support real estate valuation and other market analyses continue to be in short supply. The above results from the specificity of the real estate market which embodies various procedures and decisions, as well as the specific nature of real estate information. These shortcomings obstruct the smooth flow of comprehensive data which is required for initiating actions and making decisions regarding economic processes, business, investment, financial and promotional projects in the area of real estate.

From the analytical point of view, the solution to the problem requires the selection of appropriate methods for analyzing the available information rather than, as it is often observed in practice, the adaptation of the existing information to popular analytical methods, such as econometric models. In the era of globalization, quick and unified solutions (procedures, algorithms) are needed to enhance the objectivity and the reliability of research results. The preferred solutions should address the problem on a global scale while accounting for the local characteristics of the analyzed markets and the relevant information.

The venture point for every analysis of real estate value is the selection of adequate research methods and procedures that account for the specific attributes of the real estate market which make it different from other markets, including capital markets.

Due to the small number of observations (cases), there are limited possibilities of using statistical methods, which are generally based on the assumption of a larger number of cases compared to the data describing them. Therefore, the rough set theory was applied as a method that takes into account the small size of the data. Moreover, the assumptions of this theory are relatively simple, clear and the application is no time consuming.

Developed by Zdzislaw Pawlak a polish mathematician in two fundamental works (Pawlak 1982, 1991), Rough Set Theory is a rule-based approach that provides AVM solution in property markets with uncertain or lack of few information. It may occur in a real estate market that appraisers or real estate analyst may not have enough information to model an econometric modeling. In other case it can not be excluded that an analyst face the problem of a difficult causal relationship between price and characteristics of a property. For this reason a model providing a relationship non deterministic based on logic rules if then has been proposed (d'Amato 2002, 2004, 2007, 2008) especially for those emerging market having a lack of data or property markets with a complex urban segmentation.

The contribution is organized as follows. In the next paragraph there will be a brief theoretical introduction to Rough Set Theory. In the third paragraph the method will be applied to a small sample of commercial properties in Bari. Final remarks will be offered at the end.

2 Rough Set Theory and Valued Tolerance Relation

In the application of this method a real estate transaction is considered as an element or object to be related to a specified piece of information. Such information as property price, and technical characteristics of real estate or tenancy, are considered "attributes" of the object "property" If the real estate price is considered an object, the only available information are the specific property characteristics. The relationship between an "object" (property) and its "attributes" (characteristics) can be described by three regions of knowledge: "certainly," "possibly," and "certainly not". Assuming parking as an attribute the relationship between the object and its attribute can be defined as "certainly not" for a property (object) inside a group of property transactions (universe) without a parking (attribute). Among the property transactions (universe), properties with the same attributes can be considered indiscernible at a certain level of information. An indiscernible element is defined as an "elementary set," which cannot be confused with any other element. Two properties (objects) having similar technical features (attributes), and found in the same area, at the same price, can be considered indiscernible. This approach does not rely on any assumptions, the first stage of analysis is concerned with the relationship between objects (property transactions) and their attributes (property characteristics). The analysis starts with the so-called "informative table," whose rows represent "universe units" or objects. Considering the RST application to real estate valuation, all the attributes are listed in columns, with each measured in a different domain. Rows contain single attributes of the universe, or the real estate under consideration. This stage is very important because valuation depends on data quality and homogeneity of the data. It must be stressed that like in all the other market approach the observations should be in the same property market segment. A property would be very difficult to valuate without prior classification. It should be define precisely the type and the segment of the real estate market, and the utility function of real estate. Market type is indicative of the utility function of real estate: investment market, commercial market, industrial market, agricultural market. etc. Market segment accounts for a specific group of real estate which is identified in a given type of a market in view of its utility function. Type: investment market: residential, services, retail, etc.; type: commercial market: retail, services, offices; type: industrial market: industrial, warehouse, etc. The aim of the proposed division is to introduce a certain degree of uniformity to the valuation procedure.

Each cell contains the quantitative or qualitative description of the relationship between an object and its attribute. The presence, or the absence, of a garden (attribute) within a property (object) is marked with a dummy variable, whilst the surface (attribute) of a property (object) is expressed in square meters. It is easy to see a great similarity with the first step of regression analysis. The informative table S is expressed in formal terms, as in the following Eq. (1):

$$s = \langle U, Q, V_q, f \rangle. \tag{1}$$

In this equation, U is the universe, or a finite element set (all the property transactions in a property market segment). Q is the finite set of attributes or features (characteristics of the property). V_q is the attribute with a q domain, and f is the information function, which describes the relationship between the object and the attribute belonging to the Q set, which varies inside the V_q domain. It is possible to define an information function formally indicated as follows:

$$f : U \times Q \rightarrow V \wedge f(x, q) \in V_q \forall q \in Q \wedge \forall x \in U. \tag{2}$$

The relation indicated in the Formula (2) shows that the information function f works in a universe U of data in which information can be classified through a set of attributes Q. Each object (x) is linked to its attribute (q) through a function. This happens for each attribute inside the Q-set, and for each object (x) inside the universe U. The property transaction $x \in U$ is a line which can also be seen as a vector. Each element of this vector represents the value given to the relative attribute of the x object, and which can be defined as Des$Q$ (x). The following relationships among objects can be highlighted. There is an indiscernibleness or equivalency relationship between two objects that belong to the same universe U, when their respective attributes are identical. Two properties having a 100 sqm of garden in the same market segment, will be indiscernible with regard to this attribute. The indiscernibility relations can be expressed formally, considering a non-empty subset N of the Q-attribute set for N⊆Q:

$$IN = \{(x, y) \in U \times U : f_q(x) = f_q(y), q \in N\}. \tag{3}$$

Two objects (properties) are indiscernible if they have identical characteristics. The pair of coordinates (x, IN) defines the so-called approximative space. If $(x, y) \in IN$, then it is possible to say that x and y are N-indiscernible. In this case, the indiscernibility relation may only have two values: 1 or 0. If $N = Q$, where the Q elementary sets are known as atoms, all the elements are indiscernible. If all the X set units of the U universe are analyzed according to the N attribute set, and if they are similar (for example, where all properties are 100 sqm and are near the center), then they are indiscernible. Two real estate properties may be characterized by a single difference—but a relevant difference in price—or by two or more differences, but the same price. For this reason two important concepts must be added. Assuming U as the universe, X as a universe object set (real estate properties with known prices), Q as the attribute set (that belongs to U universe), and N as an attribute subset, the lower approximation can be defined as follows:

$$N\_(X) = \{x \in U : N(x) \subseteq X\}. \tag{4}$$

If a real property unambiguously has an attribute included in this attribute subset, then it can be defined as part of its positive or lower region, and can be defined in the following terms:

$$N-(X) = \{x \in U : N(x) \cap X = 0\}. \tag{5}$$

The upper approximation is defined by the set which shows a non-empty intersection with X. If there are some elements in set N that belong to X and others that do not, then the attribute can be described by the upper approximation. Rough Set Theory evaluates each uncertain phenomenon through these approximations. The difference between the upper or lower regions is represented by a "boundary region" of rough sets. The boundary region is expressed formally as:

$$BN_N(X) = N^-(X) - N\_(X). \tag{6}$$

The three regions described are useful to define "granular" information. Both qualitative and quantitative attributes are useful to describe an object. If the boundary region is not empty, the rough set is defined through upper approximation and lower approximation union. The granular nature of information is influenced by different aspects such as attribute characteristics, attribute numbers and each attribute domain Q. As with MRA, this procedure is heavily dependent on quality of information, ability to classify information and adequately describe single attributes, as well as levels of confidence in knowledge and problems with knowledge itself. The "informative table," is the basis to develop a "decisional table," by dividing the attributes in conditional set C and decisional set D. This distinction between conditional and decisional attributes can establish a causal relation between attributes. Assuming price as a decisional variable and attributes as a conditional variable, the RST allows us to see how conditional attributes (property characteristics) influence the decisional attribute (price). Through this procedure, an object can be valued to determine lower and upper approximation based on the relationships between the set of elements containing the price (decisional attribute) and the set containing other attributes (conditional attributes) influencing price behavior. The appraiser selects the conditional attributes in the same way he defines the independent variables that affect the valued in regression analysis. In the final step, the appraiser analyzes the relationships between conditional and decisional attributes. The relation is analyzed by taking into account the lower and the upper approximations between the decisional set D "of the price attribute" and set C of the attributes that have been selected as conditional. There are two general types of decisional rules. The former is the "exact decisional or deterministic rule," where the decisional set (price) contains the conditional attributes (area or other features). The latter is the "approximative decisional rule," in which only some conditional attributes (in our case sq. meters, date, date of construction, n. of rooms) are included in the decisional set (price). According to previous research, deterministic rules seem to be the most suitable for real estate valuation and automated valuation purposes (d'Amato 2002, 2004; Renigier-Bilozor 2010). In this case, causal relationships between property features and price can be evaluated without the problem of uncertainty. The "granularity" of the system or its uncertainty can be increased if information is based on various observations. Property valued is determined by comparing property characteristics with the rules defined for comparative properties. This phase is very important. In previous papers, classes of valued were obtained, and if-then rules applied to these classes, which then gave a specific class of valued to properties with particular characteristics. The first paper (d'Amato 2002) was based on a crisp indiscernibility relation (complete, reflexive, symmetric and transitive relation valued in the following domain {0, 1}). Since 2004 (d'Amato 2004, 2007) a valued tolerance relation as opposed to a crisp tolerance relation has been introduce in the traditional version of RST, to give an objective measure for k-threshold. Valued tolerance relation can be considered a more flexible way to deal with the indiscernibility relation. Moreover, the theory with a valued tolerance relation extension is used in many sciences, and it is often applied as the main support tool in decision-making systems (Bello and Verdegay 2012; Chi et al. 2011; Chung and Tseng 2012; Polkowski 2010; Zavadskas and Turskis 2011; Zhang 2012; Renigier-Biłozor 2011; Renigier-Biłozor and Wiśniewski 2011; Guoyin and Lihe 2012).

Classical rough set theory relies on the crucial concept of indiscernibility relation as a crisp equivalence relation. Two properties may be indiscernible only if they have the same attributes. In property markets where this is a very powerful assumption, the Value Tolerance Relation is a functional extension of Rough Set Theory and allows the appraiser to develop upper or lower approximation with different "degrees" of indiscernibility relation. The concepts of lower and upper approximation will be replaced by lower and upper approximability. The formal relation is indicated below, indicated below:

$$R_j(x,y) = \frac{\max(0, \min(c_j(x), c_j(y)) + k - \max(c_j(x), c_j(y)))}{k} \tag{7}$$

The relation R_j may assume continuous not crisp values included in the interval [0, 1]. It is a variation ratio based on sets whose membership function may express different values included in the interval [0, 1]. Value Tolerance Relation brings flexibility to traditional Rough Set Theory. In this context, the choice of the minimum in the membership function represents the intersection between two sets, while the maximum in membership function results in the union between the two sets. Two objects x and y may have different levels of indiscernibility, according to a discriminant threshold k, which measures the characteristic C_j. The k-threshold can be applied to different measures of these characteristics for all objects. For example, the indiscernibility relation between two objects (properties A and B) whose sqm area are 120 and 150 for a k-threshold of 10 sqm can be calculated as follows:

$$R(a,b) = \frac{\max(0; 120 + 10 - 150)}{10} = \frac{\max(0; -20)}{10} = \frac{0}{10} = 0 \tag{8}$$

The two objects cannot be considered similar. The result of the application of a VTR with the same k to two objects (say, property transactions) whose sqm areas are 120 and 125 is:

$$R(a,b) = \frac{\max(0; 120 + 10 - 125)}{10} = \frac{\max(0; 5)}{10} = \frac{5}{10} = 0.5 \tag{9}$$

If the valued of R_j equals 1, the k-threshold of the two objects is similar; if, as in Eq. (8), R_j is equal to 0, the k-threshold is completely different. This mathematical formula can also be used for the relationship between the object of a universe (properties) and a R_j set of rules developed for valuation purposes, where the characteristics of the object (property transaction) are compared with the conditional part of the rule considered, indicated in the following equation as $c_j(\rho)$. In this case, it is modified as follows (Stefanowski and Tsoukias 2000):

$$R_{(x,\rho)} = \frac{\max(0; \min(c_j(x), c_j(\rho)) + k - \max(c_j(x), c_j(\rho)))}{k} \tag{10}$$

In the formula output, there is a level of indiscernibility relation between the object and the rule, assuming a k level of threshold for the measure of the attribute. In the previous paper (d'Amato 2004), the measure of k-threshold was found to be subjective, due to the preferences and characteristics of the specific property market. In other contributions (d'Amato 2007, 2008; Dawidowicz et al. 2014; Renigier-Biłozor and Wiśniewski 2011; Renigier-Biłozor et al. 2014; Wang and Guan 2012) an objective measure of k-threshold is proposed as the standard deviation of each attribute belonging to the sample that originates the rule. This is an important step in defining a specific application of RST to property valuation. If rules concern properties with similar characteristics, then the threshold (standard deviation) is low. On the other hand, the threshold is high when the rules refer to a sample of elements containing properties with different features. The mathematical foundations of the method are reported in an important contribution of Stefanowski and Tsoukias (2000). It is easy to observe that:

$$\begin{aligned} &\forall x, y \in A \; R_j(x,y) = 1 \textit{ iff } c_j(x) = c_j(y) \\ &\forall x, y \in A \; R_j(x,y) \in]0,1[\textit{ iff } |c_j(x) - c_j(y)| < k \\ &\forall x, y \in A \; R_j(x,y) = 0 \textit{ iff } |c_j(x) - c_j(y)| \geq k \end{aligned} \tag{11}$$

In specific cases the k-threshold can be calculated as the standard deviation, median, dominant or arithmetic mean, in the cross-validation of the results efficiency for the very diversity database with obvious non linear the trend function (Renigier-Biłozor and Wiśniewski 2013).

The method can be applied in different steps. In particular the application of this contribution will be based on a *complete version* applied previously (d'Amato 2008, 2010). The second application of this method to real estate market of Bari is in the commercial real estate market. The sample is composed by 21 properties and it has been divided in two parts. The first one is composed by 16 commercial properties small shops. A greater sample is required to provide the rule to appraise properties. This sample will be used to generate the rules (rule sample). The second part of the sample is composed by 5 properties to apply the rule generated (appraising sample). The application provides a demonstration that this methodology may work as a further tool for limited sample. The choice of the city of Bari is related to the availability of data and the commercial sector may be interesting as a example of weak-form efficient real estate markets. That example expressed the common situation that often occurs on the market, especially in the case of commercial properties all over the world.

3 An Application of Rough Set Theory to a Small Sample of Commercial Properties in the City of Bari

Following the premise of a previous work (d'Amato 2008) this paper explore the possibility to use this method in order to provide and automatic valuation for a small sample. In particular the sample is composed by 6 property transaction obtained by a local real estate broker agency. The sample is composed by six commercial properties, small shops located in the semicentral area in Bari. The choice of this city is essentially connected with the availability of data. Two different samples have been collected. The former is a sample of 16 commercial properties (Table 1) collected to provide the rules while the latter is a sample of 5 properties (Table 2) to be appraised.

The conditional variables are DAT that means DATE. This variable is measured in moths since the moment of valuation. It is a cardinal variable. The second conditional variable is SUP or square meters of the shop. It is a cardinal variable, too. The third conditional variable is SUD and measure the square meters of store inside the shop. It is a cardinal variable. The variable VETR is a cardinal discrete variable measuring the number of showcases of the shop. The last variable APE is an energy performance indicator according to Italian law. The last two variables are cardinal discrete.

The selection of k—threshold (Table 3) is calculated in this case on the standard deviation of attributes of observations belonging to the rule sample (Table 1) in order to apply the Formulas (9) and (10).

In the next step a *comparison table* (Table 4) per each criteria will be created. In the case presented the conditional attributes are five. The comparison tables are indicated below:

Table 1 Decision table of rough set theory integrated with valued tolerance relation (rule sample)

NR	PRICE	DAT	SUP	SUD	VETR	APE
	Decisional attribute	Conditional attribute				
1	120,000	3	80	10	1	0.142857143
2	150,000	4	91	40	2	0.285714286
3	134,000	5	87	42	2	0.142857143
4	160,000	6	101	44	2	0.142857143
5	140,000	7	98	65	1	0.428571429
6	121,000	3	83	10	2	0.428571429
7	110,000	4	75	20	1	0.571428571
8	115,000	9	79	15	2	0.142857143
9	97,000	10	90	0	1	0.142857,143
10	89,000	15	65	10	1	0.142857143
11	123,000	2	83	12	1	0.428571429
12	182,000	7	120	15	3	0.428571429
13	144,000	9	97	22	2	0.142857143
14	125,000	10	85	30	2	0.142857143
15	92,000	15	67	42	1	0.142857143
16	102,000	12	78	21	1	0.285714286

Source Authors' calculation

Table 2 Decision table of rough set theory integrated with valued tolerance relation (appraising sample)

NR	DAT	SUP	SUD	VETR	APE	Price
1	7	85	12	1	0.1428571	€123,000.00
2	5	92	20	1	0.2857143	€134,000.00
3	10	100	22	2	0.1428571	€146,000.00
4	12	65	15	1	0.1428571	€105,000.00
5	13	74	27	1	0.4285714	€158,000.00

Source Authors' calculation

Table 3 k threshold

k-threshold					
	DAT	SUP	SUD	VETR	APE
Dev. st.	4.114506856	13.60989224	17.33541654	0.62915287	0.149545798

Source Authors' calculation

In all the tables indicating attributes, all the objects are compared each others through the value tolerance relation. For example according to the attribute DAT or date the comparison between the object 1 and the object 2 has a value 0.757. This value is obtained as follows:

Table 4 Comparison table

DAT	1	2	3	4	5	6	7	8	9	10	11	12	13	14	15	16
1	1.000															
2	0.757	1.000														
3	0.514	0.757	1.000													
4	0.271	0.514	0.757	1.000												
5	0.028	0.271	0.514	0.757	1.000											
6	1.000	0.757	0.514	0.271	0.028	1.000										
7	0.757	1.000	0.757	0.514	0.271	0.757	1.000									
8	0.000	0.000	0.028	0.271	0.514	0.000	0.000	1.000								
9	0.000	0.000	0.000	0.028	0.271	0.000	0.000	0.757	1.000							
10	0.000	0.000	0.000	0.000	0.000	0.000	0.000	0.000	0.000	1.000						
11	0.757	0.514	0.271	0.028	0.000	0.757	0.514	0.000	0.000	0.000	1.000					
12	0.757	0.271	0.514	0.757	1.000	0.028	0.271	0.514	0.271	0.000	0.000	1.000				
13	0.000	0.000	0.028	0.271	0.514	0.000	0.000	1.000	0.757	0.000	0.000	0.514	1.000			
14	0.000	0.000	0.000	0.028	0.271	0.000	0.000	0.757	1.000	0.000	0.000	0.271	0.757	1.000		
15	0.000	0.000	0.000	0.000	0.000	0.000	0.000	0.000	0.000	1.000	0.000	0.000	0.000	0.000	1.000	
16	0.000	0.000	0.000	0.000	0.000	0.000	0.000	0.271	0.514	0.271	0.000	0.000	0.271	0.271	0.271	1.000

(continued)

Table 4 (continued)

SUP	1	2	3	4	5	6	7	8	9	10	11	12	13	14	15	16
1	1.000															
2	0.192	1.000														
3	0.486	0.192	1.000													
4	0.000	0.486	0.192	1.000												
5	0.000	0.000	0.486	0.192	1.000											
6	0.780	0.000	0.000	0.486	0.192	1.000										
7	0.633	0.780	0.000	0.000	0.486	0.192	1.000									
8	0.927	0.633	0.780	0.000	0.000	0.486	0.192	1.000								
9	0.265	0.927	0.633	0.780	0.000	0.000	0.486	0.192	1.000							
10	0.000	0.265	0.927	0.633	0.780	0.000	0.000	0.486	0.192	1.000						
11	0.780	0.000	0.265	0.927	0.633	0.780	0.000	0.000	0.486	0.192	1.000					
12	0.000	0.780	0.000	0.265	0.927	0.633	0.780	0.000	0.000	0.486	0.192	1.000				
13	0.000	0.000	0.780	0.000	0.265	0.927	0.633	0.780	0.000	0.000	0.486	0.192	1.000			
14	0.633	0.000	0.000	0.780	0.000	0.265	0.927	0.633	0.780	0.000	0.000	0.486	0.192	1.000		
15	0.045	0.633	0.000	0.000	0.780	0.000	0.265	0.927	0.633	0.780	0.000	0.000	0.486	0.912	1.000	
16	0853	0.045	0.633	0.000	0.000	0.780	0.000	0.265	0.927	0.633	0.780	0.000	0.000	0.486	0.192	1.000

(continued)

Table 4 (continued)

SUD	1	2	3	4	5	6	7	8	9	10	11	12	13	14	15	16
1	1.000															
2	0.000	1.000														
3	0.000	0.885	1.000													
4	0.000	0.769	0.885	1.000												
5	0.000	0.000	0.000	0.000	1.000											
6	1.000	0.000	0.000	0.000	0.000	1.000										
7	0.423	0.000	0.000	0.000	0.000	0.423	1.000									
8	0.712	0.000	0.000	0.000	0.000	0.712	0.712	1.000								
9	0.423	0.000	0.000	0.000	0.000	0.423	0.000	0.135	1.000							
10	1.000	0.000	0.000	0.000	0.000	1.000	0.423	0.712	0.423	1.000						
11	0.885	0.000	0.000	0.000	0.000	0.885	0.539	0.827	0.308	0.885	1.000					
12	0.712	0.000	0.000	0.000	0.000	0.712	0.712	1.000	0.135	0.712	0.827	1.000				
13	0.308	0.000	0.000	0.000	0.000	0.308	0.885	0.596	0.000	0.308	0.423	0.596	1.000			
14	0.000	0.423	0.308	0.192	0.000	0.000	0.423	0.135	0.000	0.000	0.000	0.135	0.539	1.000		
15	0.000	0.885	1.000	0.885	0.000	0.000	0.000	0.000	0.000	0.000	0.000	0.000	0.000	0.308	1.000	
16	0.365	0.000	0.000	0.000	0.000	0.365	0.942	0.654	0.000	0.365	0.481	0.654	0.942	0.481	0.000	1.000

(continued)

Table 4 (continued)

VETR	1	2	3	4	5	6	7	8	9	10	11	12	13	14	15	16
1	1.000															
2	0.000	1.000														
3	0.000	1.000	1.000													
4	0.000	1.000	1.000	1.000												
5	1.000	0.000	0.000	0.000	1.000											
6	0.000	1.000	1.000	1.000	0.000	1.000										
7	1.000	0.000	0.000	0.000	1.000	0.000	1.000									
8	0.000	1.000	1.000	1.000	0.000	1.000	0.000	1.000								
9	1.000	0.000	0.000	0.000	1.000	0.000	1.000	0.000	1.000							
10	1.000	0.000	0.000	0.000	1.000	0.000	1.000	0.000	1.000	1.000						
11	1.000	0.000	0.000	0.000	1.000	0.000	1.000	0.000	1.000	1.000	1.000					
12	0.000	0.000	0.000	0.000	0.000	0.000	0.000	0.000	0.000	0.000	0.000	1.000				
13	0.000	1.000	1.000	1.000	0.000	1.000	0.000	1.000	0.000	0.000	0.000	0.000	1.000			
14	0.000	1.000	1.000	1.000	0.000	1.000	0.000	1.000	0.000	0.000	0.000	0.000	1.000	1.000		
15	1.000	0.000	0.000	0.000	1.000	0.000	1.000	0.000	1.000	1.000	1.000	0.000	0.000	0.000	1.000	
16	1.000	0.000	0.000	0.000	1.000	0.000	1.000	0.000	1.000	1.000	1.000	0.000	0.000	0.000	1.000	1.000

(continued)

Table 4 (continued)

APE	1	2	3	4	5	6	7	8	9	10	11	12	13	14	15	16
1	1.000															
2	0.045	1.000														
3	1.000	0.045	1.000													
4	1.000	0.045	1.000	1.000												
5	0.000	0.045	0.000	0.000	1.000											
6	0.000	0.045	0.000	0.000	1.000	1.000										
7	0.000	0.000	0.000	0.000	0.045	0.045	1.000									
8	1.000	0.045	1.000	1.000	0.000	0.000	0.000	1.000								
9	1.000	0.045	1.000	1.000	0.000	0.000	0.000	1.000	1.000							
10	1.000	0.045	1.000	1.000	0.000	0.000	0.000	1.000	1.000	1.000						
11	0.000	0.045	0.000	0.000	1.000	1.000	0.045	0.000	0.000	0.000	1.000					
12	0.000	0.045	0.000	0.000	1.000	1.000	0.045	0.000	0.000	0.000	1.000	1.000				
13	1.000	0.045	1.000	1.000	0.000	0.000	0.000	1.000	1.000	1.000	0.000	0.000	1.000			
14	1.000	0.045	1.000	1.000	0.000	0.000	0.000	1.000	1.000	1.000	0.000	0.000	1.000	1.000		
15	1.000	0.045	1.000	1.000	0.000	0.000	0.000	1.000	1.000	1.000	0.000	0.000	1.000	1.000	1.000	
16	0.045	1.000	0.045	0.045	0.045	0.045	0.000	0.045	0.045	0.045	0.045	0.045	0.045	0.045	0.045	1.000

Source Authors' calculation

Table 5 Valued tolerance table Rj

RJ	1	2	3	4	5	6	7	8	9	10	11	12	13	14	15	16
1	1.000	0.000	0.000	0.000	0.000	0.000	0.000	0.000	0.000	0.000	0.000	0.000	0.000	0.000	0.000	0.000
2	0.000	1.000	0.045	0.045	0.000	0.000	0.000	0.000	0.000	0.000	0.000	0.000	0.000	0.000	0.000	0.000
3	0.000	0.045	1.000	0.192	0.000	0.000	0.000	0.000	0.000	0.000	0.000	0.000	0.000	0.000	0.000	0.000
4	0.000	0.045	0.192	1.000	0.000	0.000	0.000	0.000	0.000	0.000	0.000	0.000	0.000	0.028	0.000	0.000
5	0.000	0.000	0.000	0.000	1.000	0.000	0.000	0.000	0.000	0.000	0.000	0.000	0.000	0.000	0.000	0.000
6	0.000	0.000	0.000	0.000	0.000	1.000	0.000	0.000	0.000	0.000	0.000	0.000	0.000	0.000	0.000	0.000
7	0.000	0.000	0.000	0.000	0.000	0.000	1.000	0.000	0.000	0.000	0.000	0.000	0.000	0.000	0.000	0.000
8	0.000	0.000	0.000	0.000	0.000	0.000	0.000	1.000	0.000	0.000	0.000	0.000	0.596	0.135	0.000	0.000
9	0.000	0.000	0.000	0.000	0.000	0.000	0.000	0.000	1.000	0.000	0.000	0.000	0.000	0.000	0.000	0.000
10	0.000	0.000	0.000	0.000	0.000	0.000	0.000	0.000	0.000	1.000	0.000	0.000	0.000	0.000	0.000	0.045
11	0.000	0.000	0.000	0.000	0.000	0.000	0.000	0.000	0.000	0.000	1.000	0.000	0.000	0.000	0.000	0.000
12	0.000	0.000	0.000	0.000	0.000	0.000	0.000	0.000	0.000	0.000	0.000	1.000	0.000	0.000	0.000	0.000
13	0.000	0.000	0.000	0.000	0.000	0.000	0.000	0.596	0.000	0.000	0.000	0.000	1.000	0.192	0.000	0.000
14	0.000	0.000	0.000	0.028	0.000	0.000	0.000	0.135	0.000	0.000	0.000	0.000	0.192	1.000	0.000	0.000
15	0.000	0.000	0.000	0.000	0.000	0.000	0.000	0.000	0.000	0.000	0.000	0.000	0.000	0.000	1.000	0.000
16	0.000	0.000	0.000	0.000	0.000	0.000	0.000	0.000	0.000	0.045	0.000	0.000	0.000	0.000	0.000	1.000

Source Authors' calculation

Table 6 Membership table

Comparables	120,000	150,000	134,000	160,000	140,000	121,000	110,000	115,000	97,000	89,000	123,000	182,000	144,000	125,000	92,000	102,000
1	1	0	0	0	0	0	0	0	0	0	0	0	0	0	0	0
2	0	1	0	0	0	0	0	0	0	0	0	0	0	0	0	0
3	0	0	1	0	0	0	0	0	0	0	0	0	0	0	0	0
4	0	0	0	1	0	0	0	0	0	0	0	0	0	0	0	0
5	0	0	0	0	1	0	0	0	0	0	0	0	0	0	0	0
6	0	0	0	0	0	1	0	0	0	0	0	0	0	0	0	0
7	0	0	0	0	0	0	1	0	0	0	0	0	0	0	0	0
8	0	0	0	0	0	0	0	1	0	0	0	0	0	0	0	0
9	0	0	0	0	0	0	0	0	1	0	0	0	0	0	0	0
10	0	0	0	0	0	0	0	0	0	1	0	0	0	0	0	0
11	0	0	0	0	0	0	0	0	0	0	1	0	0	0	0	0
12	0	0	0	0	0	0	0	0	0	0	0	1	0	0	0	0
13	0	0	0	0	0	0	0	0	0	0	0	0	1	0	0	0
14	0	0	0	0	0	0	0	0	0	0	0	0	0	1	0	0
15	0	0	0	0	0	0	0	0	0	0	0	0	0	0	1	0
16	0	0	0	0	0	0	0	0	0	0	0	0	0	0	0	1

Source Authors' calculation

Table 7 Lower approximability table

	m_{120000}	m_{150000}	m_{134000}	m_{160000}	m_{140000}	m_{121000}	m_{110000}	m_{115000}	m_{97000}	m_{39000}	m_{123000}	m_{182000}	m_{144000}	m_{125000}	m_{32000}	m_{102000}
1	1.000	0.000	0.000	0.000	0.000	0.000	0.000	0.000	0.000	0.000	0.000	0.000	0.000	0.000	0.000	0.000
2	0.000	0.955	0.000	0.000	0.000	0.000	0.000	0.000	0.000	0.000	0.000	0.000	0.000	0.000	0.000	0.000
3	0.000	0.000	0.808	0.000	0.000	0.000	0.000	0.000	0.000	0.000	0.000	0.000	0.000	0.000	0.000	0.000
4	0.000	0.000	0.000	0.808	0.000	0.000	0.000	0.000	0.000	0.000	0.000	0.000	0.000	0.000	0.000	0.000
5	0.000	0.000	0.000	0.000	1.000	0.000	0.000	0.000	0.000	0.000	0.000	0.000	0.000	0.000	0.000	0.000
6	0.000	0.000	0.000	0.000	0.000	1.000	0.000	0.000	0.000	0.000	0.000	0.000	0.000	0.000	0.000	0.000
7	0.000	0.000	0.000	0.000	0.000	0.000	1.000	0.000	0.000	0.000	0.000	0.000	0.000	0.000	0.000	0.000
8	0.000	0.000	0.000	0.000	0.000	0.000	0.000	0.404	0.000	0.000	0.000	0.000	0.000	0.000	0.000	0.000
9	0.000	0.000	0.000	0.000	0.000	0.000	0.000	0.000	1.000	0.000	0.000	0.000	0.000	0.000	0.000	0.000
10	0.000	0.000	0.000	0.000	0.000	0.000	0.000	0.000	0.000	0.955	0.000	0.000	0.000	0.000	0.000	0.000
11	0.000	0.000	0.000	0.000	0.000	0.000	0.000	0.000	0.000	0.000	1.000	0.000	0.000	0.000	0.000	0.000
12	0.000	0.000	0.000	0.000	0.000	0.000	0.000	0.000	0.000	0.000	0.000	1.000	0.000	0.000	0.000	0.000
13	0.000	0.000	0.000	0.000	0.000	0.000	0.000	0.000	0.000	0.000	0.000	0.000	0.404	0.000	0.000	0.000
14	0.000	0.000	0.000	0.000	0.000	0.000	0.000	0.000	0.000	0.000	0.000	0.000	0.000	0.808	0.000	0.000
15	0.000	0.000	0.000	0.000	0.000	0.000	0.000	0.000	0.000	0.000	0.000	0.000	0.000	0.000	1.000	0.000
16	0.000	0.000	0.000	0.000	0.000	0.000	0.000	0.000	0.000	0.000	0.000	0.000	0.000	0.000	0.000	0.955

Source Authors' calculation

Table 8 The rules

		DAT	SUP	SUD	VETR	APE		Price
1	IF	3	80	10	1	0.142857	THEN	€120,000.00
2	IF	4	91	40	2	0.285714	THEN	€150,000.00
3	IF	5	87	42	2	0.142857	THEN	€134,000.00
4	IF	6	101	44	2	0.142857	THEN	€160,000.00
5	IF	7	98	65	1	0.428571	THEN	€140,000.00
6	IF	3	83	10	2	0.428571	THEN	€121,000.00
7	IF	4	75	20	1	0.571429	THEN	€110,000.00
8	IF	9	79	15	2	0.142857	THEN	€115,000.00
9	IF	10	90	0	1	0.142857	THEN	€97,000.00
10	IF	15	65	10	1	0.142857	THEN	€89,000.00
11	IF	2	83	12	1	0.428571	THEN	€123,000.00
12	IF	7	120	15	3	0.428571	THEN	€182,000.00
13	IF	9	97	22	2	0.142857	THEN	€144000.00
14	IF	10	85	30	2	0.142857	THEN	€125,000.00
15	IF	15	67	42	1	0.142857	THEN	€92,000.00
16	IF	12	78	21	1	0.285714	THEN	€102,000.00

Source Authors' calculation

$$R(x_{1,DAT}; x_{2,DAT}) = \frac{\max(0; 3 + 4.1145 - 4)}{4.1145} = \frac{\max(0; 3.1145)}{4.1145} = \frac{3.1145}{4.1145} = 0.757 \tag{12}$$

A further step will be the calculation of the minimum of Rj as the union of all the sets and will give a general result. This is the *valued tolerance table Rj* and gives the final result of the comparison (Table 5).

For example the minimum of the comparison between the object 1 and the object 2 will be calculated (on the basis of Table 4) as follows:

$$R_j = \min(0.757; 0.192; 0; 0; 045) = 0 \tag{13}$$

In the following step each element of the rule sample is assigned to its specific class in a membership table indicated below (Table 6). In this table it is highlight the membership of each observation to a specific and crisp "class of prices".

The further stage is the construction of approximability table. It must be stressed that the concept of approximability associates a grade of membership while the concept of approximation is based on a dichotomous relationship between object and decisional class (d'Amato 2002, 2015, 2017a, b; d'Amato and Kauko 2008). Using approximability we may give a grade of membership of a property to a class of price. Two cases must be distinguished in lower approximability determination of each element z (property) for a specific decisional class. In the first case the object belongs to the decisional class. In the second case the object may not belong

Table 9 Property to be appraised

	DAT	SUP	SUD	VETR	APE	PRZ
1	7	85	12	1	0.142857143	€102,000.00
2	5	92	20	1	0.285714286	€119,000.00
3	10	100	22	2	0.142857143	€146,000.00
4	12	65	15	1	0.142857143	€94,000.00
5	13	74	27	1	0.428571429	€109,000.00

Source Authors' calculation

to the decisional class. The lower approximation of the decisional class 120,000 euro will be calculated as follows:

$$\textit{If} \quad z \in 120{,}000 \Rightarrow \mu_{120{,}000_B}(z) = \min_{x \in 120{,}000^C} (1 - R_B(z,x)) \leq \mu_{120{,}000^B(z)} = 1 \quad (14)$$

In other words the lower approximability (not approximation) is determined calculating the minimum of the complement of Rj derived from the comparison between the object and the elements that do not belong to the decisional class taken into account. In this case referring to a decisional class of 120,000:

$$\textit{If} \quad z \in 120{,}000^C \Rightarrow \mu_{120{,}000_B}(z) = 0 \leq \mu_{120{,}000^B(z)} = 1 \quad (15)$$

This calculation must be done for each class. The lower approximability can be considered determinant for the born of a deterministic rule. The property will be appraised using the lower approximability. In Table 7 the lower approximability have been calculated for each decisional class.

Finally it is possible to provide *the rule*. It consists of the rule definition based on an acceptability threshold λ which varies within the interval [0, 1]. The greater this threshold is, the lesser the accepted rule. In this case a lower lamda has been adopted equal to 0.4 (see Table 7, row m_{115000} and m_{144000}) in order to consider as much information as possible. There is a direct relation between the lamda and the dimension of the sample. The lower is the number of the property belonging to the sample and the lower will be the lambda. According to the Table 7 there are 16 rules. All of them have a lower approximability higher than the lambda term 0.4. Therefore the rules are as shown in Table 8.

For each rule can be calculated the number of objects which *support* it and the *credibility degree*. In order to define the number of objects which support the rule originated there will be a comparison among the single object and the other objects belonging to the sample. This comparison will be operated using the valued tolerance table previously indicated. The *Rj* which are greater than zero will highlight an object supporting the rule. The more objects support the rule the more the rule can be defined reliable. It seems clear that in a small sample as the one considered it is quite difficult to find a great number of objects supporting the rule. In addition to the number of objects which support the rule a *credibility degree of the rule* can be

Table 10 Rule generation process for the property n. 1

	DAT	SUP	SUD	VETR	APE	min Rj	min (Rj Cred. Rule)	
1	0.028	0.633	0.885	1.000	1.000	0.028	0.028	
2	0.271	0.559	0.000	0.000	0.045	0.000	0.000	
3	0.514	0.853	0.000	0.000	1.000	0.000	0.000	
4	0.757	0.000	0.000	0.000	1.000	0.000	0.000	
5	1.000	0.045	0.000	1.000	0.000	0.000	0.000	
6	0.028	0.853	0.885	0.000	0.000	0.000	0.000	
7	0.271	0.265	0.539	1.000	0.000	0.000	0.000	
8	0.514	0.559	0.827	0.000	1.000	0.000	0.000	
9	0.271	0.633	0.308	1.000	1.000	0.271	0.271	Rule selected
10	0.000	0.000	0.885	1.000	1.000	0.000	0.000	
11	0.000	0.853	1.000	1.000	0.000	0.000	0.000	
12	1.000	0.000	0.827	0.000	0.000	0.000	0.000	
13	0.514	0.118	0.423	0.000	1.000	0.000	0.000	
14	0.271	1.000	0.000	0.000	1.000	0.000	0.000	
15	0.000	0.000	0.000	1.000	1.000	0.000	0.000	
16	0.000	0.486	0.481	1.000	0.045	0.000	0.000	

NR	DAT	SUP	SUD	VETR	APE
1	7	85	12	1	0.142857143
k threshold	4.1145069	13.609892	17.335417	0.62915287	0.149545798

Source Authors' calculation

calculated. In formal term, after the choice of *lambda* the credibility degree for the decision class 120,000 of a rule can be indicated as follows:

$$\mu(\rho_1) = \min_{x \in s(\rho_1)}(\max(1 - R_B(x, \rho_1), \mu_{120{,}000_B}(x))) \tag{16}$$

In other terms will be calculated the maximum between two different term. The first term will be the complement of the Rj originated by the comparison between the conditional part of the rule and all the objects which support the rule. In this case only the object 2 supports the rule. The second term will be the lower

Table 11 Rule generated for all the 5 properties to appraise

Number of rule	Value	Support	Credibility degree
9	€97,000.00	1	1
11	€123,000.00	1	1
13	€144,000.00	3	1
10	€89,000.00	1	1
16	€102,000.00	2	1

Source Authors' calculation

Table 12 COD result of RST as automated valuation model

PRZ	Value	Value/PRZ	Median ratio	(Value/PRZ)-median ratio	Average absolute deviation	COD
€102,000.00	€97,000.00	0.9510	0.9510	0.0000	0.0275	0.028881035
€119,000.00	€123,000.00	1.0336		0.0826		
€146,000.00	€144,000.00	0.9863		0.0353		
€94,000.00	€89,000.00	0.9468		0.0042		
€109,000.00	€102,000.00	0.9358		0.0152		

approximation of object 2 for the specific decision class (d'Amato 2017a, b). The maximum of these two terms can be considered the credibility degree if there is one only object which support the rule. If there are two or more objects which support the rule the intersection between all these sets or the minimum of these values will be considered. The property to appraise are a group of five transactions listed below:

In the Table 9 it is possible to see the characteristics of the property from the second column to the fourth column. The last column represent the real price of these properties that will be appraised with the Rough Set Theory in order to compare price and value calculated with the proposed methodology. In order to select the right rule and therefore the right price for the property a membership grade of the object to appraise to the rule has to be calculated. The following formula indicates how to calculate the membership grade of the object to the two rules originated by the sample shown in:

$$\mu_{P_i} = \min(R_B(z, \rho_i), \mu(\rho_i)) \tag{17}$$

The first term of equation is the minimum of Rj calculated between each rule and the object while the second term is the credibility degree of the rule. Therefore the selection of the rule follows is based on the following selection criteria: the *membership degree* of the property to the rule, if this is similar therefore the rule will be selected according to the second criteria or selecting the rule with *the highest credibility degree*. If the credibility degree is the same the last criteria will be the *number of the object supporting the rule*. The more object support the rule, the stronger is the rule.

The process of rule selection is simplified in the table below (Table 10). Here it is possible to observe the Rj calculation between the single rule and the object to be appraised. For example the first value 0.028 in the first row of the first column is obtained as follows:

$$\begin{aligned} R(\rho; x_{2,DAT}) &= \frac{\max(0; \min(3;7) + 4.1145 - \max(7;3))}{4.1145} \\ &= \frac{\max(0; +3 + 4.1145 - 7)}{4.1145} = \frac{0.1145}{4.1145} = 0.028 \end{aligned} \tag{18}$$

Therefore the minimum among all the Rj in the first row will originate the result of the seventh column.

In the inferior part of Table 10 there will be the k threshold calculated from the Table 3. The single property to appraise is compared directly with the rule. This happens both in the row and in the column. In the sixth column are indicated the minimum Rj of the row. In the seventh column it is indicated the minimum between the result of the sixth column and the credibility of degree. It is easy to observe that the highest of these values is in the row number 9 that is the rule selected. This process happens for all the properties to be estimated giving the following final result in Table 11.

In the last table it is presented the COD derived by the application of Rough Set Theory as Automated Valuation Model. The Coefficient of is calculated on the five properties appraised using Rough Set Theory (Table 12).

The indicator Coefficient of Dispersion is quite encouraging. According to international standards the indicator for income producing properties should be within the interval 0.05–0.15 (IAAO (1999), Standard on Ratio Studies, p. 17)

4 Final Remarks

The lack or unavailability of data poses one of the greatest obstacles hindering the exploration of real estate valuation. Methods based on the principles of the rough set theory may constitute a valuable tool for valuation the real estate. The rough set theory was developed to analyze imprecise and vague data which is commonly found on the real estate market and accompanies decision making (fuzzy decision making) on that market (d'Amato and Siniak 2008; d'Amato and Kauko 2011, 2012; d'Amato and Anghel 2012). This solution is particularly suitable for markets that are weak-form efficient as regards information availability. According to the authors, popular analytical methods (mostly statistical) are relatively ineffective in weak-form efficient real estate markets. The preferred methods and procedures should account for the following defects in real estate data: absence of data, small number of transactions, significant variations in attribute coding, non-linear correlations between the analyzed data and the type of the underlying market. The applied methods should support market analysis at the potential (theoretical) and actual (applied) level. The work showed an application of Rough Set Theory as Automated Valuation Method. The empirical application showed that Rough Set Theory may be applied also to small sample of 21 properties that may be problematic for traditional hedonic modelling.

Some differences between Rough Set Theory and Multiple Regression Analysis must, however, be stressed. The greater difference is in the final output. Multiple regression analysis allows the appraiser to define the marginal price of each property characteristic considered in the model, while rough set theory does not give any information. In Multiple Regression Analysis the final issue is an econometric model based on continuous mathematical functions. In Rough Set

Theory the valuation is based on a discrete mathematics. In the model presented the process is based on Boolean products and the valuer arrives at the final value estimate through the right if then rule suitable for the object. Multiple Regression Analysis is based on a set of assumptions for model, error distribution and sample distribution. Furthermore Multiple Regression Analysis has a limitation in the number of observations. On the other side Rough Set Theory can work both with small and large samples without any assumption.

In the complete version presented in this work the number of the objects which support the rule, the credibility degree of the rule and the membership of the object to a specific rule play an important role. The two valuation procedures are similar from several points of view. Both the application of Rough Set Theory and Multiple Regression Analysis are based on cross sectional process. The valuation process starts with the definition of *attributes* in Rough Set and *independent variables* in Multiple Regression Analysis. In fact, a cause effect relationship is assumed in both MRA and in Rough Set Theory. MRA output is a mathematical model while in Rough Set Theory the output is a Boolean sum, or an if-then rule. Both valuation procedures give the same results starting from the same sample and the same group of attributes. Application of Rough Set Theory may be recommended for mass appraisal in those markets where the property market is not transparent or where downturn and upturn of the market diminish the quantity of reliable observations.

References

Bello, R., & Verdegay, L. (2012). Rough sets in the Soft Computing environment. *Information Sciences, 212*, 1–14.

Borst, R. A., Des Rosiers, F., Renigier, M., Kauko, T., & d'Amato, M. (2008). Technical comparison of the methods including formal testing accuracy and other modelling performance using own data sets and multiple regression analysis. In T. Kauko & M. d'Amato (Eds.), *Mass appraisal an international perspective for property valuers*. Wiley Blackwell.

Chi, D., Yeh, C., & Lai, M. (2011). A hybrid approach of dea. Rough set theory and random forests for credit rating. *International journal of innovative computing information and control, 7*(8), 4885–4897.

Chung, W., & Tseng, T. (2012). Discovering business intelligence from online product reviews: A rule-induction framework. *Expert Systems with Applications, 39*(15), 11870–11879.

d'Amato, M. (2002). Appraising properties with rough set theory. *Journal of Property Investment and Finance, 20*(4), 406–418.

d'Amato, M. (2004). A comparison between RST and MRA for mass appraisal purposes. A case in Bari. *International Journal of Strategic Property Management, 8*, 205–217.

d'Amato, M. (2007). Comparing rough set with multiple regression analysis as automated valuation methodologies. *International Real Estate Review*, 10(2), 42–64.

d'Amato, M. (2008). Rough set theory as property valuation methodology: The whole story. In T. Kauko & M. d'Amato (Eds.), *Mass Appraisal an International Perspective for Property Valuers* (Chap. 11, pp. 220–258). Wiley Blackwell.

d'Amato, M. (2010). A location value response surface model for mass appraising: An "iterative" location adjustment factor in Bari, Italy. *International Journal of Strategic Property Management, 14*(3), 231–244.

d'Amato, M. (2015). Cyclical capitalization. *International Journal of Strategic Property Management, 19*.

d'Amato, M. (2017a). Aspects of commercial property valuation and regressed DCF. In D. Lorenz (Ed.), *Behind the price: Valuation in a changing environment*. Wiley (in print).

d'Amato, M. (2017b). Cyclical capitalization. In D. Lorenz & T. Lutzkendorf (2014) (Eds.), Beyond the price: Valuation in a changing environment. Wiley Publishers, forthcoming.

d'Amato, M., & Anghel, I. (2012). *Regressed DCF, real estate value, discount rate and risk premium estimation*. A case in Bucharest, Aestimum, in print.

d'Amato, M., & Kauko, T. (2008). Property market classification and mass appraisal methodology. In T. Kauko & M. d'Amato (Eds.), *Mass appraisal an international perspective for property valuers* (Chap. 13, pp. 280–303). Wiley Blackwell.

d'Amato, M., & Kauko, T. (2011). International encyclopedia of housing and home. In S. J. Smith & M. Elsinga (Eds.), *Ong Seow Eng, Susan Watcher e Lorna Fox O'Mahoney*. Elsevier Publisher.

d'Amato, M., & Kauko, T. (2012). Sustainability and risk premium estimation in property valuation and assessment of worth. *Building Research and Information, 40*(2), March-April, 174–185.

d'Amato, M., & Siniak, N. (2008). Using fuzzy numbers in mass appraisal: The case of belorussian property market. In T. Kauko & M. d'Amato (Eds.), *Mass appraisal an international perspective for property valuers* (Chap. 5, pp. 91–107), Wiley Blackwell.

Dawidowicz, A., Renigier-Bi³ozor, M. & Radzewicz, A. (2014). An algorithm for the purposes of determining the real estate markets efficiency in Land Administration System. *Survey Review, 46*(36) (May 2014), 189–204. doi:10.1179/1752270613Y.0000000080.

Downie, M. L., & Robson, G. (2007). *Automated valuation models: an international perspective, Council of Mortgage Lenders, London* (pp.10–11). ISBN 1-905257-12-0.

IAAO (1999). *Standard on ratio studies*, Chicago, IL.

Kaklauskas, A., Daniūnas, A., Dilanthi, A., Vilius U., Lill, I., Gudauskas, R., et al. (2012). Life cycle process model of a market-oriented and student centered higher education. *International Journal of Strategic Property Management, 16*, 4, 414–430.

Kauko, T., & d'Amato, M. (2008a). Introduction: Suitability issues in mass appraisal methodology. In T. Kauko & M. d'Amato (Eds.), *Mass Appraisal an International Perspective for Property Valuers* (pp. 1–24). Wiley Blackwell.

Kauko, T., & d'Amato, M. (2008b). Preface. In T. Kauko & M. d'Amato (Eds.), *Mass appraisal an international perspective for property valuers* (p. 1). Wiley Blackwell.

Pawlak, Z. (1982). Rough sets. *International Journal of Information and Computer Sciences, 11*, 341–356.

Pawlak, Z. (1991). *Rough sets*. Theoretical Aspects of Reasoning about Data: Kluwer Academic Publisher, Dordecht.

Polkowski, L. (2010). Reductive reasoning rough and fuzzy sets as frameworks for reductive reasoning. Approximate reasoning by parts: An introduction to rough mereology. *Book Series: Intelligent Systems Reference Library, 20*, 145–190.

Renigier-Biłozor, M. (2010). Supplementing incomplete databases on the real estate market with the use of the rough set theory. *Acta Scientiarum Polonorum, Administratio Locorum*, 9(3), 2010, 107–115.

Renigier-Biłozor, M. (2011). Analysis of real estate markets with the use of the rough set theory. *Journal of the Polish Real Estate Scientific Society, 19*(3), 107–118.

Renigier-Biłozor, M., & Wiśniewski, R. (2011). The efficiency of selected real estate markets in Poland. *Acta Scientiarum Polonorum, Oeconomia, 10*(1), 83–96.

Renigier-Biłozor, M., & Wiśniewski, R. (2013). The impact of macroeconomic factors on residential property prices indices in europe. *Folia Oeconomica Stetinensia Szczecin, 12*(2), 103–125. doi:10.2478/v10031-012-0036-3.

Renigier-Biłozor, M., Wiśniewski, R., Biłozor, A., & Kaklauskas, A. (2014). Rating methodology for real estate markets—Poland case study. *International Journal of Strategic Property Management, 18*(2), 198–212. doi:10.3846/1648715X.2014.927401.

Stefanowski, J., & Tsoukias, A. (2000). Valued tolerance and decision rules. In Rough Sets and Current Trends in Computing, Vol. 2005 of the series Lecture Notes in Computer Sciences pp. 212–219.

Wang, G., & Guan, L. (2012) Data-driven valued tolerance relation. In Li et al. (Eds.), *7th International Conference in China: Rough Sets and Knowledge Technology*. RSKT 2012 (pp. 11–19). LNAI 7414. doi:10.1007/978-3-642-31900-6.

Zavadskas, E., & Turskis, Z. (2011). Multiple criteria decision making (mcdm) methods in economics: an overview. *Technological and Economic Development of Economy, 17*(2), 397–427.

Zhang, Z. (2012). A rough set approach to intuitionistic fuzzy soft set based decision making. *Applied Mathematical Modelling, 36*(10), 4605–4633.

Part V
AVM Methodological Challenges: Inputs and Models

The Theory and Practice of Comparable Selection in Real Estate Valuation

William J. McCluskey and Richard A. Borst

Abstract The article is focused n the techniques to select comparables, providing evidence that it is possible to deal with this delicate phase of the automated valuation process in a scientific way.

1 Introduction

Valuation and in particular the sales comparison approach is often described as being the application of 'art' or a 'science' or a combination of them both. The distinction is more readily applicable to the techniques employed to calculate value and not necessarily to the underlying concept of valuation. What has to be determined is how many sales constitute a sufficient number and from what specific location should the sales be drawn from (Skaff 1975). The definition of a comparable sale involves the comparison of property characteristics. Generally, a sale will be a comparable if its characteristics (used for selection purposes) agree with some specific criteria. This is particularly relevant in terms of how comparables are selected and adjusted to reflect the subject property i.e. the property being valued. The 'art' implies a certain degree of skill associated with judgment whilst a systematic approach to the application of the method may involve the science. Valuation is the process of estimating price based on market evidence. However, such an estimation can be affected by uncertainties: uncertainty in the comparable information available; uncertainty in the current and future market conditions and uncertainty in the specific inputs for the subject property. In addition, the criteria set to draw comparables from the whole set of sales transactions must be defined to ensure sufficient comparables 'similar' to the subject are extracted. The reality is

W.J. McCluskey (✉)
International Property Tax Institute, North York, Canada
e-mail: wj.mccluskey@ulster.ac.uk

R.A. Borst
Senior Research Analyst Tyler Technologies, Philadelphia, USA
e-mail: richborst@msn.com

M. d'Amato and T. Kauko (eds.), *Advances in Automated Valuation Modeling*,
Studies in Systems, Decision and Control 86, DOI 10.1007/978-3-319-49746-4_17

that with advances in research the subjectivity associated with comparable selection and determination of variable weights can be minimized. Therefore, the sales comparison approach can be viewed more as a scientific approach. In this chapter Sect. 2 will look at the issue of comparability/similarity and mathematical distance in terms of the selection of a set of 'comparable' properties; Sect. 3 will focus on comparable selection within real estate valuation; Sect. 4 provides details on the comparable weighting process; Sect. 5 provides empirical analysis using two case studies; and the final Sect. 6 provides some conclusions.

2 The Question of Similarity or Distance

To find the value of a property the valuer/appraiser normally will identify a sample of sales that are 'similar' to the 'subject property', in broad terms of location, various physical characteristics and date of sale. The records of recent sales are the basis for the 'science' with data on specific property characteristics such as size, number of bedrooms, bathrooms, age, condition etc. The data is then subjected to some scientific analysis to attempt to reduce the subjectivity of the valuation process. The difficulty can be in evaluating qualitative variables such as quality of building design, desirability of the location, condition and more intangible qualities that makes properties sell for differing prices. The sales comparison approach relies on selecting transactions of properties with similar characteristics to the subject being valued. Specific monetary or percentage adjustments are made for differences between the comparables and subject as at the valuation date (Todora and Whiterell 2002). The relationship can generally be specified as follows;

$$V = S_C + Adj_C \tag{1}$$

where V = market value, S_C = selling price of comparable property and Adj_C adjustments for differences between the subject and the comparables. Measuring similarity or 'distance' between two variables/attributes is an important component in deciding which comparables are considered as comparable to the subject. As Daiz (1990) comments when a valuer/appraiser selects comparables, he or she is choosing the most suitable alternatives from a pool of candidates. Each alternative can be represented by a set of attributes or variables, the attribute value can be considered as the degree of attractiveness.

There are several methodological approaches or rules that can be employed by the appraiser to select the most suitable comparable. These approaches according to Diaz (1990) share three characteristics; (1) the metric used to rate similarity; (2) the order of the attributes; and (3) commensurability across attributes. Methodological approaches can include statistical (Euclidean); clustering (k-means), distance weighting approaches (e.g. inverse weighting) and spatially orientated approaches such as kriging (such as universal). The concept of difference or similarity is acquiring more and more attention in the representation of data within the valuation process. It is necessary to ascertain in what manner the data are interrelated, how various data differ or agree with each other, and what the measure of their comparison should be. A metric or distance function is a function *d(x, y)* that defines the distance between attributes of a set as a non-negative real numbers. If the distance is zero, both elements are identical under that specific metric. Distance functions thus provide a way to measure how close or similar two attributes are. A small distance is therefore equivalent to a large similarity. There are multiple ways to define a distance metric on a specific parameter set. For example, a typical distance for real numbers is the their absolute difference, i.e. $d(x, y) \to |x - y|$. In a binary format the following would apply $d(x, y) = \{0, 1\}$ if $x = y$ or if $x \neq y$ are valid metrics as well. The main purpose of applying distance metrics is to evaluate quantitatively the some measure of similarity. A metric or distance function is a function which defines a distance between elements of an attribute set. Normally, the task is to define a function *Sim(X, Y)*, where X and Y are two elements or attributes of a certain class, and the value of the function represents the degree of "similarity" between the two. Formally, a distance is a function D with non-negative real values, defined on the Cartesian product $X \cdot X$ of a set X. The concept of distance can be measured in different ways. There are distances that are Euclidean (i.e. can be measured by a 'ruler') whilst others are based on other qualitative factors of similarity. For continuous data, the Minkowski Distance is a general method that is used to compute distance between two points. In applying the Minkowski measure you have two options, Manhattan and Euclidean. The equation for Euclidean distance is:

$$d(x, y) = \sum_{i=1}^{n} (x_i - y_i)^2 \tag{2}$$

Euclidean distance is the most common use of distance. It calculates distance as the square root of the sum of squared differences between corresponding elements of the two vectors. Note that the formula treats the values of X and Y as nominal as no adjustment is made for differences in scale. Euclidean distance is only appropriate for data measured on the same scale such as the size of a dwelling in square metres. (Actually we mix all sorts of number together—things measured as area, counts of things (fireplaces, bathroom fixtures) and binary (in neighborhood two or not). What makes it work is the weights assigned to each variable. We can equivalence a change in so many square feet to a change in bathroom count by the weights assigned to each). Manhattan distance or city block distance represents distance between points in a city road grid. It computes the absolute differences between coordinates of a pair of objects. The Manhattan distance function computes the distance that would be travelled to get from one data point to the other if a grid-like path is followed. The Manhattan distance between two items is the sum of the differences of their corresponding components. The formula for this distance between a point x = {X1, X2, …} and a point y = {Y1, Y2, …} is:

$$d = \sum_{i=1}^{n} |x_i - y_i| \quad (3)$$

where n is the number of variables, and x_i and y_i are the values of the *i*th variable, at points x and y respectively. The Mahalanobis Distance also called quadratic distance is a further method in determining the 'similarity' of a set of values. It accounts for the fact that the variances in each direction are different; it accounts for the covariance between variables and reduces to the familiar Euclidean distance for uncorrelated variables with unit variance. The Mahalanobis distance is a measure of the distance between a point P and a distribution D. It is a multi-dimensional generalization of the idea of measuring how many standard deviations away P is from the mean of D. This distance is zero if P is at the mean of D, and increases as P moves away from the mean. Along each principal component axis, it measures the number of standard deviations from P to the mean of D.

$$d(x, y) = \sqrt{(x - y)^t Cov(x)^{-1}(x - y)} \quad (4)$$

The *x, y* are vectors of standardized property attributes; the $Cov\ (x)^{-1}$ is a variance/covariance of the relevant attributes. The technique simply measures the nearness of the array of attributes to the subject property.

The notion of similarity for continuous data such as size, number of bedrooms, plot size is relatively well understood, but for categorical data, the similarity computation is not as straightforward (Ahmad and Dey 2007). Categorical variables can take different values and are not inherently ordered (for example if we look at VIEW, it can be categorized as NO VIEW, PARTIAL VIEW, GOOD VIEW and UNRESTRICTED VIEW). For example the simplest way to find similarity between two different categorical attributes is to assign a similarity of 1 if the values are identical and a value of 0 if the values are not identical. So you want to be able to match the subject with the comparable; i.e. to match the subject with NO VIEW to a comparable with NO VIEW. To overcome this problem, several similarity measures have been suggested for the analysis of categorical data. In the late 1800s Pearson proposed a chi-square statistic which is often used to test independence between categorical variables in a contingency table. More recently, however, the overlap measure has become one of the most commonly used similarity measure for categorical data.

3 Comparable Selection in Real Estate Appraisal—Literature

The comparable selection method is in widespread used by appraisers across the world. Representative references to its use are found in Underwood and Moesch (1982), Thompson and Gordon (1987), McCluskey and Borst (1998) and Todora and Whiterell (2002). Additional analytical foundations for the method (in particular the grid adjustment technique) are described in Colwell et al. (1983), however, they do not consider how many sales to select or which ones should be selected as comparables. They suggest that weighting schemes are a matter of judgement and experience. This is largely because they contend that the underlying probability distribution from which the comparable sales are drawn is unknown. However, to some extent this is an incorrect observation as regression analysis makes specific assumptions and uses sales data sets so there is a precedent for making operational assumptions about the underlying probability distribution. Shenkel and Eidson (1971) argue that there is not one best comparable but rather a 'best' combination of comparable sales, therefore it is necessary to select the ideal set of comparables. Vandell (1991) provides the theoretical background to provide decisions about how many comparables to select, what the comparable selection criteria should be, and how proper weights for each adjusted value estimate can be determined such that the final value estimate is both unbiased and of minimum variance. The selection of the 'best' comparable is the one with the smallest or smallest number of adjustments. Kang and Reichert (1991) compared MRA to the comparable sales method on data sets with a variety of characteristics. Gau et al. (1992) extended Vandell's method. Lai and Wang (1996) compare the accuracy of the minimum variance grid[1] method to MRA value estimates. They establish the theoretical basis upon which it is shown that the grid method should have a lower variance than MRA estimates.

Colwell et al. (1983) illustrated the mathematical similarity of hedonic regression and sales comparison adjustment grid pricing models. The regression model can be expressed as:

$$P_i = Bx + \varepsilon \tag{5}$$

where P_i = the estimated price of a subject property, B is a vector of coefficients, x is a vector of property characteristics and ε the random error term. Colwell et al. (1983) pointed out that a sales comparison adjustment grid is mathematically equivalent to an additive hedonic price model.

$$P_s = P_O + B(X_S - X_O) + \varepsilon \tag{6}$$

[1]The "grid method" or "grid estimator" is a common description given to the comparable sales analysis technique.

where P_O is the observed price of a comparable property. Krause and Kummerow (2009) used as an example of this approach the following example; consider a single comparable sale (sold two months prior) located in the same subdivision as the subject property that has identical attributes with the exception of 100 extra square feet of living space. Because of the locational proximity of these sales, the micro-spatial variations in price between them are minimal. If both are at essentially the same location, prices do not have to be adjusted for location differences. Also, because of the near identical structural features of close comparables, the only differences to be accounted for are market changes over the past two months and the premium paid for the extra living area. The righthand side of equation reduces to the sum of the adjustments for the market and for the additional square footage. Solving for the subject property value estimate simply becomes a matter of adding the adjustments to the comparable sale prices—as in a traditional sales adjustment grid.

Table 1 Chronology of research into comparable selection

Author(s)	Year	Technique	Comments
Shenkel and Eidson	1971	Sales retrieval process	Early paper on comparable selection
Skaff	1975	Standard deviation and mean functions	Applied user defined constraints. Used an iterative approach to select comparables
Tchira	1979	Weighting schemes based on Euclidean and Mahalanobis	Mahalanobis distance preferred
Isakson	1986	Mahalanobis distance	No comparable weighting scheme used
Diaz	1990	Theoretical design	Application of expert systems to comparable selection
Kang and Reichert	1991	Absolute and squared (Quadratic) weighting schemes	Comparison with OLS and ridge regression techniques
Vandell	1991	Minimum variance criteria	Comparable weighting scheme based on minimum variance
Gau et al.	1992	Coefficient of Variation	Developed the work done by Vandell
Epley	1996	Rank transformation analysis technique	Useful for small samples
Detweiler and Radigan	1996	Automated comparable selection using user defined constraints and weights	User defined iterative approach
Todora and Whiterell	2002	Minkowski metric	Weights developed by applying a correlation matrix
Krause and Kummerow	2009	Mahalanobis distance	Methodology developed on small data sets; geographic distance penalty

As shown in Table 1 research since the early 1970s has considered the development of more scientific and automated approaches to identifying and selecting comparables for the valuation of subject properties, typically unsold properties. Isakson (1986) recommended minimizing the hedonic distance between sample points when a subsample is selected for sales comparison. Isakson (1986) used Mahalanobis distance as the metric of nearness, however, did not apply any weighting to these characteristics.

4 The Weighting Process

Consider that finding the most comparable sale is equivalent to finding the least dissimilar. The actual dissimilarity measure can be based on physical separation, and differences in physical characteristics, date of sale, and the neighbourhood to which the comparable sale belongs. The expression by which "comparability" is judged was documented by Gipe (1974). Gipe (1974) recounts the early development history of the automation of the process. The main processing steps within the method include finding the n most comparable sales properties; computing an adjusted sale price for each; weighting these estimates according to their similarity to the subject; and summing the weighted comparable sales estimates to get the final estimate. Detweiler and Radigan (1996) apply a user defined approach to attribute weights and used the following expression to select comparables based on 'nearness' between the subject and the comparable.

$$Nearness_i = \frac{d_i}{S^{-1}d_i} \tag{7}$$

where

$$d_i = \sqrt{w(x_s - x_{ci})} \tag{8}$$

and.

w vector of weights decided by the user and relative to each other

x_s vector of characteristics that describe the subject

x_{ci} vector of characteristics that describes the comparable i

S covariance matrix of the characteristics used to select the comparables.

Todora and Whiterell (2002) applied the Minkowski metric to select the appropriate set of comparables. The assignment of weights for each of the variables were developed with the aid of a correlation matrix. The Minkowski metric is as follows:

$$\sum w_i \left[abs \frac{x_{si} - x_{ci}}{x_{si}} \right] \tag{9}$$

where w_i is the weight assigned to the *i*th variable; x_{si} is the value of the *i*th variable of the subject property; x_{ci} is the value of the *i*th variable of the comparable property.

Krause and Kummerow (2009) do not consider that the traditional Mahalanobis technique is not the best measure because physical distance does not offer a monotonic difference in price i.e. locations some distance away can be more similar than locations closer. They developed a dissimilarity value (DV) by adding a geographic distance penalty to the Mahalanobis distance. The DV is expressed as follows:

$$DV(x_{ij}) = \sqrt{(x_i - x_j)x \cdot S^{-1}x(x_i - x_j)} + \frac{Dist_{ij}}{k} \tag{10}$$

where;

$Dist_{ij}$ = the distance between subject property i and comparable property j

k = distance metric

The following example demonstrates the application of a weighting scheme. An important step in automating the process was the identification of a metric for measuring the comparability of one property to another. The metric takes the following form:

$$D_{jk} = \sum_{i=1}^{n} [W_i(X_{ij} - S_{ik})]^2 \tag{11}$$

where D_{jk} is the 'Distance' between subject property *j* and sale property *k*, W_i is a weighting factor assigned to characteristic *i*, X_i s the value for the subject property *k* for the *i*th variable and S_{ik} is the value of the *i*th variable for the *k*th sale. A brief narrative statement of the algorithm will help set the stage for the comparison between comparable sales and the other methods described herein. The main processing steps within the method include finding the *n* most comparable sales properties (comps); computing an adjusted sale price for each comp; weighting these estimates according to their similarity to the subject; and summing the weighted comp estimates to get the final estimate. Consider that finding the most comparable sale is equivalent to finding the least dissimilar. The actual dissimilarity measure can be based on physical separation, and differences in physical characteristics, date of sale, and the neighborhood to which the comp belongs. The weights in Eq. (10) can be determined in a number of ways. Initially consider estimating the value of a subject based on one comparable sale. Cannaday (1989) describes the method of adjustment as the difference in the MRA estimates of the subject and the comparable. This adjustment process can be expressed as:

$$Estimate\ of\ S_k\ based\ on\ C_j = ASP\left(C_j\right) + \left(ESP(S_k) - ESP\left(C_j\right)\right) \tag{12}$$

where S_k refers to subject *k*, C_j to comparable sale *j*, *ASP* (C_j) is the actual selling price of C_j, and $ESP(S_k)$ refers to the estimated selling price of S_k. Equation 12 can be rewritten as:

$$Estimate\ of\ S_k\ based\ on\ C_j = ESP(S_k) + \left(ASP\left(C_j\right) - ESP\left(C_j\right)\right) \tag{13}$$

where, the expression $(ASP\ (C_j) - ESP\ (C_j))$ is the residual error of the estimate for comparable sale *j*. Generalizing Eq. (13) to the case of valuing a subject by using several comparable sales results in:

$$CSM(S_k) = \sum_{j=1}^{n} CW_{jk}\left[ESP\left(S_k + \left(ASP(C_j) - ESP(C_j)\right)\right)\right] \tag{14}$$

With

$$\sum_{j=1}^{n} CW_{jk} = 1$$

where $CSM(S_k)$ refers to the comparable sales method value of S_k, and CW_{jk} is a "comparability weight applied to each comparable sale property. Noting that

$$\sum_{j=1}^{n} CW_{jk} ESP(S_k) = ESP(s_k) \text{ because}$$

$$\sum_{j=1}^{n} CW_{jk} = 1$$

Equation (14) can be rewritten as:

$$CSM(S_k) = ESP(S_k) + \sum_{j=1}^{n} CW_{jk}\left(ASP(C_j) - ESP(C_j)\right) \tag{15}$$

In matrix notation this expression can be rewritten as:

$$CSM(S) = X\hat{\beta} + CW\left(Y - X\hat{\beta}\right) \tag{16}$$

Table 2 Comparable weights calculation

Sale number	1	2	3	4	5
Price	135,000	105,000	109,000	110,000	168,000
Adjusted selling price	136,596	139,554	129,650	133,706	146,316
Comparability/distance (Di)	47	83	92	93	95
D_i^2	2,209	6,889	8,464	8.649	9,025
$\left(\frac{1}{2m}\right)^2$	22,500	22,500	22,500	22,500	22,500
$(2mP_i)^2$	3	38,987	11,626	16,720	6,240
D_i^2	2,209	6,889	8,464	8,649	9,025
SUM	24,712	68,376	11,626	16,720	6,240
W_i *X 10,000*	0.40	0.15	0.23	0.21	0.26
Normalized W_i	0.32	0.12	0.19	0.17	0.21
Weighted contribution	43,890	16,206	24,171	22,179	30,764
Weighted estimate	137,164				

Source Based on Thompson 1997

where $\hat{\beta}$ are the estimates of β obtained by OLS methods. The formulation of Eq. (16) can be rewritten in the compact form:

$$\hat{Y} = \hat{Y} + CW_{\hat{\varepsilon}} \tag{17}$$

where the substitution of variables is $CSM(S) = \hat{Y},\ \hat{Y} = X\hat{\beta}$, and $W_{\hat{\varepsilon}}$ represents the weighted residual errors from OLS.

Weighting mechanism

The weighting mechanism in common use for combining the n comparable sales estimates has an explicit expression.

$$W_i = \frac{1}{\left(\frac{1}{2m}\right)^2 + D_i^2 + (2mP_i)^2} \tag{18}$$

where

m = the maximum acceptable comparability

W_i = weight for the ith sale

D_i = actual comparability of the ith sale

P_i = percentage adjustment to the ith sale

Table 2 illustrates the method 5 comparables. A maximum acceptable distance (m) of 300 is assumed.

$\left(\frac{1}{2m}\right)^2$ results in $\left((^1\!/_2)300\right)^{\uparrow}2 = 22,500$

$(2mP_i)^2$ results in $2 \times 300\left[\left(\frac{136,596}{137,000}\right)\right]^2 = 3$

$$\text{SUM} = \frac{1}{\left(\frac{1}{2m}\right)^2 + D_i^2 + \cdots (2mP_i)^2}$$

Normalized W_i = 0.40/1.26 = 0.32 (normalized on 1)

The normalized W_i is multiplied by the adjusted selling price for each comparable and those results are summed to calculate the overall weighted estimate of value.

5 The Number of Comparable Sales—Case Studies

The worldwide proliferation of the use of the comparable sales method in Computer Assisted Mass Appraisal Systems (CAMA) and Automated Valuation Models (AVM) is in large part due to the continued improvements in computer hardware and software technology. The typical number of comp sales chosen for analysis ranges between three and five. This is, no doubt, due primarily to the limitations of the printed page or the computer monitor size. Higher resolution printing and display technology may influence an increase in the number of comparable sales used in valuation. If we set aside these limitations, a question arises as to the optimal number of comparable sales to be used to obtain the greatest accuracy of results. Two jurisdictions were selected as case studies to investigate study the question. The first includes a group of condominium sales from the western portion of Beijing, People's Republic of China and the second comprises residential properties (dwellings/houses) from the Town of Scarsdale, New York, United States.

Table 3 Descriptive statistics—Beijing

Variables	Mean	Median	Min	Max
Sale_price[a]	1,810,121	1,500,000	550,000	9,200,000
Age	10.41	8	1	28
Beds	2.06	2	1	6
Living Room	1.28	1	0	4
Max_floor	14.63	16	3	118
Floor	7.63	6	1	27
Size	88.79	79	30	603
Top floor	14.62	16	1	27
Wash rooms	1.20	1	1	4

[a]Chinese Yuan

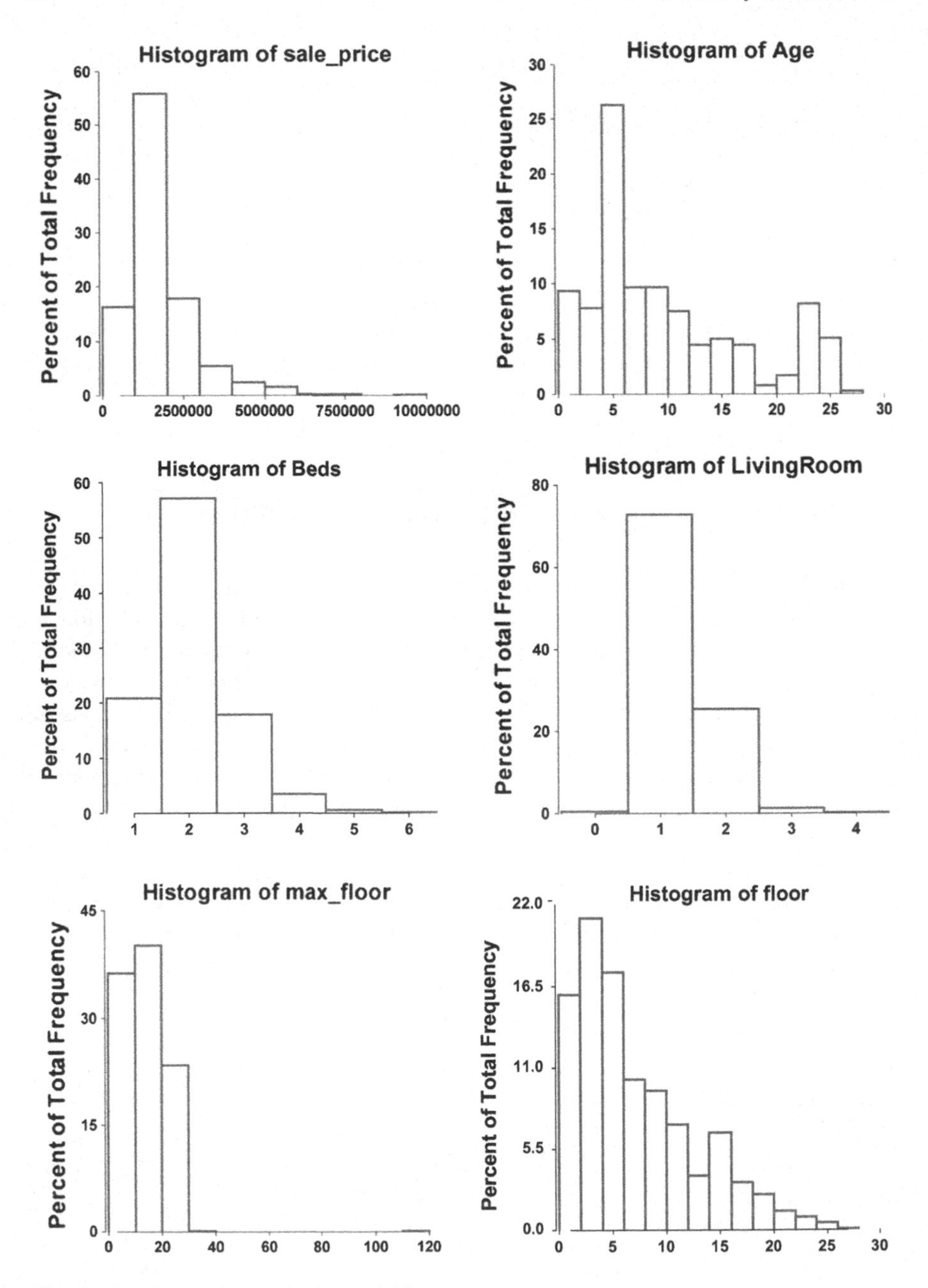

Fig. 1 Histograms of quantitative variables

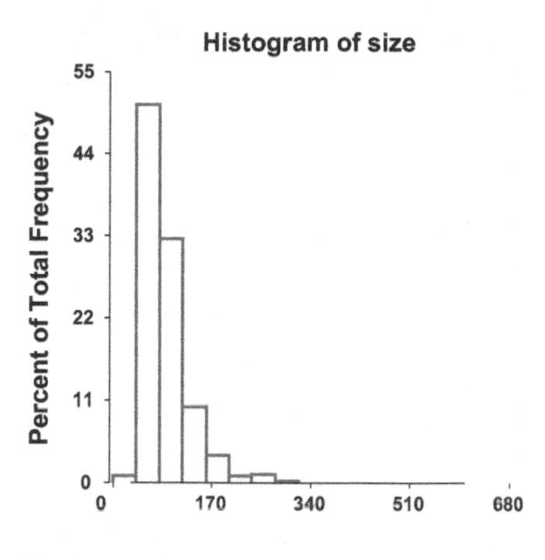

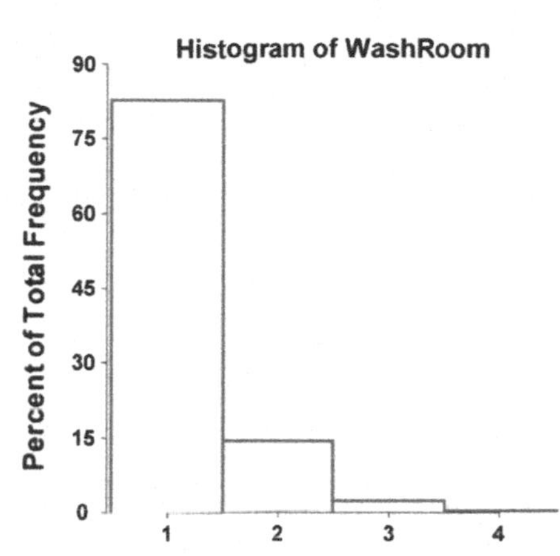

Fig. 1 (continued)

Table 4 Descriptive statistics—Scarsdale

Variables	Mean	Median	Minimum	Maximum
SALE_PRICE (US$)	1,638,597	1,325,000	485,000	10,884,000
AGE	56.62	64	1	160
NBR_BATH	3.84	3.5	1	9
NBR_BED	4.53	4	2	9
NBR_FIREPL	1.42	1	0	9
NBR_ROOMS	9.16	9	5	15
NBR_STORIES	2.01	2	1	2.7
SFLA (Sq. Ft.)	3341.74	3088	1232	8452

6 Case Studies

The data consists of 2,180 valid sales of condominium flats in Beijing, Peoples Republic of China. he properties are located due west of Beijing center. A summary of selected data elements is provided in the Table 3.

Figure 1 show a set of histograms which provides more insight into the distribution of each data element. For example, for the MAX_FLOOR (the number of floors in the block) there is really only one at 118 stories. The rest are in the range of 3–30. Similarly the size (in sq. meters) is generally under 340 m^2.

A second sample data consists of 366 valid sales of single family homes in Scarsdale, United States The properties are located in an exclusive suburb north of New York City. A summary of selected data elements is provided in the Table 4.

As with the Beijing data, Fig. 2 provides more insight into the distribution of each data element. For example, for Age, it can be seen that there are two concentrations, a group under 20 years old, and another in the 80–100 years age range. It is also clear that Scarsdale is a "two-story" jurisdiction

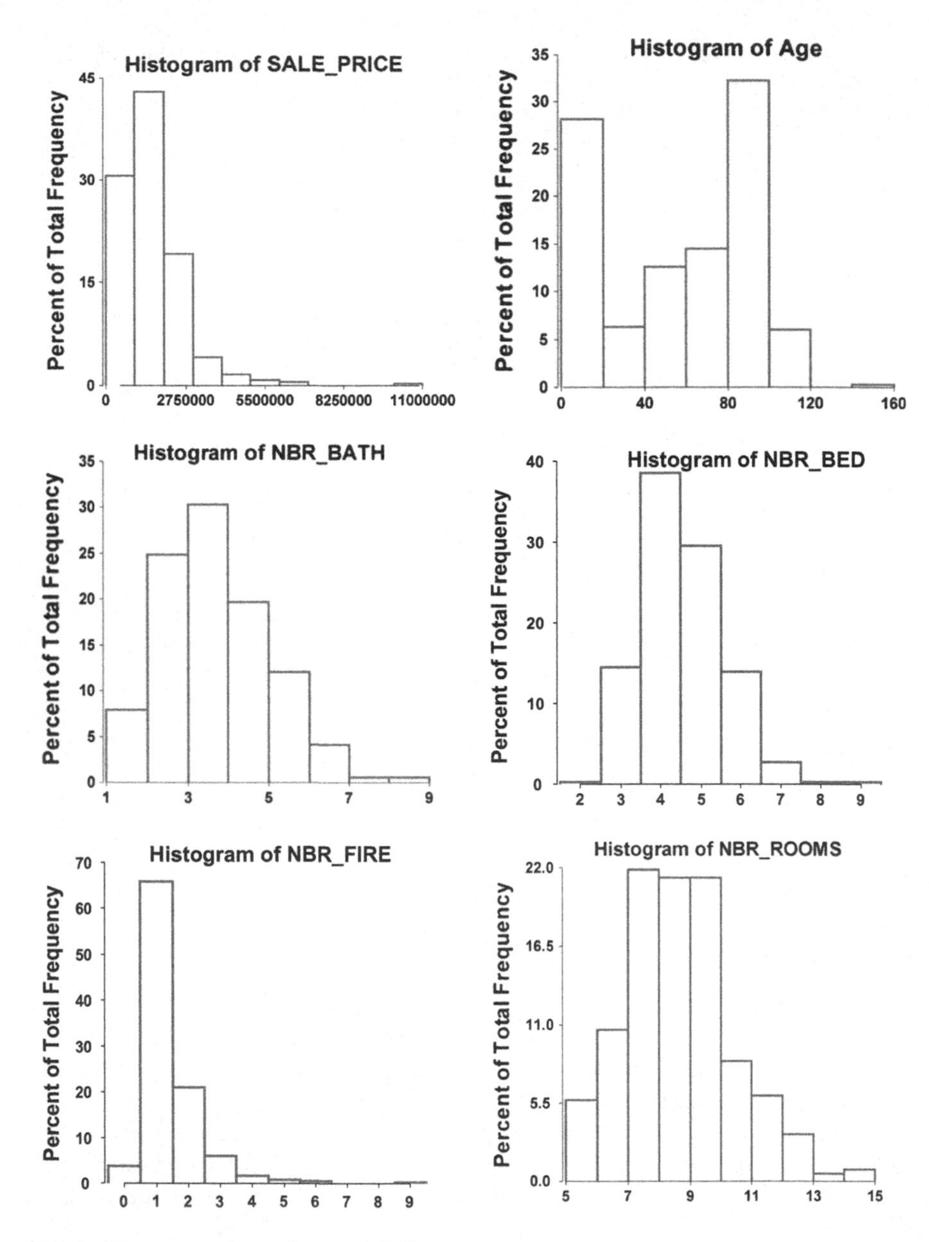

Fig. 2 Histograms of quantitative variables

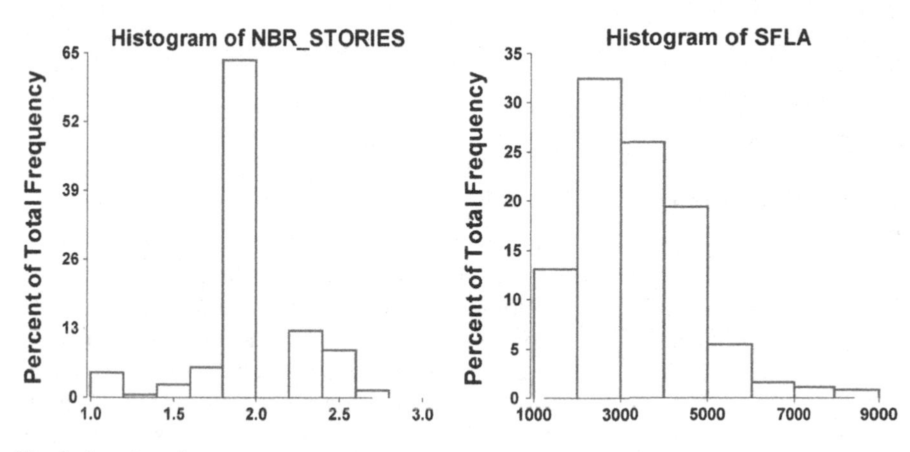

Fig. 2 (continued)

7 Analysis

A commercial-off-the shelf (COTS) MRA/Comparable sales program was used to evaluate a varying number of comp sales for accuracy of prediction as measured by the Coefficient of Dispersion (COD) and Moran's I for spatial autocorrelation in the ratio of the estimated value to the actual sales value. Among other measures, the model with the lowest COD is a candidate for the best model from among competing models. Borst (2014) states that "Moran's I is the most commonly known single measure of spatial autocorrelation in a dataset. Properly specified parameters of the calculation yield value ranges between −1 and 1. Associated with the value is a measure of significance, namely a z-score, a p-value or both. The Moran's I results are viewed in terms of a null hypothesis that the distribution of errors is random. If the z-score is high enough and the p-value is low enough, then the hypothesis is rejected. Assuming the null hypothesis is rejected, it can be said that when Moran's I is closer to 1 there is a high positive autocorrelation—high values near high values and low values near low values. When it is closer to zero, there is there is little overall spatial autocorrelation in the data. When closer to −1 means high values will be near low neighbor values and low values near high neighbor values consistently." The definition of Moran's I is given by Eq. (19):

$$I = \frac{N}{\sum_i \sum_j w_{ij}} \frac{\sum_i \sum_j w_{ij}(X_i - \bar{X})(X_j - \bar{X})}{\sum_i (X_i - \bar{X})^2} \tag{19}$$

where N is the number of spatial units indexed by i and j
X is the variable of interest,
$\bar{X}$ is the mean of X,
and w_{ij} is an element of a row standardized spatial weights matrix, i.e

$$\sum_j w_{ij} = 1, and\, w_{ii} = 0$$

8 Multiple Regression Models

It is helpful to begin by describing the comparable sales model in narrative form. The steps in the process are as follows: find the sales properties which are most comparable to the subject property to be valued; adjust the sale price for each comparable to account for differences between it and the subject's characteristics and also for the date of sale; weight these adjusted comparable sales estimates according to their similarity to the subject; sum the weighted comparable sales estimates to get the final estimate

In terms that are more formal:

$$ESP = SP_{comp} + ADJ_{difference} \tag{20}$$

where:

ESP = Estimated selling price

SP_{comp} = Selling price of the comp sale

$ADJ_{differences}$ = An adjustment for the differences in characteristics between the subject and the comp sale

From Borst (2014), this can be rewritten as:

$$ESP = MRA_{subj} + \mathrm{Res}idError_{compESP} \tag{21}$$

Generalizing for more than one comp:

$$ESP = MRA_{subj} + w_1\mathrm{Res}id_1 + w_2\mathrm{Res}id_2 + w_3\mathrm{Res}id_3 + \ldots w_n\mathrm{Res}id_n \tag{22}$$

where $\sum w_i = 1$

Therefore, the process of computing comparable sales values first starts with calibrating an MRA model that is used as the basis for adjusting the comparable sale to the subject.

In Beijing, the model and its performance statistics is shown in Table 5.

The dependent variable is Price. The term LNAGE is the natural log of age plus "e", the base of the natural logarithm. So, for example if age = 0, the value of the term is 1.0. The term LOCF represents a location factor derived by a separate process, not a topic of this chapter. See Borst (2014) for a full explanation of how it is derived. The Model for Scarsdale is given by Table 6.

The variable names have been expanded to make them self-explanatory. SFLA refers to square foot of living area. The term SQRTSFLA_GRADE ADJUST is the

Table 5 MRA model for Beijing data

Beijing Model			
Dependent	SALE_PRICE		
Std Error for estimate	0.1249		
Constant:	18,070.26		
Attribute	Coeff.	Std. Error	t value
SIZE	1.0043	0.0104	96.5297
WASHROOM	0.1045	0.013	8.0072
LOCF	0.8441	0.0379	22.2413
FLOOR	0.0131	0.0041	3.2023
MAX FLOOR	0.0233	0.0064	3.6658
LNAGE (ln(age+e))	−0.1291	0.0069	−18.8349
Model statistics			
Total valued	2180		
R squared	0.93		
Adjusted R squared	0.9298		
COD	9.5738		
COV median	12.5697		
COV Mean	12.4926		
Median	1.0023		
Mean	1.0078		
Weighted mean ratio	0.9911		

Table 6 MRA model for Scarsdale data

Scarsdale Model			
Dependent	SALE_PRICE		
Std error for estimate	233,697.77		
Constant:	−615,309.99		
Attribute	Coeff.	Std. error	t value
LAND VALUE	0.9082	0.0356	25.5259
FINISHED BASEMENT AREA	95.8451	23.8355	4.0211
SFLA_A_GRADE	125.3039	10.7238	11.6846
SFLA_B_GRADE	38.1035	9.1235	4.1764
AGE*SFLA	−1.2668	0.1216	−10.4217
SQRTSFLA*GRDADJPCT	228.2003	12.0093	19.0019
Model statistics			
Total valued	366		
R squared	0.9529		
Adjusted R squared	0.9521		
COD	10.0248		
COV median	12.7431		
COV mean	12.7502		
Median	0.9999		
Mean	1.0007		
Weighted mean ratio	1		

square root of square foot of living area multiplied by a quality grade adjustment factor.

9 Comparable Sales Results

The COD and Moran's I results are presented for both jurisdictions in tabular and graphic from. In the tables, the values for the typical five comps are highlighted along with the optimal outcome. The results are discussed immediately following the presentation of the data.

10 Beijing

Comparability measures were defined within the COTS and trials were run from two to twenty comparable sales. The results for COD are presented in Table 7 and Fig. 3.

The results for Moran's I are presented in tabular and graphic from (Table 8).

Table 7 Beijing results

Beijing comp Est/Sale Price*100				
No Comps	Median	Mean	PRD	COD
2	100.14	100.79	1.01	6.34
3	100.17	100.74	1.00	6.11
4	100.16	100.70	1.01	6.04
5	*100.24*	*100.72*	*1.00*	*6.02*
6	100.24	100.70	1.00	5.98
7	100.22	100.71	1.00	5.98
8	100.26	100.71	1.00	5.97
9	100.26	100.70	1.00	5.96
10	**100.25**	**100.70**	**1.00**	**5.95**
11	100.24	100.68	1.00	5.99
12	100.23	100.68	1.00	5.98
13	100.22	100.67	1.00	5.97
14	100.21	100.67	1.00	5.99
15	100.21	100.67	1.00	5.98
16	100.22	100.67	1.00	5.98
17	100.22	100.66	1.00	5.99
18	100.24	100.65	1.00	5.99
19	100.25	100.66	1.00	6.02
20	100.25	100.65	1.00	6.01

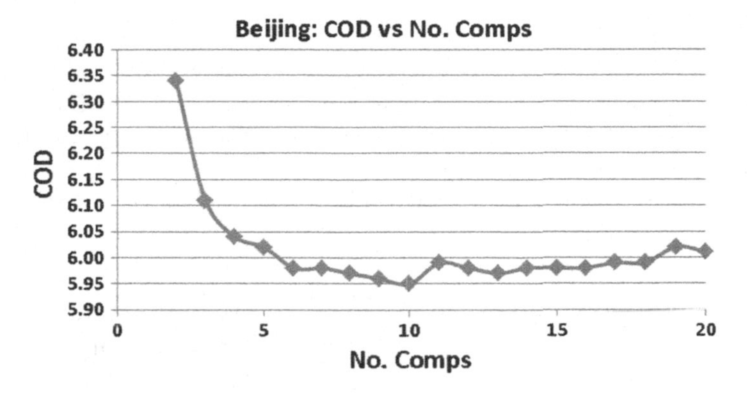

Fig. 3 Beijing COD versus number of comps

Table 8 Moran's I—Beijing

No comps	Moran's I	P value
2	−0.01589	0.033
3	−0.01885	0.014
4	−0.01517	0.028
5	*−0.01508*	*0.046*
6	−0.01337	0.068
7	**−0.01209**	**0.085**
8	−0.01249	0.066
9	−0.01227	0.078
10	−0.01238	0.091
11	−0.01779	0.015
12	−0.01862	0.015
13	−0.01946	0.012
14	−0.02427	0.001
15	−0.02420	0.004
16	−0.02492	0.003
17	−0.02606	0.002
18	−0.02497	0.001
19	−0.02693	0.001
20	−0.02391	0.003

11 Scarsdale

Similar to Beijing, comparability measures were defined within the COTS and trials were run from two to sixteen and twenty comparable sales. The results for COD are presented in Table 9 and Fig. 4.

The results for Moran's I are presented in tabular and graphic from (Table 10).

Table 9 Scarsdale results

Comps.	Median	Mean	PRD	COD
2	100.32	101.56	1.01	10.43
3	101.03	101.63	1.01	10.07
4	101.17	101.65	1.01	9.88
5	*101.23*	*101.61*	*1.01*	*9.73*
6	100.96	101.55	1.01	9.71
7	100.98	101.42	1.01	9.62
8	100.78	101.22	1.01	9.54
9	100.79	101.26	1.01	9.45
10	100.82	101.19	1.01	9.52
11	101.01	101.14	1.01	9.44
12	100.79	101.08	1.01	9.41
13	100.67	101.15	1.01	9.45
14	100.86	101.10	1.01	9.40
15	**100.56**	**101.10**	**1.01**	**9.38**
16	100.57	101.11	1.01	9.41
20	100.65	101.05	1.01	9.46

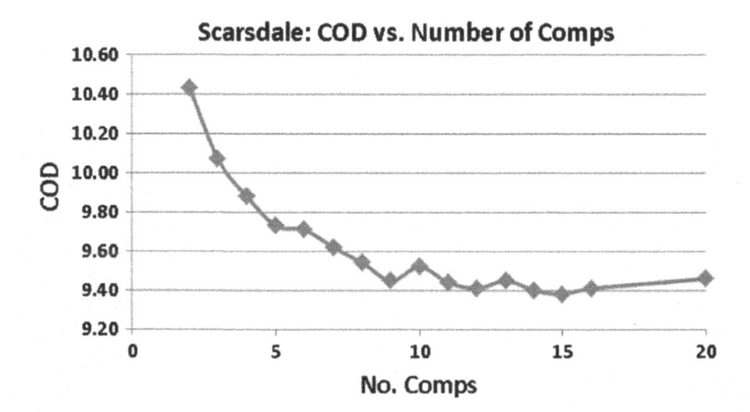

Fig. 4 Scarsdale COD versus number of comps

Table 10 Moran's I—Scarsdale

Num comps	Moran's I	P value
2	−0.10631	0.001
3	−0.10455	0.001
4	−0.07811	0.001
5	**−0.05712**	**0.006**
6	−0.04112	0.040
7	−0.03242	0.089
8	−0.02227	0.175
9	−0.02524	0.176
10	−0.01733	0.280
11	−0.00712	0.433
12	−0.00468	0.472
13	−0.00248	0.491
14	**−0.00047**	**0.452**
15	0.00228	0.398
16	−0.00065	0.482
20	0.01190	0.262

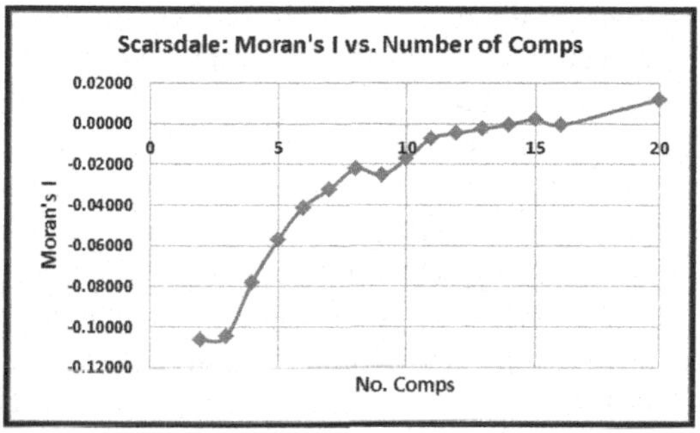

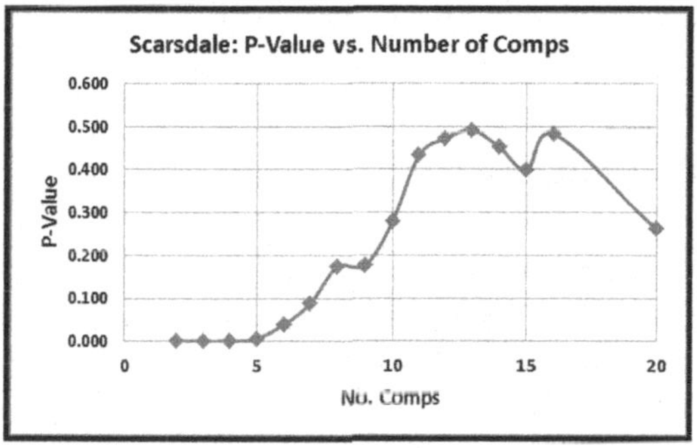

12 Discussion of Results

12.1 Beijing

The optimal COD of 5.95 for the Beijing data occurs at 10 comps. It should be noted that the difference among the comp sales results (low of 5.95 to high of 6.34) is smaller than the difference between any result and the MRA COD of 9.57. In other words the comp sales method is far superior to MRA and the difference among the choices of say 5 or 10 comps may be a matter of individual preference. The optimal Moran's I (the one closest to 0.0) occurs at 7 comps. However, the p-value has importance in the comparison. The Moran's I is really a hypothesis test, with the null hypothesis, H_0 that the value of Moran's I = 0. The p-value is the probability of rejecting H_0 when it is true. So, when we have a small p-value, we are more confident that Moran's I is not 0.0, although it can be very close to 0.0. At 7 comps the p-value = 0.085, or an 8.5 % chance of incorrectly rejecting the null hypothesis. In the case of Beijing, all computed values are quite close to 0.0 anyway.

13 Scarsdale

The optimal COD of 9.38 occurs at a value of 15 comps. In this case, the differences among the various numbers of comps are noteworthy, ranging from a low of 9.38 to a high of 10.43. The MRA model has a COD of 10.02. This is quite different than the Beijing case. The comparable sales method only performs better than MRA after the number of comps reaches 4. From there on, it is always better than MRA. The Moran's I results are quite interesting. At 14 comps, the value is closest to zero. However the p-value is 0.452. The decision would be to not reject the null hypothesis. The practical interpretation is that the comps sales method has removed virtually all spatial autocorrelation in the errors of the model. In either case, COD is best at 15 comps, and Moran's I is best at 14. The agreement is quite good and at a COD of 9.38, versus 9.73 for 5 comps, the difference in accuracy might well be worth considering the higher number.

14 Conclusions

This chapter has sought to outline the theory and practice of the application of the comparable sales method. Beginning in the early 1970s there have been several papers that have attempted to place the selection of comparables on a scientific basis. A key development has been the use of distance metrics drawing on the notion that one can measure similarity between two properties based on the differences of their respective attributes or characteristics. Within the traditional grid adjustment technique adjustments tended to rely largely on the appraisers subjective opinion. However, correlation and co-variance measures have refined the ability to more objectively measure variations between attributes within the same set. Developments in Computer Assisted Mass Appraisal applied within the property tax environment has led to a scientific approach to comparable selection. Given the scale (i.e. numbers) of properties to be valued an automated approach was essential. Following on from these developments the finance/banking industry has recognized an opportunity to utilize the theory of comparable sales in terms of the development of Automated Valuation Models (AVMs). The intention of this chapter has been to link the theory and practice of comparable sales. The case studies provide empirical evidence as to the age old problem of 'how many comparables should be used?' The industry mantra of using three-to-five has to some extent been refined to select the 'best' number by reference to diagnostic statistics such as Coefficient of Dispersion (COD) and Moran's I. If pure accuracy is the desired goal, it is best to test a variety of numbers of comp sales to see which performs best in a given setting. No universal statement about the best number of comps to use is evident from this admittedly limited case study.

References

Ahmad, A., & Dey, L. (2007). A method to compute distance between two categorical values of same attribute in unsupervised learning for categorical data set. *Pattern Recognition Letters, 28*(1), 110–118.

Borst, R. A. (2014). *Improving mass appraisal valuation models using spatio-temporal methods*. Toronto, Ontario, Canada: International Property Tax Institute.

Borst, R. A., & McCluskey, W. J. (2007). Comparative evaluation of the comparable sales method with geostatistical valuation models. *Pacific Rim Property Research Journal, 13*(1), 106–129.

Cannaday, R.E. (1989). How should you estimate and provide market support for adjustments in single family appraisals. *Real Estate Appraiser and Analyst, 55*(4), 43–54

Colwell, P., Cannaday, R., & Wu, C. (1983). The analytical foundations of adjustment grid methods. *Journal of the American Real Estate and Urban Economics Association, 11*(1), 11–29.

Detweiler, J., & Radigan, R. (1996). Computer-assisted real estate appraisal. *Appraisal Journal January* 91–101.

Diaz, J. (1990, October). The process of selecting comparable sales. *The Appraisal Journal*, 533–540.

Epley, D. R. (1996). Note on the optimal selection and weighting of comparable properties. *Journal of Real Estate Research, 14*(1/2), 175–182.

Gau, G. W., Lai, T.-Y., & Wang, K. (1992). Optimal comparable selection and weighting in real property valuation: An extension. *AREUEA Journal, 20*(1), 107–123.

Gibson, D., Kleinberg, J., & Raghavan, P. (2000). Clustering categorical data: An approach based on dynamical systems. *The VLDB Journal, 8*(3), 222–236.

Gipe, G. W. (1974). Analysis of residuals to improve multiple regression equations. *Assessors Journal, 9*(1), 9–16.

Guha, S., Rastogi, R., & Shim, K. (2000). ROCK: A robust clustering algorithm for categorical attributes. *Information Systems, 25*(5), 345–366.

Huang, Z. (1998). Extensions to the k-means algorithm for clustering large data sets with categorical values. *Data Mining and Knowledge Discovery, 2*(3), 283–304.

Isakson, H. (1986). The nearest neighbors appraisal technique: An alternative to the adjustment grid methods. *Journal of the American Real Estate and Urban Economics Association, 14*, 274–286.

Kang, H.-B., & Reichert, A. K. (1991). An empirical analysis of hedonic regression and grid-adjustment techniques in real estate appraisal. *AREUEA Journal, 19*(1), 70–91.

Krause, A., & Kummerow, M. (2009). An iterative approach to minimizing valuation Errors using an automated comparables sales model. *Journal of Property Tax Assessment & Administration, 8*(2), 39–52.

Lai, T.-Y., & Wang, K. (1996). Comparing the accuracy of the minimum-variance grid method to multiple regression in appraised value estimates. *Real Estate Economics, 24*(4), 531–549.

McCluskey, W.J., & Borst, R.A. (1998). *Application of hybrid intelligent appraisal techniques within the field of comparable sales analysis*. Paper presented at IAAO Annual Conference: Today's Vision Tomorrow's Reality, Orlando, Florida.

Moore, J. (2009). A history of appraisal theory and practice: Looking back from IAAO's 75th year. *Journal of Property Tax Assessment and Administration, 6*(3), 23–49.

Pace, R., & Gilley, O. (1998). Generalizing the OLS and grid estimators. *Real Estate Economics, 1*, 331–346.

Shenkel, W.M., & Eidson, A.S. (1971, October). Comparable sales retrieval systems. *Appraisal Journal*, 540–555.

Skaff, M.S. (1975, April). The search for comparable sales: A new approach. *Assessors Journal*, 7–16.

Sneath, P.H.A., & Sokal, R.R. (1973). *Numerical taxonomy: The principles and practice of numerical classification*. San Francisco: W. H. Freeman and Company.

Tchira, A. (1979, January). Comparable sales selection—A computer approach. *The Appraisal Journal*, 86 – 98.
Thompson, J.F. (1997). *Comparable sales as a computer assisted valuation (Mass Appraisal) Technique—Designing efficient programs for comparable sales selection and valuation*. Paper presented at IAAO Annual Conference (pp 476—485).
Thompson, J.F., & Gordon J.F. (1987). Constrained regression modeling and the multiple regression analysis - comparable sales approach. *Property Tax Journal*, *6*(4), 251–262
Todora, J., & Whiterell, D. (2002, January, February). Automating the sales comparison approach. *Assessment Journal*, 25–33.
Underwood, W. E., & Moesch, J.R. (1982). *The second generation of CAMA in New York State*. Paper presented at the 1st World Congress on Computer Assisted Valuation Cambridge MA
Vandell, K. D. (1991). Optimal comparables selection and weighting in real property valuation. *Real Estate Economics, 19*(2), 213–239.
Wang, X., De Baets, B., & Kerre, E. (1995). A comparative study of similarity measures. *Fuzzy Sets and Systems, 73*(2), 259–268.

Reducing the Appraisal Bias in Manual Valuations with Decision Support Systems

Carsten Lausberg and Anja Dust

Abstract Any appraiser is subject to many biasing influences which compromise the accuracy of the appraisal. One of the most prominent biases is the anchoring heuristic: appraisers involuntarily anchor to reference points such as their previous valuation, the value opinion of the seller, or the last transaction price. While many studies have proven the importance of the anchoring effect, very few studies have suggested practical means to counter it. In this chapter we demonstrate that the effect can be reduced with a tool helping the valuer to make better decisions. In our experiment probands appraised an office building with the help of a self-written valuation software. The software came in three versions with different features for debiasing in order to test its influence on the appraised values. It turned out that the probands who used the decision support version of the software produced significantly less dispersed market values than the others.

Keywords Appraisal bias · Anchoring heuristic · Debiasing · Decision support system · Valuation variation

1 Introduction

There are many ways to improve property valuations. One way is to improve the *automated valuations*, which is the topic of this book. Another way is to improve the *manual valuations* by technical means, which is the topic of this chapter. We claim that supporting the valuer's decision-making with appropriate decision-making aids is a key to better manual valuations. Here we see great potential for improvement because many insights from modern decision theory and

C. Lausberg (✉) · A. Dust
Institute for Real Estate Information Technology, Nürtingen-Geislingen University, Geislingen/Steige, Germany
e-mail: carsten.lausberg@hfwu.de

A. Dust
e-mail: anja.dust@googlemail.com

M. d'Amato and T. Kauko (eds.), *Advances in Automated Valuation Modeling*,
Studies in Systems, Decision and Control 86, DOI 10.1007/978-3-319-49746-4_18

computer science have not yet been applied to the practice of real estate appraisal. These insights center on the cognitive and judgmental limitations of human beings and how machines can help to counteract and possibly overcome them.

Our work was triggered by a survey which we carried out in 2013 among German valuers. It turned out that (1) many valuers do not realize that they make decisions, and (2) the tools they use do not support decision-making. This was somewhat surprising as it is not a new insight that appraisal requires judgment (i.e., making a decision after careful thinking), and valuation software has been on the market for decades. When we looked deeper into the matter we discovered some more phenomena, for instance, that valuation textbooks often ignore the issue of decision-making and that the technological gap between software to support manual valuations and their "cousins" AVMs has widened dramatically in recent years. None of the usual valuation tools contains more than basic decision support functionalities, not to mention any form of artificial intelligence. Leading software packages like Argus or Cougar possess sophisticated information handling abilities, e.g., financial modelling, data analysis, and plausibility checks, but they do not help the valuer to decide which data source to trust or which comparables to use.

To test our assumption that a decision support system (DSS) can improve valuations we developed a software and set up an experiment which we will describe after a literature survey in part 3 and 4 of this chapter. It should be regarded as work in progress since we intend to repeat the experiment in two more settings and combine it with our other research on DSS.

2 Literature Review

Our work is based on three distinct streams within the real estate valuation literature:

1. In 1985 Hager and Lord (1985) started a discussion on valuation accuracy and variance which led to the concept of the "margin of error".[1] It is generally accepted that different appraisers come to different results, but that the variance should be kept to a minimum.
2. Later many studies were undertaken to identify the reasons for valuation variation. Many of them looked into behavioral issues of valuation and discovered the prominent role of cognitive biases such as the anchoring-and-adjustment heuristic, which was first studied in a real estate context by Northcraft and Neale (1987).[2]

[1]See Crosby et al. (1998) and Babawale and Omirin (2012) for a review of the literature in this field.

[2]For a literature review see, for example, Amidu (2011).

3. In recent years many authors addressed the question how technology can help in debiasing. Drawing on findings from computer science, psychology, and other fields, some of them—for instance Tidwell (2011)—suggest the use of decision support systems.[3]

2.1 Valuation Variance

Real estate valuation is often considered to be both a science and an art. For a long time this view implied that a certain degree of variation between two valuations of a property is natural. This hindered an objective analysis of the driving forces of valuation variance. It was not before the mid-1980s that researchers started to quantify the level of variation (Crosby et al. 1998, p. 318). In the beginning this was done with fairly simple descriptive statistics, later with more sophisticated measures (Shiller and Weiss 1999).

Babawale and Omirin (2012) classify the reasons for valuation variance into seven groups: characteristics of the property market, data problems, different value definitions, integrity issues, client pressure, absence of valuation standards, different market expertise, and finally individual disparities in skills, experience, and judgment. In our article we focus on the latter, being aware that improving the ability of valuers to judge can only have a limited effect on the total valuation variance.

2.2 Anchoring

Cognitive biases are systematic deviations in judgment. They often arise from heuristics which human beings use to compensate their limited information processing capabilities. The anchoring-and-adjustment heuristic, which was introduced by Tversky and Kahnemann (1974), is particularly well researched. The term describes the effect that people—consciously or not—use a starting point when making estimates and then adjust this number for the final answer. Numerous studies have proven the importance of the anchoring effect for many different domains.[4] Real estate-related decisions seem particularly prone to the anchoring effect because here, (1) a market value is not objectively determinable and (2) high information requirements encounter low information availability (Northcraft and Neale 1987; Tidwell 2011).

In real estate the research started with Northcraft and Neale (1987). Other influential research was conducted by Diaz, Gallimore, Hansz, Havard, and Wolverton—to name but a few (Diaz 1997; Diaz and Hansz 1997; Diaz and Hansz

[3]For a literature review see Tidwell (2011) and Bhandari et al. (2008).

[4]For a general literature survey see Furnham and Boo (2011).

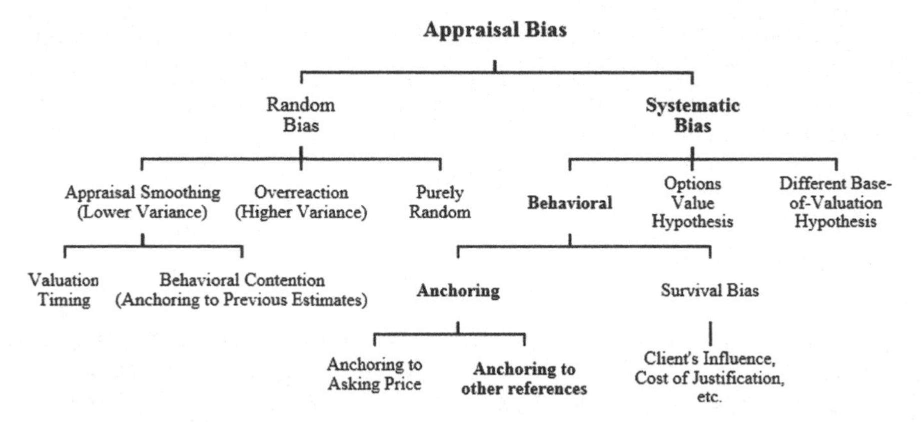

Fig. 1 Appraisal bias according to Yiu et al. (2006)

2001; Havard 1999; Gallimore 1994; Gallimore and Wolverton 1997). They found evidence for several reference points that may anchor property appraisals. Some of them, such as the transaction price of the property or comparable properties, are acceptable because they contain relevant information, whereas others, such as the value opinion of the owner, are not (Diaz and Hansz 2001). Figure 1 shows that the anchoring effect is one of many appraisal biases. Other biases arise, for example, from client pressure or the tendency of appraisers to smooth their valuations.

In the Northcraft-Neale-Experiment both experts and novices anchored their value opinion to the listing price of the property. Since then it has been debated whether there is a link between the professional experience and the influence of anchors. Cypher and Hansz (2003), for instance, showed that experts differ between meaningful anchors, such as the opinions of other experts in case they are not familiar with the local market, and anchors which are fundamentally inappropriate.

Another area for ongoing research is debiasing, i.e., the measures that can be taken to minimize the unsolicited effects of the appraisal bias. The basic options are to modify the environment, for instance by improving the information situation, or the decision maker, for instance by training, feedback, warnings, or technical support (Fischhoff 1982; Larrick 2004; Soll et al. 2015). Debiasing strategies can further be divided into …

- motivational strategies which try to stimulate the decision maker to put more effort into the decision-making process; (Lopes 1987)
- cognitive strategies which aspire to increase the cognitive skills of the subjects; (Larrick 2009)
- technological strategies which try to improve decisions by technical aids, regardless of the cognitive ability of decision makers (Larrick 2004).

DSSs are an example for a technical aid. To our knowledge they are not available for property valuation yet, with the exception of tools for the analysis of market and property information (Tidwell 2011, pp. 3–4).

2.3 Decision Support Systems

A DSS "is a computerized aid designed to enhance the outcomes of an individual's decision-making activities" (Singh 1998, p. 145). There are many varieties, ranging from simple tools which facilitate the entry, analysis, or interpretation of data to complex systems of artificial intelligence, such as expert systems and artificial neural networks. They have a long tradition in industries such as banking (Arnott and Pervan 2005; Bhandari and Hassanein 2004), and even in real estate DSS are used, for instance for management tasks (business intelligence systems) and location analysis (spatial DSS).

Most systems have been created to counteract the human information processing deficits, only a few to correct the decision-making biases (George et al. 2000, pp. 195, 26). Bhandari et al. (2008) gave one example for the latter. They demonstrated that their simple investment DSS can minimize occurring heuristics. Tidwell followed a different path and used the existing information service CoStar. In his experiment he was able to show that the availability of meaningful data may result in reduced cognitive effort and thus less reliance on cognitive simplification mechanisms (Tidwell 2011, p. 4).

In our experiment we transformed a standard Excel spreadsheet into a DSS by integrating several features that—according to the literature, e.g., (George et al. 2000; Kleinmuntz and Schkade 1993)—are useful to reduce the anchoring effect and other biases: input assistance, process orientation, verbal and quantitative data analysis, plausibility checks, help texts, and different kinds of information display (graphs, tables, emoticons).

3 Research Hypotheses and Research Method

Our main research question was whether the appraisal bias can be reduced with a decision support system. For this purpose we set up an experiment in which probands were given the task to determine the market value of an office building, based on a set of documents and with the help of a self-written valuation software.

3.1 Construction of the Experiment

We chose to invent a fictitious building so it could perfectly correspond to a set of criteria for an average property and would be fairly easy to appraise. It was important to minimize the influence of the test persons' individual knowledge differences on the results, but still leave enough leeway to justify dispersed market values.

- Type of use: office
- Moderate size: five units with a total rentable area of 1,400 m^2
- Tenants with high creditworthiness: four law firms
- Average lease expiry: one unit vacant, four units with expiries between one and nine years
- Average age: ten years old, very well-kept, high standard in terms of energy efficiency and other technologies
- Average location: large city (Nuremberg), but not one of the "Big 7" German real estate markets; central, but not prime location
- A few special features that may separate inexperienced from seasoned appraisers: for instance proximity to the courthouse (=perfect location for law firms) and small outdoor facilities (=lower than usual maintenance costs)

The probands received an information package which consisted of six parts:

- General information on the property (year of construction, rentable area, rent roll, book value, picture of the building, floor plan)
- A table with distances to the courthouse and other important locations
- A map extract with average land values
- Handwritten notes from the site inspection with the owner
- Market reports from different sources
- Benchmarks for maintenance costs, useful life, and other data

As the anchor we chose the number of inhabitants of Nuremberg and the book value of the property. Both figures did not contain any relevant information for the market value. We deliberately chose low anchors because Hansz and Diaz found out that appraisers fear "too low" valuation judgments more than "too high" judgments. In other words, there may be a natural tendency towards higher values, which we could not control in our experiment Hansz and Diaz (2001, p. 563).

The valuation software is a Microsoft Excel file consisting of three sheets: a cover page with some basic explanations, the data entry and calculation sheet, and a questionnaire to collect some personal data. The calculation is limited to the "Ertragswertverfahren", the German version of the income approach, which is suitable for our case and known to any German valuer. We developed three versions of the software that differ in their degree of decision support:

- Standard (= no support for identifying anchors): In this version the appraiser transfers the figures from the information package to the calculation sheet, either directly or after some mental arithmetics. No hints are given to the nature of anchoring or the possible anchors.
- Modified (= little support for identifying anchors): This version contains a written warning that the appraised value may be too close to the book value and a short explanation of the anchoring bias. After that the test person has the opportunity to adjust the market value with a sliding switch.
- Decision support system (= all-round support for performing the appraisal task): This version has several features typical for a DSS. Example: Before the user

can type in the estimated market rent she first has to evaluate the quality of the various market reports. Furthermore she is offered a graphical comparison of the market rents and a written recommendation, made up of text blocks.

3.2 Hypotheses

To answer our research question we hypothesized that if the appraiser is debiased and supported in his decisions by an appropriate valuation system the anchoring effect and the valuation variation are reduced. In a more formal way our hypotheses can be written as:

H_1 On average the unadjusted market values are smaller (= closer to the anchor) than the adjusted market values after the warning message.

H_1 The variation of the market values decreases with increasing decision support.

It should be mentioned that we did not intend to measure the anchoring effect itself because that was convincingly done in previous experiments by other authors. Instead, we regard it as an indicator for the anchoring effect if an appraiser adjusts the market value after reading the warning message.

4 Results

4.1 Description of the Sample

For meaningful results we aimed at a sample size of 90, i.e., 30 probands per software version and 45 experts/students. We recruited the experts via random sampling from the membership rosters of RICS (Royal Institution of Chartered Surveyors) and BIIS (Association of Investment Property Valuers). To enhance the response rate we also used personal contacts. The student sample was mainly recruited in a fourth-semester bachelor course in property valuation at Nürtingen-Geislingen University.

A total of 289 real estate experts were contacted in June and July 2014. The email included a cover letter, the information package, and—randomly chosen, but equally distributed—one of the three versions of the valuation tool. After a follow-up by e-mail and telephone 44 experts had participated in the experiment (=15.2 % response rate). In addition 54 students were contacted of which 47 participated. The high response rate of 87 % in this sub-sample was achieved because the experiment was partly conducted during a lecture. After eliminating two incomplete questionnaires a total of 89 data sets remained which were roughly equally distributed over the sub-samples (see Table 1).

The sample of the experts can be described as follows: About half of the test persons are under 30 years old and have less than five years of professional

Table 1 Number of participants in the sub-samples

	Total	Experts	Students
Standard version	28	13	15
Modified version	28	13	15
DSS version	33	17	16
Total	89	43	46

experience. 41 % of the respondents have completed property-related academic studies. This points to a potential bias in the data which can be attributed to the recruitment of some of the experts from the group of former students of our university. We assume that the probands in our sample are *on average* significantly younger than the basic population of German property appraisers, have a higher affinity to information technology, greater academic knowledge, and lower market expertise. However, in the data we could not find any evidence that this could have distorted the findings.

The sample of students includes only participants under the age of 30 who have mostly no work experience. Accordingly, they possess a low knowledge of the Nuremberg real estate market and the office market in general. In contrast to the students most experts have a high to very high knowledge of the office markets and a low to mediocre knowledge of the Nuremberg real estate market.

4.2 Testing the Hypotheses

Hypothesis 1 Table 2 shows that 3 experts and 9 students adjusted the market value after reading the warning notice and that the adjustment was equally used in both versions. However, in four cases the direction of the adjustment was negative, i.e., towards the anchor, which is illogical. Maybe the warning message was not explicit enough or was misunderstood. Table 2 further shows that the average adjustment was fairly low: the mean of the adjusted market values of MOD and DSS was only 0.86 % higher than the unadjusted market values.
Hence, the first null hypothesis

$$H_0 : \text{Mean}(\text{UnadjMV}_{\text{MOD}} + \text{UnadjMV}_{\text{DSS}}) \geq \text{Mean}(\text{AdjMV}_{\text{MOD}} + \text{AdjMV}_{\text{DSS}}) \quad (1)$$

is accepted if all valuations are counted and rejected if the invalid valuations are eliminated. Although the effect is small, we conclude that at least some test persons were prone to the anchoring and adjustment effect.

Hypothesis 2 In the next step we analyze the effect of the decision support functionalities. As variation can be measured with different ratios, the null

Table 2 Adjustments of market values

		Mean market values		Adjustments	
		Unadjusted (EUR)	Adjusted (EUR)	Magnitude (% of MV)	Number (and %) of all valuations
All valuations	Experts	2,595,573	2,611,865	0.63	3 (7)
	Students	2,661,213	2,643,890	−0.65	9 (20)
	Modified	2,587,998	2,566,537	−0.83	6 (21)
	DSS	2,663,662	2,680,409	0.63	6 (18)
	Mean	2,625,830	2,623,473	−0.09	
Only valuations with correct adjustment	Modified	2,521,724	2,543,115	0.85	3 (12)
	DSS	2,671,711	2,694,997	0.87	5 (18)
	Mean	2,596,718	2,619,056	0.86	

Table 3 Variation measures

	Standard	Modified	DSS
Range (min/max/%)	1.62/4.22/161	1.62/3.6/123	2.21/3.52/59
Standard deviation	0.50	0.50	0.32
Variation coefficient (%)	18.7	19.5	12.0

hypothesis shall be rejected if the *majority* of the measures show the lowest variation for STD and the highest variation for DSS.

$$H_0 : \text{Variation}\,\text{MV}_{\text{STD}} \leq \text{Variation}\,\text{MV}_{\text{MOD}} \leq \text{Variation}\,\text{MV}_{\text{DSS}} \quad (2)$$

We used three simple variation measures: range, standard deviation and variation coefficient. Table 3 shows that the market values in the STD version range between EUR 1.62 million to EUR 4.22 million (=161 %), which is much higher than in both the MOD version (123 %) and the DSS version (59 %). The standard deviation is almost the same in STD and MOD (EUR 0.5 million), and significantly lower in DSS (EUR 0.32 million). The variation coefficient confirms these findings, although it is slightly lower in STD than in MOD.

Figure 2 displays the results of the total sample in the form of a frequency distribution of variations from the mean. It is easy to see that the test persons using the DSS version produced more accurate market values than the two other versions, i.e., the values were more often close to the mean and less frequently extreme outliers. We conclude that a simple warning message is not sufficient to improve the valuation quality; however, there is clear evidence to support our assumption that a decision support system can effectively reduce valuation variation.

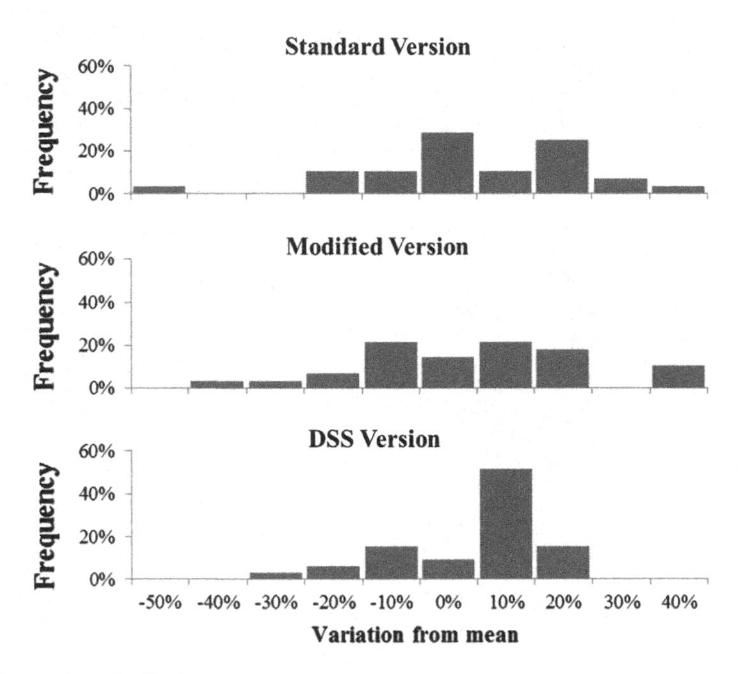

Fig. 2 Frequency distributions of outcomes

Table 4 Significance test

Hypothesis	Test-statistic	p-value
H_0: $Var_{STD} = Var_{MOD} = Var_{DSS}$	2.8799	0.0618*
H_0: $Var_{STD} = Var_{DSS}$	3.3743	0.0714*

Note: * Significant at the 5% level.

To evaluate the significance of the results we used the Brown-Forsythe-Test (1974).[5] The null hypothesis that the variances are homogeneous cannot be rejected at the 0.05 level of significance (Table 4).

A closer look at the data yielded some more interesting results. For example we tested whether the decision support is more beneficial for novices than for experienced valuers. To our surprise this was not the case. In the student sample the MOD version produced the most dispersed results, while the STD and the DSS versions were almost equal. In the expert sample the results were similar to the total sample, i.e., the DSS clearly had the lowest variation.

Two more hypotheses aimed at explanations for the reduced variation. Firstly, the DSS version required more reading and more data inputs, thus forcing the valuer to spend more time on decision-making. Indeed the average processing time increased from 18:25 min (STD) to 24:22 min (DSS). Secondly, the DSS required the proband to compare the market data sources and make an explicit judgment.

[5]We adjusted the test for small and different sized samples with the method of Noguchi and Gel (2009).

This significantly reduced the variation of the market rent in the DSS version (standard deviation EUR 0.87 compared to EUR 1.58 for the STD and MOD versions).

5 Conclusions

Our study demonstrated that a DSS can help to reduce valuation variation. This was expected on grounds of previous research from different sides: valuation, decision-making, and informatics. But the results did not meet all our expectations; the warning message, for instance, was not very effective. For more robust results the experiment should be replicated.

- with other properties in other countries,
- using other valuation methods and an improved software,
- covering other biases and incorporating other debiasing features,
- with a larger sample of valuers from more diverse backgrounds,
- employing more sophisticated measures of valuation variance and accuracy.

It can be expected that further research in this direction will yield more interesting results because the combination of behavioral real estate research and computer science is still in its infancy. To get an impression of what is possible if contemporary technology is used we refer to Argiolas et al. (2010) who give an example of how different technologies can work together to support decision-making in real estate.

Furthermore we believe that traditional appraisals would greatly profit from integrating techniques that have proven successful in AVMs, such as statistical analysis of market data, and risk valuation, such as Monte Carlo simulation, because they provide valuable decision support for the knowledgeable user (Downie and Robson 2008; Lausberg 2013). As Amidu (2011) points out, the crucial point is that not every valuer is knowledgeable; in the future valuation training should include more statistical training and cover behavioral issues.

References

Amidu, A. R. (2011). Research in valuation decision making processes: Educational insights and perspectives. *Journal of Real Estate Practice and Education, 147*(1), 19–30.

Argiolas, M., Dessì, N., Marchi, G., & Pes, B. (2010). Real estate decision making processes and web-based applications: An integrated approach. In G. Phillips-Wren et al. (Eds.), *Advances in intelligent decision technologies* (Vol. 4, pp. 329–338). Berlin, Heidelberg: Springer.

Arnott, D., & Pervan, G. (2005). A critical analysis of decision support systems research. *Journal Information Technology, 20*(2), 67–87.

Babawale, G., & Omirin, M. (2012). An assessment of the relative impact of factors influencing inaccuracy in valuation. *International Journal of Housing Markets and Analysis, 5*(2), 145–160.

Bhandari, G., & Hassanein, K. (2004). A cognitive DSS for investment decision making: Challenges & opportunities. McMaster University Working Paper. http://hdl.handle.net/11375/5329

Bhandari, G., Hassanein, K., & Deaves, R. (2008). Debiasing investors with decision support systems: An experimental investigation. *Decision Support Systems, 46*(1), 300–410.

Brown, M. B., & Forsythe, A. B. (1974). Robust tests for equality of variances. *Journal of the American Statistical Association, 69*(1), 364–367.

Crosby, N., Lavers, A., & Murdoch, J. (1998). Property valuation variation and the 'margin of error' in the UK. *Journal of Property Research, 15*(4), 305–330.

Cypher, M. L., & Hansz, J. A. (2003). Does assessed value influence market value judgments? *Journal of Property Research, 20*(4), 305–318.

Diaz, J., III. (1997). An investigation into the impact of previous expert value estimates on appraisal judgment. *Journal of Real Estate Research, 13*(1), 57–66.

Diaz, J., III, & Hansz, J. A. (1997). How valuers use the value opinions of others. *Journal of Property Valuation & Investment, 15*(3), 256–260.

Diaz, J., III, & Hansz, J. A. (2001). The use of reference points in valuation judgment. *Journal of Property Research, 18*(2), 141–148.

Downie, M. L., & Robson, G. (2008). Automated valuation models: An international perspective. In: RICS Automated Valuation Models Conference: AVMs Today and Tomorrow, 4 November 2008, London. http://nrl.northumbria.ac.uk/1683/

Fischhoff, B. (1982). Debiasing. In D. Kahnemann, P. Slovic, & A. Tversky (Eds.), *Judgement under uncertainty: Heuristics and biases* (pp. 422–444). Cambridge: Cambridge University Press.

Furnham, A., & Boo, H. (2011). A literature review of the anchoring effect. *The Journal of Socio-Economics, 40*(1), 35–42.

Gallimore, P. (1994). Aspects of information processing in valuation judgment and choice. *Journal of Property Research, 11*(2), 97–110.

Gallimore, P., & Wolverton, M. L. (1997). Price knowledge induced bias: A cross-cultural comparison. *Journal of Property Valuation and Investment, 15*(3), 261–273.

George, J., Duffy, K., & Ahuja, M. (2000). Countering the anchoring and adjustment bias with decision support systems. *Decision Support Systems, 29*(2), 195–206.

Hager, D. P., & Lord, D. J. (1985). The property market, property valuations and property performance measurement. *Journal Institute Actuar, 112*(01), 19–60.

Hansz, J. A., & Diaz, J., III. (2001). Valuation bias in commercial appraisal: A transaction price feedback experiment. *Real Estate Economics, 29*(4), 553–565.

Havard, T. (1999). Do valuers have a greater tendency to adjust a previous valuation upwards or downwards? *Journal of Property Investment & Finance, 17*(4), 365–373.

Kleinmuntz, D., & Schkade, D. (1993). Information displays and decision processes. *Psychological Science, 4*(4), 221–227.

Larrick, R. (2004). Debiasing. In D. Koehler & N. Harvey (Eds.), *Blackwell handbook of judgment and decision making* (pp. 316–337). Malden: Blackwell Publishing.

Larrick, R. (2009). Broaden the decision frame to make effective decisions. In E. Locke (Eds.), *Handbook of principles of organizational behavior. Indispensable knowledge for evidence-based management* (pp. 461–480). Hoboken: Wiley.

Lausberg, C. (2013). Economic scenarios for the real estate market: Incorporating uncertainty and risk in real estate appraisals. Aestimum, August, pp. 427–442.

Lopes, L. (1987). Procedural debiasing. *Acta Psychologica, 64*(2), 167–185.

Noguchi, K., & Gel, Y. R. (2009). *Combination of Levene-type tests and a finite-intersection method for testing equality of variances against ordered alternatives.* Working paper, Waterloo University of Waterloo–Department of Statistics and Actuarial Science.

Northcraft, G. B., & Neale, M. (1987). Experts, amateurs, and real estate: An anchoring-and-adjustment perspective on property pricing decisions. *Organizational Behavior and Humans Decision Processes, 39*(1), 84–97.

Shiller, R. J., & Weiss, A. N. (1999). Evaluating real estate valuation systems. *The Journal of Real Estate Finance and Economics, 18*(2), 147–161.

Soll, J. B., Milkman, K. L., & Payne, J. W. (2015). A user's guide to debiasing. In G. Keren & G. Wu (Eds.), *Wiley-Blackwell handbook of judgment and decision making. Chichester: John Wiley & Sons.*

Thomassin Singh, D. (1998). Incorporating cognitive aids into decision support systems: The case of the strategy execution process. *Decision Support Systems, 24*, 145–163.

Tidwell, O. A. (2011). *An investigation into appraisal bias: The role of decision support tools in debiasing valuation judgments*. Georgia State University–Department of Real Estate–Real Estate Dissertations. Atlanta.

Tversky, A., & Kahnemann, D. (1974). Judgment under uncertainty: Heuristics and biases. *Science, 185*(4157), 1124–1131.

Yiu, C. Y., Thang, B., Chiang, H., & Choy, T. (2006). Alternative theories of appraisal bias. *Journal of Real Estate Literature, 14*(3), 321–344.

An Application of Regressed Discounted Cash Flow as an Automated Valuation Method: A Case in Bari

Maurizio d'Amato and Yener Coskun

Abstract The application of automated valuation methodology (AVM) procedure to income approach normally deals with direct capitalization. This happens although the great diffusion of discounted cash flow (DCF) analysis. The main objectives of paper are twofold: first, we aim to propose an AVM procedure based on the relationship between the DCF inputs and outputs. Second, we seek to determine discount rate and local risk premium in the case of Bari commercial market The study also refines discussions on risk premium factor in the regressed DCF application. The study also and identifies the room for enhancing the suggested methodology. The solution proposed is the model A of Regressed DCF (d'Amato and Kauko 2012).

Keywords Automated valuation methodology · Discounted cash flow analysis · Regressed DCF · Commercial property

1 Introduction

The work provides the application of the regressed discounted cash flow (DCF) analysis (d'Amato and Kauko 2012) to a small sample of commercial properties in Bari real estate market. This methodology can be used in different ways. The first one is a further proposed automated valuation method (AVM) based on discounted cash flow (DCF) analysis instead of direct capitalization. The

---

The paper has been written in strict cooperation between the two authors. Therefore the credit of the article should be equally divided between them.

M. d'Amato (✉)
Technical University Politecnico Di Bari, Bari, Italy
e-mail: madamato@fastwebnet.it

Y. Coskun
Capital Markets Board of Turkey & Izmir University of Economics, İzmir, Turkey
e-mail: ycoskun@spk.gov.tr

M. d'Amato and T. Kauko (eds.), *Advances in Automated Valuation Modeling*,
Studies in Systems, Decision and Control 86, DOI 10.1007/978-3-319-49746-4_19

application of the method may be also helpful for discount rate and risk premium determination. In this case, the valuer will be able to calculate both discount rate and risk premium using this methodology. The choice of the city of Bari is motivated by the availability of data and the desire to test this kind of method in a small property market with evident data constraints. In this way the method provides an estimation of risk premium and discount rate based on few data. This paper is organized as follows. In the next section, we analyze theoretical background. Following section involves data construction, methodology, and outcomes of regressed DCF application. And the last section is reserved the final remarks.

2 Theoretical Background

2.1 AVM, DCF, and Regressed DCF

Automated valuation methods are an important and growing field in literature and professional activity. In the RICS Information Paper, AVM is defined as "*...one or more mathematical techniques to provide an estimate of value of a specified property at a specified date, accompanied by a measure of confidence in the accuracy of the result, without human intervention post-initiation*" (RICS 2013, p. 5). This mathematical modeling can be used as a valuation tool in the field of mortgage origination. According to Basel II they may represent a method to periodically appraise properties used as bank guarantees. It is also possible to use AVM in the valuation of real estate portfolio. The International Assessing Officers Association (IAAO) has provided a *Standard on Automated Valuation Models*. The methods exposed in these standards are based on the international framework of real estate appraisal techniques: market, income and cost approaches. Income approach relies on the traditional direct capitalization formula (IAAO 2003, pp. 9–10). The formulas (1) and (2) illustrated the AVM model included in the IAAO standards:

$$MV = NOI/R \tag{1}$$

where MV is the market value, NOI is the Net Operate Income and the R is the overall capitalization rent. The second formula indicated in the IAAO standards is the following number 2:

$$MV = GI \cdot GIM \tag{2}$$

where GI is the gross income and the GIM is the gross income multiplier. In this chapter a further approach is proposed in order to create AVM models based on DCF analysis, called Regressed DCF. Before discussing Regressed DCF, it may be useful to recall the concept of DCF "*...where a discount rate is applied to a series of cash flows for future periods to discount them to a present value...*" (IVSC 2013,

IVS Framework para 60). DCF can be considered as a methodology to estimate the valuation of a property and a real estate investment project assessment (Baum et al. 1996). The method consists in discounting several cash flows related to an interval of time called holding period. After this period a direct capitalization (scrap value, going out value) will be summed up to the previous value. Several contributions analysed different aspects of this method like: reliability of cash-flow analysis (Willison 1999); the influence of vacancy and market analysis (Rabianski 2002); the sensitivity of inputs on the final output (Taylor and Rubin 2002; Wheaton et al. 2001; Hendershott and Hendershott 2002). This methodology is also widely used for assessment of worth. In this case the application of Discounted Cash Flow Analysis is useful to determine the value created by the real estate investment project. Regressed DCF explores a deterministic relationship between inputs and outputs of DCF in order to propose an automatic valuation method and a tool to determine a "local" discount rate and risk premium. In this model the dependent variable will be the prices of comparable and the financial characteristics, such as net operate income, going in cap rate, discount rate and going out cap rate used for reversion calculation taken from previous valuations or real transaction data, will be the independent variables. The group of models has been recently proposed and tested (d'Amato and Kauko 2012; d'Amato and Anghel 2012; d'Amato 2017) using a limited number of comparables. In the real estate market the lack of data is one of the problem often raised when applying methodologies, this is particularly true either in emerging markets (Renigier-Biłozor et al. 2014a, b) or in property markets having a limited number of transactions.

3 Defining Discount Rate and Risk Premium

Defining optimal level of risk premium is critical in determining the value and hence valuation of both financial and non-financial assets. However it is out of the scope of this paper, we may note that finance theory reveals several discussions on the determination of risk premium. In this respect, final output of Capital Asset Price Method (Sharpe 1964) or Arbitrage Price Theory (Ross 1976) is a measure of risk premium for areas with limited property market segmentation. There are also several attempts in the market practice to define local risk premium. For example, recently the Asset Quality Review manual required discount rates used in hope value calculations should reflect *local market risk premia* (Asset Quality Review Phase II Manual Paragraph 5.6). Therefore the role of analysis improving the efficiency of local risk premium determination can be considered strategic for valuation purposes. The local dimension of valuation is essential in order to consider property market segmentation (d'Amato 2010)

For the purpose of our study, the application of regressed DCF may be helpful to provide a single point estimate but also in the determination of discount and risk premium rate. According to International Valuation Standards 2013 (IVS 230 Valuation Real Property Interest), "*C.19 The yield or discount rate discussed above will be determined by the objective of the valuation. If this is to establish the value to a particular owner or potential owner based on their own investment criteria, the rate used may reflect their required rate of return or the weighted average cost of capital. If it is to establish the market value, the rate will be derived from observation of the returns implicit in the price paid for real property interests traded in the market between market participants*" (IVSC 2013, p. 67). The same standard defines (IVS 230 Valuation Real Property Interest) "…C.20 *The appropriate discount rate should be determined from analysis of the rates implicit in transactions in the market. Where this is not possible an appropriate discount rate may be built up from a typical risk free return adjusted for the additional risks and opportunities specific to the particular real property interest…*" (IVSC 2013, p. 67). It is possible to distinguish two possible alternatives to determine discount rate for real estate valuation purposes. The former is the selection of internal rate of return related to "*traded in the market between participants*" or "*transactions in the market*". In this case the selection of an appropriate comparable is essential. The latter will be the sum between a risk free return and additional risk premium (Hoesli and Macgregor 2000). In the second case risk premium can be determined in a subjective way or using and additive model to take into account different kinds of risk.

4 Data and Methodology

The sample consists in a small number of offers and transactions because of the difficulties in collecting data (Ciuna et al. 2014a, b). The data are real estate price and offers collected in the commercial real estate market of Bari. The data refers to 14 commercial shops whose surface varied between the 200 and 400 m^2. The data have been collected among three real estate broker agencies that made available some real transactions and asking prices in a peripherical area of Santa Caterina in Bari, south east of Italy. They also provide a forecast in time on market discussing about a possible variability of price. This variability of price is caused by the economic crisis. The relation between DCF outputs and inputs will be tested on this small sample in order to see how it works verifying it also with mean absolute percentage error and the coefficient of dispersion.

The proposed methodology is composed by three groups, previously defined A, B and C (d'Amato and Kauko 2012). In this work, we will apply only the "Model A". In a further application of this model it will be possible to realize risk premium maps (see, d'Amato 2010, 2017; d'Amato 2015, 2004; d'Amato and Siniak 2008; d'Amato and Kauko 2008). In this contribution the Regressed Model A will be applied to a small sample in Italy in the city of Bari. In particular the model A provides the following mathematical relationship between inputs and outputs of a DCF.:

$$PPRICE = DCF(Y) = LOC \cdot NOI^{b_1} \cdot \frac{1}{(GICR \cdot DR)^{b_2}} \cdot DAT^{b_3} \quad (3)$$

In the Formula (3) it is possible to see the dependent variable the PRICE or the value (DCF(Y)), the independent variables are LOC as the constant term or location, NOI as Net Operate Income, GICR is the going in capitalization rate while DR is the discount rate and finally the DAT is the data. The introduction of the variable DAT is new, because in the previous samples used for the application this kind of information was not available. The terms b_1, b_2, and b_3, are the coefficients. In log terms the equation is the following:

$$\ln(PRICE) = \ln(LOC) + b_1 \ln(NOI) - b_2 \ln(GICR \cdot DR) + b_3 \ln(DAT) \quad (4)$$

5 An Application of Model a in Bari

As indicated, the observations are located in an urban area of Bari, an area in the south east of Italy, Santa Caterina. The sample list involves 14 observations on commercial properties consisting of both final transaction price and asking price (offers) (Table 1). Because of the great difficulty to find the data it was not possible to select a greater number of comparables data. The final transaction prices and the asking price in euro are indicated with the acronym PRICE (DCF) in the first line and Price in the second line of the second column. The contractual rent is indicated in the Table 1 as Net Operate Income in the first line of the third column and NOI in the second line of the third column. The going in cap rate that can be used in a direct capitalization, is indicate with the term Going in Cap Rate in the first line of the fourth column and the acronym GICR in the second line of the fourth column. The cap rate has been calculated on the gross income. The discount rate is indicated in first line of the fifth column and as DR in the second line of the fifth column. The last sixth column indicates the DATE calculated in number of days to the moment of valuation. This is the first case in which it is possible to introduce this variable in the model.

In the Table 2 it is possible to observe descriptive statistics of the sample.

In the Table 3 the data are organized in order to apply the model A of regressed DCF.

We collect both final transaction price and asking prices of a specific property to understand the volatility of the final transaction price. As a consequence, a further step on the variability of the price has been applied. Consulting three agencies in the area the price included in the model has been adjusted for almost all the asking

Table 1 A sample of 14 DCF inputs in Bari

	Price (DCF)	Net operate income	Going in cap rate	Discount rate	Date
Nr.	Price	NOI	GICR	DR	DAT
1	€2,370,000.00	€133,000.00	0.056	0.069	2
2	€2,000,000.00	€100,000.00	0.050	0.062	3
3	€3,650,000.00	€225,000.00	0.062	0.070	100
4	€2,550,000.00	€178,000.00	0.070	0.090	20
5	€1,000,000.00	€65,000.00	0.065	0.072	22
6	€1,250,000.00	€92,000.00	0.074	0.078	50
7	€8,560,000.00	€570,000.00	0.067	0.072	100
8	€4,120,000.00	€250,000.00	0.061	0.065	15
9	€2,228,000.00	€162,000.00	0.073	0.078	20
10	€3,850,000.00	€290,000.00	0.075	0.092	30
11	€320,000.00	€22,000.00	0.069	0.075	25
12	€1,150,000.00	€90,000.00	0.078	0.080	60
13	€700,000.00	€48,000.00	0.069	0.075	10
14	€520,000.00	€37,000.00	0.071	0.076	14

Table 2 Descriptive statistics of the sample

	Mean	St. Dev	Min	Max
Price	€2,447,714.29	€2,154,528.52	€320,000.00	€8,560,000.00
NOI/year	€161,571.43	€143,369.73	€22,000.00	€570,000.00
GICR	0.07	0.01	0.05	0.08
DR	0.08	0.01	0.06	0.09
DAT	33.64	32.34	2.00	100.00

Table 3 Variables of the model

	Variables			
Nr.	Price	NOI	1/(GICR.DR)	DAT
1	€2,370,000.00	€133,000.00	258.25433148	2
2	€2,000,000.00	€100,000.00	322.58064516	3
3	€3,650,000.00	€225,000.00	231.74603175	100
4	€2,550,000.00	€178,000.00	159.17602996	20
5	€1,000,000.00	€65,000.00	213.67521368	22
6	€1,250,000.00	€92,000.00	174.19175028	50
7	€8,560,000.00	€570,000.00	208.57699805	100
8	€4,120,000.00	€250,000.00	253.53846154	15
9	€2,228,000.00	€162,000.00	176.32162077	20
10	€3,850,000.00	€290,000.00	144.30284858	30
11	€320,000.00	€22,000.00	193.93939394	25
12	€1,150,000.00	€90,000.00	159.72222222	60
13	€700,000.00	€48,000.00	194.44444444	10
14	€520,000.00	€37,000.00	184.92176387	14

Table 4 Adjustment for asking price

	Price—asking price variability	**Price adjusted**
Price	0	**€2,370,000.00**
Price	0	**€2,000,000.00**
Price	0	**€3,650,000.00**
Asking price	**−0.12**	**€2,244,000.00**
Asking price	**−0.23**	**€770,000.00**
Asking price	**−0.3**	**€875,000.00**
Price	0	**€8,560,000.00**
Asking price	**−0.24**	**€3,131,200.00**
Asking price	**−0.3**	**€1,559,600.00**
Price	0	**€3,850,000.00**
Price	0	**€320,000.00**
Price	0	**€1,150,000.00**
Price	0	**€700,000.00**
Price	0	**€520,000.00**

prices in negative way (Kaklauskas 2012; Kauko and d'Amato 2008; Kauko and d'Amato 2011). This is an important step because it allows the valuer to create a more realistic relationship between dependent and independent variables. The Table 4 shows these adjustments.

Price variation is higher for asking price than for final transaction price. They reflect the situation at the moment requiring the application of AVM. These adjustments have been done on the price and are related to the variability of selling price of the property. The Pearson correlation coefficient shows a strong correlation between the location (constant term) and the NOI (Net Operate Income) as indicated in the Table 5.

In the Table 6 there are the log of the variables including mean and standard deviation.

At the end of the Table 6, there are also the calculation of mean and standard deviation that may be helpful for the standardization, a statistical technique to give equal weight to all the variables of the model (Gelman 2007). The new variable z is calculated as in the Formula (5) below

$$z = \frac{x - \mu}{\sigma} \tag{5}$$

In the Formula (5), z is the transformed variable, x is the original variable (log of original data as in Table 4), μ is the arithmetic mean and σ is the standard deviation. The Table 7 shows the standardized variables.

Table 5 Pearson correlation coefficient

	LOC	NOI	1/(GICR*DR)	DAT
LOC	1	0.971	0.198	0.186
NOI	0.971	1	0.035	0.294
1/(GICR*DR)	0.198	0.035	1	−0.525
DAT	0.186	0.294	−0.525	1

Table 6 Log calculation

Nr.	PRZ	NOI/year	1/(GICR*DR)	DAT
1	14.6784	11.7981	5.55394488	0.693147181
2	14.5087	11.5129	5.776353167	1.098612289
3	15.1102	12.3239	5.445642081	4.605170186
4	14.6238	12.0895	5.070010697	2.995732274
5	13.5541	11.0821	5.364457169	3.091042453
6	13.6820	11.4295	5.160156706	3.912023005
7	15.9626	13.2534	5.340308268	4.605170186
8	14.9569	12.4292	5.535515534	2.708050201
9	14.2599	11.9954	5.172309718	2.995732274
10	15.1636	12.5776	4.971914206	3.401197382
11	12.6761	9.9988	5.267545708	3.218875825
12	13.9553	11.4076	5.073436195	4.094344562
13	13.4588	10.7790	5.27014649	2.302585093
14	13.1616	10.5187	5.219932838	2.63905733
Mean	€14.27	€11.66	5.30	3.03
St. Dev	€0.90	€0.88	0.22	1.14

Table 7 Standardized variables

Nr.	PRZ	NOI	1/GICR*DR	DAT
1	0.455	0.161	1.147	−2.038
2	0.267	−0.164	2.157	−1.684
3	0.934	0.759	0.655	1.380
4	0.394	0.492	−1.052	−0.026
5	−0.791	−0.654	0.286	0.057
6	−0.650	−0.259	−0.642	0.774
7	1.879	1.817	0.176	1.380
8	0.764	0.879	1.063	−0.278
9	−0.009	0.385	−0.587	−0.026
10	0.993	1.048	−1.497	0.328
11	−1.765	−1.887	−0.154	0.169
12	−0.347	−0.284	−1.036	0.934
13	−0.897	−0.999	−0.143	−0.632
14	−1.227	−1.295	−0.371	−0.338

Table 8 Regressed DCF as valuation of discount rate

Nr.	LOC	NOI	1/(GICR*DR)	DAT	ln (NOI)	Ln (PRICE)	Ln (GICR)	Ln (DAT)
1	−0.00000000000000031	0.97884628	0.1293149	−0.03413660	0.1607	0.46	−1.398	−2.038
2					−0.1638	0.27	−2.342	1.684
3					0.7590	0.93	−0.629	1.380
4					0.4924	0.39	0.388	−0.026
5					−0.6539	−0.79	−0.195	0.057
6					−0.2586	−0.65	0.822	0.774
7					1.8167	1.88	0.002	1.380
8					0.8789	0.76	−0.758	−0.278
9					0.3852	−0.01	0.722	−0.026
10					1.0478	0.99	1.011	0.328
11					−1.8867	−1.77	0.264	0.169
12					−0.2836	−0.35	1.324	0.934
13					−0.9989	−0.90	0.243	−0.632
14					−1.2951	−1.23	0.545	−0.338

A regression analysis has been carried out to define the AVM model. In the Table 8, the output of regression model is indicated.

	LOC	NOI	1/(GICR*DR)	DAT
Coefficients	−0.0000000000000031	0.9788462840856390	0.1293148585868640	−0.034136601
Adj R2	0.960868371			
F	99.21910178			
t—student	−0.00000000000006	16.57310737	1.949872188	−0.49221910342019
Eigenvalue	1.587	1.03	1	0.383
Conditional index	1	1.241	1.26	2.037

Therefore the model has been verified though the Mean Absolute Percentage Error indicated in below formula:

$$MAPE = \sum_{i=1}^{n} \frac{\left|\frac{PS_i - AS_i}{AS_i}\right| \cdot 100}{n} \tag{6}$$

In the above formula, MAPE is the mean absolute percentage error, PS means the predicted selling while the AS means the actual selling.

	PRZ	Application of the model	Destandardization	Predicted PRZ	Percentage error	Mean absolute percentage error
1	€2,370,000.00	0.375195723	14.6063932	€2,205,342.11	0.0695	0.1377
2	€2,000,000.00	0.176119498	14.42684494	€1,842,888.89	0.0786	
3	€3,650,000.00	0.780487092	14.97192838	€3,178,526.81	0.1292	
4	€2,244,000.00	0.346831976	14.58081174	€2,149,641.72	0.0420	
5	€770,000.00	−0.605098202	13.72225914	€910,964.44	0.1831	
6	€875,000.00	−0.362659379	13.94091644	€1,133,608.47	0.2956	
7	€8,560,000.00	1.753948626	15.84990026	€7,647,583.43	0.1066	
8	€3,131,200.00	1.007221367	15.17642164	€3,899,744.65	0.2454	
9	€1,559,600.00	0.302020823	14.54039625	€2,064,495.09	0.3237	
10	€3,850,000.00	0.82076596	15.00825618	€3,296,118.68	0.1439	
11	€320,000.00	−1.87250069	12.57917982	€290,447.99	0.0924	
12	€1,150,000.00	−0.443523188	13.8679848	€1,053,875.41	0.0836	
13	€700,000.00	−0.974682574	13.38892837	€652,736.21	0.0675	
14	€520,000.00	−1.304127031	13.09180007	€484,949.51	0.0674	

According to IAAO Standard on Ratio Study, the table below shows the calculation of Coefficient of Dispersion a ratio that helps to verify the approximation between the calculation of actual data and the output of the method.

Predicted Price (1)	Predicted Price/Actual Price (2)	Median Ratio (3)	(Predicted Price/Actual price) − Median Ratio (2–3)	Sum ((Predicted Price/Actual price) − Median Ratio) (4)	Average Absolute Deviation Sum ((Predicted Price/Actual price) − Median Ratio)/Number of Observation (5)	COD (5/3)
€2,205,342.11	0.930524099	0.931502202	0.000978103	1.574425456	0.112458961	0.120728605
€1,842,888.89	0.921444443		0.01005776			
€3,178,526.81	0.870829262		0.06067294			
€2,149,641.72	0.957950855		0.026448653			
€910,964.44	1.183070701		0.251568499			
€1,133,608.47	1.29555254		0.364050338			
€7,647,583.43	0.893409279		0.038092923			
€3,899,744.65	1.245447321		0.313945119			
€2,064,495.09	1.323733707		0.392231504			
€3,296,118.68	0.856134722		0.07536748			
€290,447.99	0.907649982		0.02385222			
€1,053,875.41	0.916413402		0.0150888			
€652,736.21	0.932480306		0.000978103			
€484,949.51	0.932595216		0.001093013			

It is possible to calculate also the discount rate for the area and the relative risk premium. Starting from the relationship

$$PRICE = DCF(Y) = LOC \cdot NOI^{b_1} \cdot \frac{1}{(GICR \cdot DR)^{b_2}} \cdot DAT^{b_3} \tag{7}$$

It is possible to rewrite the Formula (7) as in the Formula (8) below:

$$\ln(DR) = \frac{\ln(LOC) + 0.9712870 \ln(DAT) - \ln(PRICE) - 0.1532660 \ln(GICR)}{0.1532660} \tag{8}$$

In the Formula (8), DR means Discount Rate while LOC means Location Variable or constant term and NOI indicates Net Operate Income, PRICE the price of the property (but could be used also the value) and GICR the going in cap rate. Following previous works, the variable DAT has been introduced for the date in this paper. The Table 9 shows the application of the Regressed DCF to the model.

In the Table 9 the first column is the number of the observation, the second column represents the coefficient of instrumental variable LOC, the third column is the coefficient of variable NOI while the fourth column is related to the 1/(GICRxDR) variable the term for going in cap rate and discount rate proposed like as the other terms in the Formula (8). The fifth column indicates the coefficient of variable DAT. The remaining four columns indicates the log for several all the

Table 9 Regressed DCF as valuation of DR: empirical results

Nr.	DR standardized	DR destandardize	Predicted DR	Actual DR	Percentage error	Mean absolute percentage Error
1	−0.366310511	−2.632	0.071940936	0.069	0.042622259	0.15425928
2	−0.515974324	−2.648	0.070778136	0.062	0.141582837	
3	−1.211423974	−2.724	0.065616683	0.070	0.06261881	
4	0.295105873	−2.560	0.077312843	0.090	0.140968416	
5	1.35080333	−2.445	0.086730344	0.072	0.20458811	
6	2.040666933	−2.370	0.093495729	0.078	0.198663198	
7	−1.144955812	−2.717	0.066093275	0.072	0.082037851	
8	1.577179453	−2.420	0.08889461	0.065	0.367609386	
9	2.26943029	−2.345	0.09585373	0.078	0.228893973	
10	−0.84569991	−2.684	0.068282247	0.092	0.257801662	
11	−0.940229195	−2.694	0.06758307	0.075	0.098892395	
12	−1.036497237	−2.705	0.066878391	0.080	0.164020118	
13	−0.699349343	−2.668	0.06937901	0.075	0.074946533	
14	−0.77274557	−2.676	0.068826788	0.076	0.094384365	

variables excluding the location term. Therefore the mean absolute percentage error between actual discount rate and predicted discount rate is indicated in the Table 10.

Technical Information Paper and the International Valuation Standards indirectly defines the risk premium as the difference between the discount rate determined through the Regressed DCF and at the moment of valuation the risk free which has been assumed as 0.0195 (August 2015 emission of Italian BTP)[1] rent of the Italian gilt.

$$\ln(DR) = \frac{\ln(LOC) + 0.9712870\ln(NOI) - 0.0192925\ln(DAT) - \ln(PRICE) - 0.1532660\ln(GICR)}{0.1532660} \tag{9}$$

In the Table 11 it is possible to observe a comparison between the risk premium obtained by the 14 comparables and the risk premium predicted through Regressed DCF. The comparison is proposed assuming the following relationship:

Both the risk premium and the discount rate in a small are of a little town in the south east of Italy may be calculated as the mean of the data indicated in the previous Table 11.

[1]The emission is available at the following internet address http://www.dt.mef.gov.it/export/sites/sitodt/modules/documenti_it/debito_pubblico/risultati_aste/risultati_aste_btp_10_anni/BTP_10_Anni_Risultati_Asta_del_28-31.08.2015.pdf last contact 11.09.2015.

Table 10 Regressed DCF model A as valuation of risk premium

Nr.	Risk free (Italian gilt)	Predicted DR	**Predicted risk premium**	Selected DR	**Risk premium**	Percentage error	Mean absolute percentage error
1	0.0195	0.072	**0.052**	0.069	**0.050**	0.05941284	0.22562
2	0.0195	0.071	**0.051**	0.062	**0.043**	0.20654437	
3	0.0195	0.066	**0.046**	0.070	**0.051**	0.08679835	
4	0.0195	0.077	**0.058**	0.090	**0.071**	0.17995968	
5	0.0195	0.087	**0.067**	0.072	**0.053**	0.28057798	
6	0.0195	0.093	**0.074**	0.078	**0.059**	0.26488426	
7	0.0195	0.066	**0.047**	0.072	**0.053**	0.11250905	
8	0.0195	0.089	**0.069**	0.065	**0.046**	0.52515626	
9	0.0195	0.096	**0.076**	0.078	**0.059**	0.30519196	
10	0.0195	0.068	**0.049**	0.092	**0.073**	0.32714142	
11	0.0195	0.068	**0.048**	0.075	**0.056**	0.13363837	
12	0.0195	0.067	**0.047**	0.080	**0.061**	0.21688610	
13	0.0195	0.069	**0.050**	0.075	**0.056**	0.10127909	
14	0.0195	0.069	**0.049**	0.076	**0.057**	0.1269595	

Table 11 Risk premium and discount rate

	Risk premium	Discount rate
Minimum	0.0461	0.0656
Maximum	0.0764	0.0959
Mean	0.0560	0.0755
Standard dev.	0.0109	0.0656

6 Conclusion

Commercial real estate market in Bari is the second area in which has been applied the Regressed DCF. The method potentially may be applied as a further possible AVM method in those contexts in which there are valuation and transaction based on DCF. The application with few possible transactions showed the possibility to determine a "local risk premium" and a "local discount rate" based on a limited number of transactions and strictly related to a specific urban segment. In this case the risk premium was determined for a peripheral place of the city of Bari with several commercial properties. More contribution are required to increase the number of variables involved in this AVM method. In the present application an adjustment of price based on the different variability of selling price has been proposed.

References

Baum, A., Crosby, N., & MacGregor, B. D. (1996). Price Formation, Mispricing and Investment Analysis in the Property Market. *Journal of Property Valuation and Investment, 10*, 709–726.
Ciuna, M., Salvo, F., & De Ruggiero, M. (2014a). Property prices index numbers and derived indices. *Property Management, 32*(2), 139—153. doi: 10.1108/PM-03-2013-0021.
Ciuna, M., Salvo, F., & Simonotti, M. (2014b). Multilevel methodology approach for the construction of real estate monthly index numbers. *Journal of Real Estate Literature, 22*(2), 281–302.
d'Amato. (2017). Aspects of commercial property valuation and regressed DCF. In D. Lorenz (Ed.), *Behind the price: Valuation in a changing environment*. Wiley, in print.
d'Amato, M. (2015). Income approach and property market cycle. *International Journal of Strategic Property Management, 29*(3), 207–219.
d'Amato, M. (2004). A comparison between RST and MRA for mass appraisal purposes. A case in Bari. *International Journal of Strategic Property Management, 8*, 205–217.
d'Amato, M. (2008). Rough set theory as property valuation methodology: The whole story, Chap. 11. In T. Kauko & M. d'Amato (Eds.), *Mass* Appraisal an International Perspective for Property V*aluers* (pp. 220–258). Wiley Blackwell.
d'Amato, M. (2010). A location value response surface model for mass appraising: An "Iterative" location adjustment factor in Bari, Italy. *International Journal of Strategic Property Management, 14*, 231–244.
d'Amato, M., & Anghel, I. (2012): Regressed DCF, real estate value, discount rate and risk premium estimation. A case in Bucharest. In Saverio Miccoli (Ed.), *Appraisals Evolving Proceedings in Global Change* (Vol. I, pp. 765–776). Real Estate Market and Information Systems, Italian Association of Appraisers and Land Economists, CeSET.
d'Amato, M., & Kauko, T. (2012). Sustainability and risk premium estimation in property valuation and assessment of worth. *Building Research and Information, 40*(2), 174–185.
d'Amato, M., & Kauko, T., (2008). Property market classification and mass appraisal methodology, Chap. 13, In T. Kauko, & M. d'Amato (Eds.), *Mass appraisal an international perspective for property valuers* (pp. 280–303), Wiley Blackwell.
d'Amato, M., & Siniak, N. (2008). Using Fuzzy numbers in mass appraisal: The case of belorussian property market, Chap. 5. In T. Kauko, M. d'Amato (Eds.), *Mass appraisal an international perspective for property valuers* (pp. 91–107). Wiley Blackwell.
European Central Bank. (2014). Asset Quality Review phase 2, Manual, March.
Gelman, A. (2007). Scaling regression inputs by dividing by two standard deviations. Technical Report, Department of Statistics, Columbia University.
Hendershott, P. H., & Hendershott, R. J. (2002). *On Measuring real estate risk* (pp. 35–40). Winter: Real Estate Finance.
Hoesli, M., & MacGregor, B. (2000). *Property investments. principles and practice of portfolio management*. Edinburgh: Longman.
IAAO. (2003). *Standard on Automated Valuation Models (AVM)*. Chicago, IL: International Association of Assessing Officers. Retrieved August 25, 2015 from http://docs.iaao.org/media/standards/AVM_STANDARD.pdf
IVSC. (2013). International Valuation Standards, London.
IVSC. (2011). Technical information paper No. 1, The Discounted Cash Flow (DCF) Method – Real Property and Business Valuations, Exposure Draft. Retrieved August 25, 2015 from http://www.ivsc.org/sites/default/files/DCF%20Exposure%20Draft.pdf
Kaklauskas, A., Daniūnas, A., Dilanthi, A., Vilius, U., Lill Irene, Gudauskas, R., et al. (2012). Life cycle process model of a market-oriented and student centered higher education. *International Journal of Strategic Property Management, 16*(4), 414–430.
Kauko, T., & d'Amato, M. (2011). Neighbourhood effect. *International encyclopedia of housing and home*. Edited by Elsevier Publisher.

Kauko, T., & d'Amato, M. (2008), Introduction: Suitability issues in mass appraisal methodology In T. kauko, & M. d'Amato (Eds.), *Mass appraisal an international perspective for property valuers* (pp. 1–24). Wiley Blackwell.

Kauko, T., & d'Amato, M. (2008). Preface In T. Kauko, & M. d'Amato (Eds.), *Mass appraisal an international perspective for property valuers* (p. 1). Wiley Blackwell.

Rabianski, J.G. (2002). Vacancy in market analysis and valuation. *The Appraisal Journal*, 191–199.

Renigier-Biłozor, M., Wiśniewski, R., Biłozor, A., & Kaklauskas, A. (2014a). Rating methodology for real estate markets—Poland case study. *Pub. International Journal of Strategic Property Management, 18*(2). 198–212. ISNN. 1648-715X.

Renigier-Biłozor, M., Dawidowicz, A., & Radzewicz, A. (2014b). An algorithm for the purposes of determining the real estate markets efficiency in Land Administration System. *Pub. Survey Review, 46*(336), 189–204.

RICS. (2013). Automated Valuation Models. Information Paper.

Ross, S. (1976). The Arbitrage price theory of capital asset pricing. *Journal of Economic Theory, 13*, 341–360.

Sharpe, W. F. (1964). Capital asset price: A theory of market equilibrium under condition of risk. *Journal of Finance, 19*(3), 425–442.

Taylor, M. A., & Rubin, G. (2002). *Raising the bar: Simulation as a tool for assessing risk for real estate properties and portfolios* (pp. 18–34). Winter: Real Estate Finance.

Wheaton, W. C., Torto, R. G., Sivitanides, P. S., Southard, J. A., et al. (2001). *Real estate risk: A forward-looking approach* (pp. 20–28). Fall: Real Estate Finance.

Willison, D.L. (1999). Towards a more reliable cash flow analysis. *The Appraisal Journal*, 75–82.

Automatic Research of the Capitalization Rate for the Residential Automated Valuation: An Experimental Study in Cosenza (Italy)

Marina Ciuna, Manuela De Ruggiero, Francesca Salvo and Marco Simonotti

Abstract When a market oriented approach is not applicable because of the lack of comparable sales data (Ciuna et al. 2014a), International Valuation Standards suggest the use of the Income Approach, which has the task of simulating the market through an economic and financial scheme. The central question of this approach is due to the quantification of the capitalization rate, whereas it is generally extracted in a market segment different from that of the subject. This paper intends to propose an automated valuation model of the capitalization rate based on a real estate database built through a computerized geocoding automatic procedure, starting from the application of statistical models of multiple regression analysis, performed to separately build the prediction function of the price and the prediction function of the income.

Keywords Automated valuation models · Income approach · GIS

1 Introduction

International valuation standards (IAAO 2003) suggest the use of the income approach in all circumstances in which there are not enough real estate data to apply the market approach or to appraise properties that are frequently sold based on their income streams. The financial method has the task of simulating the market through an economic and financial scheme for which it has to foresee the series of future

M. Ciuna (✉) · M. Simonotti
University of Palermo, Palermo, Italy
e-mail: marina.ciuna@unipa.it

M. Simonotti
e-mail: marco.simonotti@unipa.it

M. De Ruggiero · F. Salvo
University of Calabria, Rende, Italy
e-mail: manueladeruggiero@virgilio.it

F. Salvo
e-mail: francesca.salvo@unical.it

M. d'Amato and T. Kauko (eds.), *Advances in Automated Valuation Modeling*,
Studies in Systems, Decision and Control 86, DOI 10.1007/978-3-319-49746-4_20

income and costs of the property, the capitalization rate, and duration. The central problem of this approach is due to the quantification of the capitalization rate, whereas it is generally extracted in a segment different from that of the subject. The real estate appraisal literature indicates different methods to estimate the above-mentioned cap rate: the direct research, which considers prices of the properties and income in the same market segment of the property to be valued, and the remote search, which considers the differences of the parameters of market segments that conceptually stops when it intercepts a comparable market segment for which prices and rents are available. According to international valuation standards, the search for the capitalization rate should reflect the data and market information relating to prices and rates, and their trend as well as the practice and expectations of the operators. Direct capitalization for the mass appraisal involves developing an Over All Capitalization Rate (OAR). Directly from the market place. The OAR is then used with the estimated net income to estimate value by the income capitalization. A contribution to the transparency of market data and, in operational terms, to the applicability of the Income Approach can be gained through automated GIS techniques in order to research the capitalization rate. Where a computerized database of real estate prices and income is available, in fact, you can implement models for calculating the capitalization rate which, using geo-referenced data, can return a prediction function of the rate itself. This paper intends to propose an automated valuation model of the capitalization rate based on a real estate database built through a computerized geocoding automatic procedure of real estate data, tested on the city of Cosenza in southern Italy. The method adopted aims at carrying out the analysis by market segment area, starting from the application of statistical models of multiple regression analysis, performed to separately build the prediction function of the price and the prediction function of income. The regression was carried out in a GIS environment to draw directly from the database real estate data that meet specific space requirements (area), to return in the first instance a function of the appraised rate obtained as the ratio between the functions of the appraised income and that of the price, and then an algorithm, through the remote search of the capitalization rate, able to appraise the capitalization rate in segments where this information is lacking.

The study addresses two case studies: the first regards segments in which both income and market prices are available, and in which it is possible to estimate the capitalization rate in a direct way; the second concerns segments in which it is possible to detect only market prices and appraised income (for example, for tax purposes) by interpolation with a function that uses the capitalization rate of the next segments and segments in which it is possible to detect only real estate income and appraised values. In the case study, the prediction function of the capitalization rate leads to point values for every investigated area and market segment, once the ordinary property characterizing the sub-market has been defined. The spatial mapping of capitalization rates through the automatic functions, as well as providing the data necessary for the application of the Income Approach, is a useful tool for investment choices.

2 Background and Literature Review

The income approach can be found in the Old Testament. The ancient Jewish law stated that the price of land could be appraised as the sum of annual crops equal to the missing dated to sabbatical (Simonotti 1997). The Spaniards of the sixteenth century tasadores esteemed the taxable value of agricultural land by measuring the annual harvest piled in storage by farms and calculating the value patrimoniale with the capitalization of the harvest rate based on percentage fixed by the central tax (Simonotti 1997). In the fifties and sixties, the evaluators noted that the relationship between the prices of housing units and their rents for offices located in buildings in Manhattan, was constant and equal to 7. A number of studies have sought to explain the rent paid on the market for rental units using hedonic price theory. This hedonic approach was developed by Lancaster (1971), Rosen (1974) and others. This literature has been reviewed by Jud et al. (1996). A recent study of the relationship between rents and distance is available in Soderberg and Janssen (2001). In addition, Frew and Wilson (2000) examined the rent-distance relationship in the Portland area. Property rents and values are related through the capitalization (cap) rate. The cap rate as used in the real estate literature refers to the ratio of Net Operating Income (NOI) to property value. This rate has a particularly important role in property valuation, because the income capitalization method converts the expected NOI stream from commercial property into an estimate of asset value by dividing the net NOI stream by the capitalization rate (Brueggeman and Fisher 1993). Because NOI is the difference between effective rents and operating expenses, property value is fundamentally related to rents and the cap rate, if expenses are a constant fraction of rents. The cap rate bears a close relation to the Weighted Average Cost of Capital (WACC) as defined in the corporate finance literature (Copeland and Weston 1988). The WACC is the rate of discount that reflects the average costs of debt and equity capital employed by a firm. Empirical work in the real estate literature seeks to explain the cap rate relative to other rates and macroeconomic factors (Froland 1987; Evans 1990). Ambrose and Nourse (1993) developed an investment approach based on the WACC; however, they do not incorporate the capital asset pricing model (CAPM) in their model. Instead, they rely on the intuitive argument that debt rates on mortgages should be related to government debt rates and that the cap rate should be related to the earnings-price ratio. Jud and Winkler (1995) drew on the theoretical underpinnings of the WACC and the CAPM models in the corporate finance and investment literature to develop a theoretic model of the capitalization rate for real estate properties. Recognizing the imperfect market conditions relevant to real estate transactions, they use a lag component process for market variables as suggested by Evans (1990). The resulting empirical model explains a substantial portion of the variation in cap rates. Simulation of real estate return and risk analysis using discounted cash flow (DCF) valuation dates from the late 1960s and continues today. Pioneers such as Graaskamp (1969), Pyhrr (1973) and Woffard (1978) were instrumental in producing much of the initial research in the field. A historical review of the literature reveals that although technology advancements have alleviated much of the initial impediments to wide spread adoption, real estate investors still limit their use of probabilistic analysis when evaluating real estate investments (Farragher and Savage 2008). One major limitation appears to be a lack of statistical knowledge and training needed by the analyst to conduct such analysis (Amidu 2011; Foster and Lee 2009; Farragher and Savage 2008). Stephen Pyhrr (1973) was one of the first researchers to study real estate return simulation and risk during the late 1960s. Very few, if any, real estate investment decision makers were using computerized decision support systems (DSS) during this time. Phyrr advocated that better alternatives were available than the subjective approaches being practiced, which he stated included "hunches, intuition, judgment and instinct" (Pyhrr 1973). Computer technology was at its infancy and simulations required access to a mainframe computer. Likewise, obtaining the data needed for the simulation was cost prohibitive. Fortunately, many of the computing obstacles faced by Phyrr have been addressed today.

Throughout the late 70s and early 80s computerized decision support system (DSS) continued to advance, especially those focused on financial modeling. Woffard's (1978) work called Computerized Real Estate Appraisal System or CREAS, addressed some of the concerns during that time, such as single point DCF valuation estimates used by real estate appraisers. Continuing earlier technological advancements in DCF modeling by Graaskamp (1969) and Pyhrr (1973), Woffard's 32-variable model allowed appraisers to use computerized models to simulate market-based appraisals using probabilistic inputs. Although successful in its overall objective, Woffard acknowledged that input parameter distributions inherently reflect the subjective views of the appraiser. The outcome of such modeling culminated in a range of possible valuation metrics reflecting the appraiser's perception of the market. Keliher and Mahoney (2000) created a DCF cash flow modeling tool to assist analysts when valuing property using three distinctive approaches: deterministic point estimates, sensitivity analysis, and probabilistic modeling. They intended to show how more advanced Monte Carlo simulation (MCS) techniques could improve point estimate modeling and reveal risk when appraising real estate assets. Weaver and Michelson (2004) created a simple Excel based model that used discounted cash flow point estimates to produce a plus or minus range of internal rate of return (IRR) equity values. Hoesli et al. (2006) incorporated a variation of the traditional DCF analysis by using the Adjusted Present Value valuation technique (Myers 1974) on 30 Swiss properties. Their research focused on several DCF limitations, such as determining the present value of the reversion price, variations in the discount rate risk premium over the holding period and the absence of risk metrics in single point values using traditional DCF analysis. One of the more recent research endeavors to investigate the use of pro forma DCF analysis and simulation was produced by Foster and Lee (2009). The authors reviewed multiple real estate development projects in the New England area from 2001–2007 and compared ex post returns to proposed ex ante returns. Farragher and Savage (2008) conducted a recent survey of real estate institutional and private investors to determine what types of decision-making processes they used when making real estate investment decisions.

3 Methods

The present work aims at determining the capitalization rate in an computerized environment using the Income Approach and multiple regression models, according to the International Valuation Standards. An brief illustration of the above-mentioned approaches follows, in order to explicate the appraisal methodology applied in this work.

4 Income Approach

The Income Approach provides an indication of value by converting future cash flows to a single current capital value. This approach considers the income that an asset will generate over its useful life and indicates value through a capitalization process (d'Amato 2015). Capitalization involves the conversion of income into a capital sum through the application of an appropriate discount rate. The income stream may be derived under a contract or contracts, or be non-contractuals, e.g., the anticipated profit generated from either the use of holding of the asset. Methods that fall under the income approach include:

- Income capitalization, when an all-risks or overall capitalization rate is applied to a representative single period income,
- Discounted cash flow where a discount rate is applied to a series of cash flows for future periods to discount them to a present value,
- Various option pricing models.

The Income approach can be applied to liabilities by considering the cash flows required to service a liability until it is discharged (IVS 2013). Generally, the Income Approach is to be used when sales data belonging to the same market segment of that to be appraised are absent. Consequently, the detection of rents and sales takes place in one or more segments of similar markets, comparable with the segment of the property to be appraised according to their parameters. The income capitalization approach is based on three methods: the Direct Capitalization, the Yield Capitalization and the Discounted Cash Flow Analysis (Appraisal Institute 2002). The Direct Capitalization (Formula 1) is a method used to convert an estimate of a single year's income expectancy, or annual average of several years' income expectancies, into an indication of value in one direct step—either by dividing the income estimate by an appropriate income rate or by multiplying the income estimate by an appropriate factor called gross rent multiplier (GRM) (Formula 2):

$$V = \frac{R}{i} \tag{1}$$

$$V = R \cdot GRM \tag{2}$$

where V is the property value, i is the capitalization rate, R is the annual income and GRM is the gross rent multiplier and is the inverse of the price in a specific market segment. Direct capitalization may be applied to every kind of income. The income selected for capitalization is influenced by the data available. The Yield Capitalization is based on the conversion of future benefits into present value by applying appropriate yield rate. The Discounted Cash Flow Analysis (DCFA) is a valuation method used to estimate the attractiveness of an investment opportunity. Discounted cash flow (DCF) analysis uses future free cash flow projections and discounts them (most often using the weighted average cost of capital) to arrive at a present value, which is used to evaluate the potential for investment. The selection of the capitalization rate may be carried out by a direct survey, based on the survey of prices and rents set by the market, and with an indirect approach aimed at analyzing the circumstances that affect quantitatively on it. The direct research of the capitalization rate appeals to the inverse formulas of the Income Approach, if income, sales data and their functional relationship are known. In this paper, the determination of the capitalization rate is based on the direct research and on the survey of incomes and sales data.

5 The Multiple Regression Models

The multiple regression model is a statistical tool that is of wide spread application in the economic and appraisal scope (d'Amato 2004, 2008, Kauko and d'Amato 2008). The regression model establishes a correlation between the market value, or income, and the intrinsic and extrinsic characteristics that most significantly affect them, in order to express the mutual correlations between the characteristics and the quantitative effect produced by each characteristic on the price or income. In the appraisal analysis the multiple regression model provides the equation of price or income and identifies a hyperplane (regression plane) (Formulas 3 and 4, Ciuna 2010):

$$P_j = p_0 + p_1 \cdot x_{j1} + p_2 \cdot x_{j2} + \cdots + p_n \cdot x_{jn} + e_j \tag{3}$$

$$R_j = r_0 + r_1 \cdot x_{j1} + r_2 \cdot x_{j2} + \cdots + r_n \cdot x_{jn} + e_j \tag{4}$$

where j = 1, 2, …, m is the generic observation in a sample of m property trades, P_j is the sale price of the generic j property (or the income R_j), x_{ij} (with i = 1, 2, …, n) the property features. In the appraisal context, the equation is usually presented in a deterministic form, where the stochastic variable e_j is omitted. There are numerous indices and criteria to verify the multiple regression model, the most important for estimation purposes are:

(1) the standard error s^2, which expresses the deviation between the observed data and the data interpolated with the model, it represents a first measure of the closeness of the relationship between the variables;
(2) the coefficient of determination R^2, which measures the goodness of the proximity of the model to the original data;
(3) the t-test on the statistical significance of the individual variables;
(4) the F-test, derived from the corresponding analysis of variance, which allows to decide if the variance induced by regression is statistically significant or may be assigned to the case to a predetermined level of confidence;
(5) the residual analysis, which may assess the appropriateness of the model by defining residuals and examining residual plots. Residuals (e) are the difference between the observed value of the dependent variable and the predicted value.

A series of practical criteria led to suggest the following acceptable values of the previous statistical tests and statistical measures percentages in the real estate appraisal context (Ciuna 2011):

a. Value of the coefficients of determination equal to or greater than 0.95, less restrictively are proposed values close to 0.90;
b. The average error rate should not be higher than 10 %,and errors for each observation must not be greater than 15 %, or the standard error should not be greater than 5 % of the average price;
c. the limit values of the t and F test are based on the confidence interval that generally is equal to 0.95. In the regression analysis, the condition about the number of observations in relation to the number of variables becomes obligatory for the appraisal purposes. In most situations, there is only a reduced availability of observations, while a higher number of explanatory variables is required. Practical criteria suggest that the relation between the number of observations and the number of variables, expressed as inequality, is to act in terms of $m > (n + 1)^{1/2}$, o $m > 10$ n, o $m > n + 30$, or less restrictively $m > 4n$ at least up to $n = 10$ in respect of the previous inequality, or even $m > 5n$. The choice of the criterion is related to the actual availability of the data.

6 Capitalization Rate Determination

The methodology proposed in this paper is based on the implementation of an algorithm designed to determine the estimating function of e the capitalization rate using the disaggregated estimate of the function of price and the function of income, drawing on data from a real estate database available in a GIS environment, specially constructed for pilot study area. The method aims to support the principles of the Income Approach with the techniques related to statistical models of multiple regression, acquiring data from a complex and computerized real estate database on the basis of spatial (location) and attribute (amount of real estate features) criteria of selection. International standards require that capitalization rates are derived solely from the real estate market. To this end, the study proposes the research of the capitalization rate in market segments other than that of the property to be assessed, according to the general rules of the market-oriented methods applied to the parameters of the segments. In the Income Approach, international valuation standards make constant reference to the need for comparative data (IVS 2013). It should be remembered that the capitalization rate is not directly expressed s by the market, such as the rate of interest on a loan operation, but is, in the first instance, a quantity derived from the ratio between the income and the market price of a property: the first is a rent stated in the rental market, the second is a sale data recorded in the real estate trades market (Renigier-Biłozor ed al 2014, a). Generally, the Income Approach is applied when sales data of similar properties are not available. Consequently, the detection of rents and trades takes place in one or more market segments comparable with that of the property to be assessed according to their parameters. This detection process in the next market segments can be indicated as "remote research process of the capitalization rate". In operational terms, we decided to divide the territory as indicated by the Land Registry and illustrated in Table 1:

Once the database has been split in real estate sub-samples with the same locational characteristics (d'Amato and Kauko 2008), corresponding to the homogeneous areas

Table 1 Subdivision of the territory as indicated by the land registry (OMI)

OMI homogeneous area
B1/Central/CENTRAL—VIA ROMA, MONTESANTO, XXIV MAGGIO, PARISIO, C.SO D ITALIA, V.LE TRIESTE, P.ZZA LORETO
B2/Central/CENTRAL—C.SO MAZZINI, V.LE ALIMENA, P.ZZA FERA
C1/Semi-central/SEMI-CENTRAL—V.LE DELLA REPUBBLICA, VIA ARNONE, C.SO UMBERTO, VIA VENETO, P.ZZA EUROPA, VIA RIVOCATI
C2/Semi-central/SEMI-CENTRAL—V.LE DELLA REPUBBLICA, VIA ARNONE, C.SO UMBERTO, VIA VENETO, P.ZZA EUROPA, VIA RIVOCATI
D1/Peripheral/PERIPHERAL EST—VIA POPILIA, REGGIO CALABRIA
D2/Peripheral/PERIPHERAL NORTH—CITTA 2000, VIA PANEBIANCO, V.LE COSMAI
D3/Peripheral/PERIPEHRAL (HISTORICAL CENTER)—VIA PAPARELLE, C.SO TELESIO, COLLE TRIGLIO, PORTAPIANA, P.ZZA SPIRITO SANTO
D4/Peripheral/PERIPHERAL (STADIUM AREA)—SAN VITO, SERRA SPIGA

indicated by the Land Registry, the survey starts with an analysis on the amount of data available. In particular, for each OMI zone: data of trades and rents contained in the database are counted; average unit prices and average unit rents are calculated (where data are available); missing unit prices and/or income are acquired from the OMI quotations (Salvo et al. 2013a). All these data are summarized in tabular form showing for each OMI, which represent the conditions of applicability of the regression models.

For areas that do not contain any information on prices and/or income, missing information is integrated with the real estate quotation provided by the Land Registry, already available in the database in a polygonal layer, designed to store the information provided by OMI. Once available information about number of data and amount of unit price/unit income are obtained, it has to assess whether there are the conditions necessary to the applicability of the multiple regression models. In particular, in this work it was considered essential to establish a minimum number of observations necessary to the application of statistical models of multiple regression, fixed in the condition $m > n + 30$, which in the case study becomes $m > 42$. If the obligatory is not satisfied in some areas, there is the necessity to turn to similar areas, referring to the problem of the "remote search of the capitalization rate." The condition of similarity is expressed in terms of closeness between the amounts of the average unit price/average unit income between the deficit area (indicated by the subscript 0) and the others in which the information is available, assessed as the percentage difference between the quotation of the area investigated and that of other areas. In particular, the similar area is characterized by a lesser percentage divergence, identified through a similarity index as formula 5 below

$$\gamma_o = \min\left\{ \left| \frac{\bar{p}_0 - \bar{p}_k}{\min(\bar{p}_o; \bar{p}_k)} \right| \middle| k \neq 0 \right\}. \tag{5}$$

Once the analysis of the data has been done, for each area if its data are sufficient they are directly regressed, otherwise, if the area is instead a deficit of these information, data related to the most similar area are regressed and corrected through the index. More specifically, for areas with sufficient data, the functions of the price and income are separately appraised, regressing the data from the real estate database which meet the requirement of the chosen location, once they have been exported from the GIS system in a electronic computing spreadsheet. The price and income of properties depend on many variables. The explanatory variables x_i ($i = 1, 2, \ldots, n$) that affect the price or income are numerous, so the linear multiple regression model is presented as follows in the Formulas (6) and (7) below:

$$P_j = p_0 + p_1 \cdot x_{j1} + p_2 \cdot x_{j2} + \cdots + p_n \cdot x_{jn} \tag{6}$$

$$R_j = r_0 + r_1 \cdot x_{j1} + r_2 \cdot x_{j2} + \cdots + r_n \cdot x_{jn} \tag{7}$$

where P_j and R_j respectively indicate the generic price and the generic income of the j detected data, p_i and r_i are the hedonic prices and hedonic income of the generic i feature, x_{ij} is the i generic feature of the j property purchase. Once the prediction function of the price and the prediction function of income are known, it is possible to derive the prediction function of the capitalization rate as set by the International Standards in the application of the Direct Capitalization as in the Formula (8):

$$i = \frac{R_j}{P_j} = \frac{r_0 + r_1 \cdot x_{j1} + r_2 \cdot x_{j2} + \cdots + r_n \cdot x_{jn}}{p_0 + p_1 \cdot x_{j1} + p_2 \cdot x_{j2} + \cdots + p_n \cdot x_{jn}} \tag{8}$$

For areas in which data of price and/or income are not sufficient, the functions are obtained by regressing the data of the more similar area. The similarity factor is then applied in order to correct the calculated functions like in the Formula (9) and (10) below:

$$P_j^0 = [\gamma(p]_0 + p_1 \cdot x_{j1} + p_2 \cdot x_{j2} + \cdots + p_n \cdot x_{jn}); \tag{9}$$

$$R_j^0 = \gamma(r_0 + r_1 \cdot x_{j1} + r_2 \cdot x_{j2} + \cdots + r_n \cdot x_{jn}) \tag{10}$$

The capitalization rate of the 0 area is then calculated in the following Formula (11):

$$i = \frac{R_j^0}{P_j^0} = \frac{r_0 + r_1 \cdot x_{j1} + r_2 \cdot x_{j2} + \cdots + r_n \cdot x_{jn}}{p_0 + p_1 \cdot x_{j1} + p_2 \cdot x_{j2} + \cdots + p_n \cdot x_{jn}} \tag{11}$$

In operational terms, the methodology is set up as an assisted automatic valuation model which draws data from a real estate information system by selecting the data in each zone through a spatial selection of intersection between the real estate data layer and the zoning one. The data are processed in an automatic procedure implemented by the Model Builder tool of Arc GIS which in an early stage needs to export the information to an excel spreadsheet and to then re-import the results. Model builder tool of Arc GIS, in fact, while still allowing many operations on the data, does not allow the implementation of multiple regression models. The modeling is so made operational in excel, through the instrument data analysis, estimating the parameters of the regression model once defined the input data corresponding to the dependent variable and the independent ones and specified confidence intervals for the coefficients of the model (also indicating the confidence level) and the type of the residues to be analyzed. In this paper, we propose the linear regression model with a zero location blind (in order to delete the location factor); the results validation has been assigned to the main statistics indices as described in Sect. 3.2. Although the procedure is fully automated, the model obviously requires the supervision of an experienced operator who is able to assess the reliability of the results returned in output (regression equation and tests). The function used to estimate the capitalization rate, zone by zone, is then used to calculate the capitalization rate of a virtual ordinary property, whose features are considered as averages of that of the data of sales and income in the area investigated, respectively, in the function of price and in the function of income.

7 International Valuation Standards in the Case Study

The modeling must comply with the requirements of international standards of automated valuation methods even with appropriate adaptations and specifications to the specific real estate market and the aims of the project.

(1) Property identification

The first step in any appraisal problem is to identify the property to be appraised and to acquire information about property to be appraised. In this work, the identification concerns the definition of the ordinary virtual property representative of each zone. The identification of the aforementioned virtual property is automatically generated by calculating the average of each feature of all real estate data in the investigated area.

(2) Assumptions

The key assumption is the intended use. The sale price (or rent) usually reflects the highest and best use, potentially different from the actual use at the time of the contract, with variations that may also be significant.

The real estate data in the pilot database relates only to residential properties, and the model works in the estimation of properties having the same destination. In the area under investigation, the most frequent destination is undoubtedly the housing one, and therefore there should not be inconsistencies. It is clear that who uses the tool should still check every possibility, taking duly into account any differences between the current use and the most profitable one.

(3) Data Management and Quality Analysis

The quality of the comparison data, which, however, has been extensively tested in the preliminary phase of the creation of the computerized database, has to be verified downstream of the regression through the main statistical test (Borst et al. 2008).

(4) Stratification

Stratification is the process of grouping properties for modelling and analysis. Stratification begins with property type. Residential properties in urban areas are generally stratified into "market areas." Market areas are broad, somewhat homogeneous geoeconomic areas that appeal to buyers in similar economic brackets.

In this real estate database, all real estate data contained fall within the same market segment (that is residential and used flats in condominiums located in multi-storey buildings), except for size and location.

The incidence associated with the location of the property is controlled by the operations of intersection with the polygon layer representative of the land zoning.

The size factor is managed by subtypes, which in principle limit the search of comparable to those that fall in a specific segment. It has to be specified that in the case study the size factor is not taken into account for the small availability of segmented data that would reduce the possibility to apply the regression models.

(5) Model specification and calibration

The specification is to identify, describe and implement the model configuration and define the variables to be included in the system.

Calibration is the process of determining the adjustments or the coefficients of the variables used in the AVM through market analysis.

For the proposed model, we have chosen to refer to the multiple linear regression, considering a set of independent variables which coincide with the characteristics specified in paragraph 4 while the dependent variable represents the market value or the income (respectively for the prediction function of the price and that for income).

The calibration is represented by the coefficients of the regression designed to simulate the hedonic prices of real estate features.

(6) Value defense

The model results are verified through the use of statistical tests.

8 Case Study

The methodology proposed in this paper has been tested on a pilot real estate information system built for the city of Cosenza, Italy. The real estate data were acquired from the Observatory of Real Estate Market, a regional real estate market observatory instituted by University of Calabria, one of the rare examples of databases containing information on the real market prices, although on a local scale, initially for teaching purposes. It has been surveyed a sample of approximately 866 data relating to real estate sale transactions and 573 pertaining to rental, all data related to segment of the used market for condominiums in multistory buildings, in a period between 1993 and 2013. Properties are physically, technically, economically and functionally similar, the use is exclusively the housing one, and motivations of purchase and sale are usually attributable to transfer. The shape of the market is restricted due to monopolistic competition, in which franchise agents are the most common form of intermediation. Specifically, we created two different layers, one containing data relating to sale one containing data relating to rent. In particular, we have archived all of the property by associating a cardinal number to them in order to use the sample as a basis for the implementation of a GIS. Specifically, the features detected and stored for each property included:

Address (ADD): indicates where the apartment is located—street name and number;

Age of building (ABU): indicates the period when the property was built;

Sale contract Date (SCD): measured with a cardinal scale, indicates the year when the property was sold;

Rent contract Date (RCD): measured with a cardinal scale, indicates the year when the rent contact was done;

Total Surface Area (TSA): measured in square meters, is the sum of the principal surface and the secondary ones, annexed (balconies) and associates (attic, basement, garage), each in respect of its commercial ratio (Ciuna and Simonotti 2014; d'Amato 2010);

Car Box (BOX): measured with a cardinal scale, indicates the number of exclusive property's car box;

Parking place (PAP): measured with a cardinal scale, indicates the number of assigned external car parking spaces (each one means a parking space of 12.5 m^2);

External surface condominium in portions (ESCP) measured in square meters, indicating the portion of the external communal surface attributable to property;

External surface in exclusive property (ESEP): measured in square meters, indicates the exclusive external surface;

Restrooms (RES): n° measured, indicates the number of toilets;

Floor Level (LEV): indicates the apartment floor level, measured with an ordinal scale;

Elevator (ELE): expressed as a dichotomous variable, with 0 indicating the absence of the lift, with 1 its presence;

Heating (HEA): expressed as a dichotomous variable, with 0 indicating the absence of heating, with 1 its presence;

Views (VIE): measured with a cardinal scale, indicate the number of sides on which there are windows;

Maintenance and conservation status (MSC): measured with a scale score, 0 was attributed to indicate a state of poor maintenance, 1 for mediocre maintenance, 2 for adequate maintenance, 3 for a discrete state, 4 if the maintenance status is good, 5 if the maintenance condition is excellent;

Orientation (ORI): measured with a scale score, indicates the predominant exposure of the property. The value is 2 if the exposure is east-west, 1 if the exposure is north-east or south-east, 0 otherwise.

Sale Price (SPR): measured in €, indicates the true price paid in the real estate transaction;

Sale Average Unit Price (SAUP): measured in €/sqm, is the average unit price on the retail space;

Monthly rent (MOR): expressed in €/month, indicating the monthly fee paid for the property. In the procedure, the tool provides for the conversion from monthly to annual fee (respectively Monthly and Annual Net Rent—MOR and ANR) as in the Formula (12):

$$ANR = MOR \cdot \frac{(1 + i_k)^{n \cdot k} - 1}{i_k} \tag{12}$$

where n represents the number of years, k represents the number of times that fee is paid in a year, i_k represents the perioral assay obtained by the annual rate as in the Formula (13) (Ciuna et al. 2014b; Ciuna et al. 2014c; Salvo et al. 2013):

$$i_k = (i + 1)^{\frac{1}{k}} - 1 \tag{13}$$

From the gross annual rent, the net annual rent is calculated by deducting all costs borne by the owner to manage the property that contribute to reducing the positive components of income. Table 2 shows number of data and unit average prices. Costs and relative percentages are listed in Table 3.

Table 2 Number of data, unit average price

Area		B1	B2	C1	...	R1	R2
Prices	Number of data	$n_{PB1} = \ldots$	$n_{PB2} = \ldots$	$n_{PC1} = 0$	...	$n_{PR1} = 0$	$n_{PR2} = \ldots$
	Unit average price	$\bar{p}_{B1}$	$\bar{p}_{B2}$	?	...	?	$\bar{p}_{R2}$
Income	Number of data	$n_{RB1} = \ldots$	$n_{RB2} = 0$	$n_{RC1} = \ldots$	...	$n_{RR1} = 0$	$n_{RR2} = \ldots$
	Unit average income	$\bar{r}_{B1}$	?	$\bar{r}_{C1}$	...	?	$\bar{r}_{R2}$

Table 3 Managing costs

Costs	Percentage of the gross annual income (%)
Maintenance costs	$2 \div 4$
Amortisation charge	0.5
Insurance costs	0.5
Administration costs	3
Charges for services	$2 \div 5$
Allowances for vacancy and uncollectable	$0 \div 5$
Tax expense	$10 \div 22$

Looking at all managing costs, the annual net gun is obtained by deducting the annual gross fee in reason of 27 %. The real estate database also provides for the presence of a layer of polygonal geometry, representative of the zoning of the territory in homogeneous areas proposed by the Land Registry; there is also a layer of linear geometry, meant to represent the urban road network, as well as orthophotos over the study area. As for more details on database design, data structure, spatial and attribute validation rules, see the contribution "Automated procedures based on Market Comparison Approach. A pilot study in the city of Cosenza" which refers to the same GIS structure.

9 Results

Hedonic regression fits observed transaction prices on house characteristics. Once fitted, the estimated hedonic function can be used to predict both the market value of a property given its characteristics and its annual income, in order to determine a function for the capitalization rate, obtained as a ratio between the regression equation for the market value and that for the income one. Considering the necessity to have enough data to implement the regression models, as described in paragraph 3.2., a similarity index has been calculated in order to relate areas without enough data with others provided with a superior number of information (remote research). Results are reported in Table 4.

Table 4 Area data sample

OMI Area	Prices			Income		
	Number of data	Average price (€/ mq)	Prices similarity index	Number of data	Average Income (€/ mq anno)	Income similarity index
B1	197	980.43	γ = 0 (enough data)	85	34.12	γ = 0 (enough data)
B2	104	1,073.42	γ = 0 (enough data)	78	38.73	γ = 0 (enough data)
C1	182	878.50	γ = 0 (enough data)	89	31.17	γ = 0 (enough data)
C2	51	893.50	γ = 0.016 (related to C1)	97	32.41	γ = 0 (enough data)
D1	48	715.20	γ = −0.23 (related to D2)	52	28.33	γ = -0.10 (related to C1)
D2	196	935.70	γ = 0 (enough data)	73	33.87	γ = 0 (enough data)
D3	42	650.15	γ = −0.30 (related to D2)	51	24.15	γ = −0.29 (related to C1)
D4	46	638.72	γ = −0.32 (related to D2)	48	22.31	γ = −0.40 (related to C1)

In our specification exercise, we have considered two separate specifications of the hedonic regression model, the first related to the sales data, the second one based on income data.

Both these equations are expressed as in the Formulas (14) and (15):

$$P_j = [(\gamma + 1)(p_0 + p_1 \cdot x_{j1} + p_2 \cdot x_{j2} + \cdots + p_n \cdot x_{jn})]; \tag{14}$$

$$R_j = (\gamma + 1)(r_0 + r_1 \cdot x_{j1} + r_2 \cdot x_{j2} + \cdots + r_n \cdot x_{jn}) \tag{15}$$

where the term $(\gamma + 1)$ means to correct the equation for the area where data are not enough, the location blind are considered zero, $p_1 \dots p_n$ and $r_1 \dots r_n$ are the hedonic prices and rates, $x_{j1} \dots x_{jn}$, are the amounts of the real estate features. Each regression equation appraised for every single area has been reported in Table 5 (Functions of price) and Table 6 (Functions of Income):

In order to verify the significance of the obtained results, we calculated:

a. the value of the coefficient of determination, which must be equal to or greater than 0.95, less restrictively are proposed values close to 0.90;

b. the average percentage error, which should not be higher than 10 %, and the relative errors at each observation, which must not be more than 15 %;

c. the standard error, which should not be greater than 5 % of the average price;

Table 5 Function of price

OMI Area	Function of price
B1	$PRS_{B1} = 1.398.68 \frac{€}{mq} \cdot TSA - 4.801.32 \frac{€}{anno} \cdot DAP + 6.817.30 \frac{€}{n°} \cdot RES + 8.210.88 \frac{€}{livello} \cdot LEF + 6.812.41 \frac{€}{punto} \cdot HEA + 3.870.30 \frac{€}{punto} \cdot ELE + 3.141.28 \frac{€}{n°} \cdot VIE + 2.987.16 \frac{€}{punto} \cdot MSC + 1.610.27 \frac{€}{punto} \cdot O$
B2	$PRS_{B2} = 1.418.31 \frac{€}{mq} \cdot TSA - 5.101.32 \frac{€}{anno} \cdot DAP + 6.418.21 \frac{€}{n°} \cdot RES + 5.121.32 \frac{€}{livello} \cdot LEF + 6.847.11 \frac{€}{punto} HEA + 3.442.23 \frac{€}{punto} \cdot ELE + 3.743.22 \frac{€}{n°} \cdot VIE + 3.131.67 \frac{€}{punto} \cdot MSC + 1.841.23 \frac{€}{punto} \cdot O$
C1	$PRS_{C1} = 1.180.22 \frac{€}{mq} \cdot TSA - 4.129.12 \frac{€}{anno} \cdot DAP + 6.123.45 \frac{€}{n°} \cdot RES + 3.373.48 \frac{€}{livello} \cdot LEF + 8.481.14 \frac{€}{punto} \cdot HEA + 3.870.46 \frac{€}{punto} \cdot ELE + 3.008.27 \frac{€}{n°} \cdot VIE + 2.622.84 \frac{€}{punto} \cdot MSC + 1.128.95 \frac{€}{punto} \cdot O$
C2	$PRS_{C2} = 1.169.53 \frac{€}{mq} \cdot TSA - 4.198.44 \frac{€}{anno} \cdot DAP + 6.235.401 \frac{€}{n°} \cdot RES + 3.938.51 \frac{€}{livello} \cdot LEF + 5.222,32 \frac{€}{punto} \cdot HEA + 3.935.44 \frac{€}{punto} \cdot ELE + 3.053.69 \frac{€}{n°} \cdot VIE + 2.666.87 \frac{€}{punto} \cdot MSC + 1.145.87 \frac{€}{punto} \cdot$
D1	$PRS_{D1} = 1.298.68 \frac{€}{mq} \cdot TSA - 4.301.87 \frac{€}{anno} \cdot DAP + 6.318.42 \frac{€}{n°} \cdot RES + 4.827.42 \frac{€}{livello} \cdot LEF + 8.641.18 \frac{€}{punto} \cdot HEA + 3.178.42 \frac{€}{punto} \cdot ELE + 3.181.82 \frac{€}{n°} \cdot VIE + 2.844.31 \frac{€}{punto} \cdot MSC + 1.608.11 \frac{€}{punto} \cdot O$
D2	$PRS_{D2} = 992.64 \frac{€}{mq} \cdot TSA - 3.288.12 \frac{€}{anno} \cdot DAP + 4.829.47 \frac{€}{n°} \cdot RES + 3.689.83 \frac{€}{livello} \cdot LEF + 5.076.17 \frac{€}{punto} \cdot HEA + 2.429.42 \frac{€}{punto} \cdot ELE + 2.409.09 \frac{€}{n°} \cdot VIE + 2.174.04 \frac{€}{punto} \cdot MSC + 1.225.33 \frac{€}{punto} \cdot OR$
D3	$PRS_{D3} = 902.35 \frac{€}{mq} \cdot TSA - 3.622.88 \frac{€}{anno} \cdot DAP + 4.390.11 \frac{€}{n°} \cdot RES + 3.354.22 \frac{€}{livello} \cdot LEF + 4.614.47 \frac{€}{punto} \cdot HEA + 2.208.45 \frac{€}{punto} \cdot ELE + 2.188.97 \frac{€}{n°} \cdot VIE + 1.976.30 \frac{€}{punto} \cdot MSC + 1.118.28 \frac{€}{punto} \cdot OR$
D4	$PRS_{D4} = 886.49 \frac{€}{mq} \cdot TSA - 2.936.51 \frac{€}{anno} \cdot DAP + 4.313.03 \frac{€}{n°} \cdot RES + 3.295.25 \frac{€}{livello} \cdot LEF + 4.533.35 \frac{€}{punto} \cdot HEA + 2.169.63 \frac{€}{punto} \cdot ELE + 2.151.47 \frac{€}{n°} \cdot VIE + 1.941.56 \frac{€}{punto} \cdot MSC + 1.094.30 \frac{€}{punto} \cdot OR$

Table 6 Function of income

OMI Area	Function of income
B1	$ANR_{B1} = 35.14 \frac{€}{mq \cdot anno} \cdot TSA + 138.81 \frac{€}{n° \cdot anno} \cdot RES + 107.14 \frac{€}{livello \cdot anno} \cdot LEF + 141.17 \frac{€}{punto \cdot anno} \cdot HEA + 68.23 \frac{€}{punto \cdot anno} \cdot ELE + 67.21 \frac{€}{n° \cdot anno} \cdot VIE + 62.15 \frac{€}{punto \cdot anno} \cdot M$
B2	$ANR_{B2} = 41.17 \frac{€}{mq \cdot anno} \cdot TSA + 138.63 \frac{€}{n° \cdot anno} \cdot RES + 108.42 \frac{€}{livello \cdot anno} \cdot LEF + 142.27 \frac{€}{punto \cdot anno} \cdot HEA + 71.41 \frac{€}{punto \cdot anno} \cdot ELE + 68.31 \frac{€}{n° \cdot anno} \cdot VIE + 63.67 \frac{€}{punto \cdot anno} \cdot M$
C1	$ANR_{C1} = 33.21 \frac{€}{mq \cdot anno} \cdot TSA + 129.47 \frac{€}{n° \cdot anno} \cdot RES + 106.45 \frac{€}{livello \cdot anno} \cdot LEF + 129.41 \frac{€}{punto \cdot anno} \cdot HEA + 64.81 \frac{€}{punto \cdot anno} \cdot ELE + 61.15 \frac{€}{n° \cdot anno} \cdot VIE + 33.15 \frac{€}{punto \cdot anno} \cdot M$
C2	$ANR_{C1} = 35.43 \frac{€}{mq \cdot anno} \cdot TSA + 130.12 \frac{€}{n° \cdot anno} \cdot RES + 104.32 \frac{€}{livello \cdot anno} \cdot LEF + 141.43 \frac{€}{punto \cdot anno} \cdot HEA + 71.12 \frac{€}{punto \cdot anno} \cdot ELE + 60.42 \frac{€}{n° \cdot anno} \cdot VIE + 60.03 \frac{€}{punto \cdot anno} \cdot M$
D1	$ANR_{D1} = 35.43 \frac{€}{mq \cdot anno} \cdot TSA + 130.12 \frac{€}{n° \cdot anno} \cdot RES + 104.32 \frac{€}{livello \cdot anno} \cdot LEF + 141.43 \frac{€}{punto \cdot anno} \cdot HEA + 71.12 \frac{€}{punto \cdot anno} \cdot ELE + 60.42 \frac{€}{n° \cdot anno} \cdot VIE + 60.03 \frac{€}{punto \cdot anno} \cdot M$
D2	$ANR_{D2} = 33.65 \frac{€}{mq \cdot anno} \cdot TSA + 130.61 \frac{€}{n° \cdot anno} \cdot RES + 110.41 \frac{€}{livello \cdot anno} \cdot LEF + 131.01 \frac{€}{punto \cdot anno} \cdot HEA + 62.44 \frac{€}{punto \cdot anno} \cdot ELE + 60.87 \frac{€}{n° \cdot anno} \cdot VIE + 59.88 \frac{€}{punto \cdot anno} \cdot M$
D3	$ANR_{D3} = 24.02 \frac{€}{mq \cdot anno} \cdot TSA + 90.46 \frac{€}{n° \cdot anno} \cdot RES + 77.68 \frac{€}{livello \cdot anno} \cdot LEF + 92.69 \frac{€}{punto \cdot anno} \cdot HEA + 44.10 \frac{€}{punto \cdot anno} \cdot ELE + 43.86 \frac{€}{n° \cdot anno} \cdot VIE + 39.35 \frac{€}{punto \cdot anno} \cdot MSC$
D4	$ANR_{D4} = 20.42 \frac{€}{mq \cdot anno} \cdot TSA + 76.88 \frac{€}{n° \cdot anno} \cdot RES + 66.02 \frac{€}{livello \cdot anno} \cdot LEF + 78.78 \frac{€}{punto \cdot anno} \cdot HEA + 37.48 \frac{€}{punto \cdot anno} \cdot ELE + 38.72 \frac{€}{n° \cdot anno} \cdot VIE + 33.87 \frac{€}{punto \cdot anno} \cdot MSC$

d. the thresholds of the t and F test, based on the interval of confidence that arises generally equal to 0.95. Results have been shown in Table 7.

Analyzing the values in the table, it can be seen that the checks tend to be met, any distortion having to be attributable not to a lack of skills in interpretation of the model, rather to the quality of the data contained in the database, in line with the viscosity that characterizes some real estate market, like the Italian one, which are often characterized by anomalies. The coefficients of determination are met in terms of statistics. The regression equations are, in fact, been set for homogeneous areas, deleting the location effect (no intercept), and being therefore characterized by R^2 coefficients never less than 0.998. Even the t and F test tends to be statistically verified, being able also in this case to attribute the discrepancy to the nature of the used data. Once calculated, area by area, the functions of the capitalization rate as the ratio of the corresponding functions of the income and the price, the punctual capitalization rate attributable to each homogeneous area was calculated, by identifying a virtual ordinary property having features whose values correspond to the average characteristics of the properties in the selected area (Table 8).

Table 7 Results

OMI Area	Sales Prices			Income		
	Average percentage error (%)	Error in each observation	Standard error (%)	Average percentage error (%)	Error in each observation	Standard error (%)
B1	8.23	<15 % in 95 % cases	4.38	9.50	<15 % in 92 % cases	5.15
B2	7.51	<15 % in 96 % cases	5.03	8.32	<15 % in 91 % cases	5.17
C1	8.18	<15 % in 91 % cases	4.18	9.31	<15 % in 87 % cases	4.92
C2	9.78	<15 % in 87 % cases	4.71	12.42	<15 % in 82 % cases	5.03
D1	11.35	<15 % in 82 % cases	5.36	13.57	<15 % in 79 % cases	5.61
D2	12.56	<15 % in 87 % cases	5.10	11.42	<15 % in 80 % cases	4.31
D3	9.38	<15 % in 80 % cases	4.93	10.06	<15 % in 91 % cases	4.65
D4	10.72	<15 % in 81 % cases	5.65	9.87	<15 % in 88 % cases	5.27

Table 8 Capitalization rate for area OMI

OMI homogeneous Area	Capitalization Rate (%)
B1/Central/CENTRAL—VIA ROMA, MONTESANTO, XXIV MAGGIO, PARISIO, C.SO D ITALIA, V.LE TRIESTE, P.ZZA LORETO	2.33
B2/Central/CENTRAL—C.SO MAZZINI, V.LE ALIMENA, P.ZZA FERA	2.07
C1/Semi-central/SEMI-CENTRAL—V.LE DELLA REPUBBLICA, VIA ARNONE, C.SO UMBERTO, VIA VENETO, P.ZZA EUROPA, VIA RIVOCATI	2.71
C2/Semi-central/SEMI-CENTRAL—V.LE DELLA REPUBBLICA, VIA ARNONE, C.SO UMBERTO, VIA VENETO, P.ZZA EUROPA, VIA RIVOCATI	2.82
D1/Peripheral/PERIPHERAL EST—VIA POPILIA, REGGIO CALABRIA	3.01
D2/Peripheral/PERIPHERAL NORTH—CITTA 2000, VIA PANEBIANCO, V.LE COSMAI	2.65
D3/Peripheral/PERIPEHRAL (HISTORICAL CENTER)—VIA PAPARELLE, C.SO TELESIO, COLLE TRIGLIO, PORTAPIANA, P. ZZA SPIRITO SANTO	3.17
D4/Peripheral/PERIPHERAL (STADIUM AREA)—SAN VITO, SERRA SPIGA	3.08

10 Conclusions

Predicting market values of residential properties is important for investment decisions and the risk management of households, banks and real estate developers. The increased access to electronically stored information has spurred the development and implementation of AVMs, which provide appraisals at low cost. The key challenge of AVM development is finding a workable statistical approach that uses the data and provides reasonably reliable appraisals. A large body of academic literature has developed concerning the statistical modelling of real estate prices. This literature offers a variety of approaches, but does not aim at prediction or, if it does, eschews the constraints and trade-offs faced by AVM developers. In this paper, we have implemented a model which attempts to define a function in order to appraise the capitalization rate, using multiple regression models to determine on the basis of a double sample of sales and rent, respectively, the function of price and the function of income, to obtain finally the function of the capitalization rate as the ratio between the two. The model has been implemented using the tool Model Builder in ArcGIS, in some phases integrated by an excel spreadsheet, to get in the end a very flexible tool which, while using data contained in the specific real estate GIS, is still separate from it and may be used in contexts other than pilot database. The results have been checked through the main statistical tests, which provided acceptable values, where any distortive element has to be related to quality of the input data and not to the predictive ability of the model. The tool proposed in this paper does not intend to propose a numerical result, but to provide an investigation methodology, certainly perfectible in terms of computer implementation of the algorithm, but no doubt innovative in the panorama of automated valuation models in viscous real estate markets.

References

Ambrose, B. W., & Nourse, H. O. (1993). Factors influencing capitalization rates. *Journal of Real Estate Research, 8*(2), 221–237.

Amidu, A. R. (2011). Research in valuation decision making processes: Educational insights and perspectives. *Journal of Real Estate Practice and Education, 14*(1), 19–34.

Appraisal Institute. (2002). *The Appraisal Journal.* Chicago: The Appraisal Institute

Borst, R. A., Des Rosiers, F, Renigier, M., Kauko, T., & d'Amato, M. (2008). Technical comparison of the methods including formal testing accuracy and other modelling performance using own data sets and multiple regression analysis. In T. Kauko, & M. d'Amato (Eds.), *Mass appraisal an international perspective for property valuers*. Wiley Blackwell.

Brueggeman, W. B., & Fisher, J. D. (1993). *Real estate finance and investments*, Ninth edn. Homewood, IL: Irwin.

Ciuna, M., Salvo, F., & Simonotti, M. (2014, November, 22/24). The expertise in the real estate appraisal in Italy. In *Proceedings of the 5th European Conference of Civil Engineering (ECCIE '14).* Mathematics and Computers in Science and Engineering Series p. 120–129, North Atlantic University Union, ISSN: 2227-4588, Firenze.

Ciuna, M. (2010). L'Allocation Method per la stima delle aree edificabili. *AESTIMUM, 57*, 171–18.

Ciuna M. (2011). The valuation error in the compound values. AESTIMUM [S.l.], (pp. 569–583). August 2013. ISSN 1724-2118.

Ciuna, M., Salvo, F., De Ruggiero, M. (2014). *Property prices index numbers and derived indices*. Property management (Vol. 32, Iss. 2, pp. 139–153). doi:10.1108/PM-03–2013-0021.

Ciuna, M., Salvo, F., & Simonotti, M. (2014). Multilevel methodology approach for the construction of real estate monthly index numbers. *Journal of Real Estate Literature, 22*(2), 281–302.

Ciuna, M., & Simonotti, M. (2014). Real estate surfaces appraisal. *AESTIMUM 64, Giugno, 2014*, 1–13.

Copeland, T. E., & Weston, J. F. (1988). *Financial theory and corporate policy*, Third edn. Reading, MA: Addison-Wesley.

d'Amato. (2010). A location value response surface model for mass appraising: an "iterative" location adjustment factor in Bari, Italy. *International Journal of Strategic Property Management, 14*, 231–244.

d'Amato, M. (2015). Income approach and property market cycle. *International Journal of Strategic Property Management, 29*(3), 207–219.

d'Amato, M. (2004). A comparison between RST and MRA for mass appraisal purposes. *A Case in Bari, International Journal of Strategic Property Management, 8*, 205–217.

d'Amato, M. (2008). Rough set theory as property valuation methodology: The whole story, Chapter 11. In T. Kauko, & M. d'Amato (Eds.), *Mass appraisal an international perspective for property valuers* (pp. 220–258). Wiley Blackwell.

d'Amato, M. & Kauko, T. (2008). Property market classification and mass appraisal methodology, Chapter 13. In T. Kauko, & M. d'Amato (Eds.), *Mass appraisal an international perspective for property valuers* (pp. 280–303), Wiley Blackwell.

Evans, R. D. (1990). A transfer function analysis of real estate capitalization rates. *Journal of Real Estate Research, 5*(3), 371–380.

Farragher, E., & Savage, A. (2008). An investigation of real estate investment decision making Processes. *Journal of Real Estate Practice and Education, 11*(1), 29–40.

Foster, J. J. & Lee, B. D. (2009). Sophisticated sensitivity: Can developers guess smarter? (Master's thesis). Retrieved from MIT University, MIT Libraries Web site. http://dspace.mit.edu/handle/1721.1/54853.

Frew, J., & Wilson, B. (2000). *Estimating the connection between location and property value, essays in honor of James A.* Graaskamp, Boston, MA: Kluwer Academic Publishers.

Froland, C. (1987). What determines cap rates on real estate. *Journal of Portfolio Management, 13*, 77–83.

Graaskamp, J. A. (1969). A practical computer service for income approach. *Appraisal Journal, 37*, 50–57.

Hoesli, M., Jani, E., & Bender, A. (2006). Monte carlo simulations for real estate valuation. *Journal of Property Investment and Finance, 24*(2), 102–122.

International Association Assessing Officers. (2003). *Standard on automated valuation models (AVM)*. Chicago IL: IAAO.

Jud, G. D., & Winkler, D. T. (1995). The capitalization rate of commercial properties and market returns. *Journal of Real Estate Research, 10*(5), 509–518.

Jud, G. D., Benjamin, J. D., & Sirmans, G. S. (1996). What do we know about apartments and their markets? *Journal of Real Estate Research, 11*(3), 243–257.

Kauko, T., & d'Amato, M. (2008). Preface. In T. Kauko, & M. d'Amato (Eds.), *Mass appraisal an international perspective for property valuers* (p. 1). Wiley Blackwell.

Kelliher, C. F., & Mahoney, L. S. (2000). Using monte carlo simulation to improve long-term investment decisions. *The Appraisal Journal, 68*(1), 44–56.

Lancaster, K. (1971). *Consumer demand theory: A new approach*. New York, NY: Columbia University Press.

Myers, S. (1974). Interactions in corporate financing and investment decisions– Implications for capital budgeting. *Journal of Finance, 29*(1), 1–25.
Pyhrr, S. A. (1973). A computer simulation model to measure risk in real estate investment. *Journal of the American Real Estate and Urban Economics Association, 1*(1), 48–78.
Renigier-Biłozor, M., Dawidowicz, A., & Radzewicz, A. (2014a). An algorithm for the purposes of determining the real estate markets efficiency in Land Administration System. *Publication Survey Review, 46*(336), 189–204.
Renigier-Biłozor, M., Wiśniewski, R., Biłozor, A., & Kaklauskas, A. (2014). Rating methodology for real estate markets—Poland case study. *Publication International Journal of Strategic Property Management. 18*(2). 198–212. ISNN. 1648-715X.
Rosen, S. (1974). Hedonic prices and implicit markets: Product differentiation in price competition. *Journal of Political Economy, 82*, 35–55.
Salvo, F., Ciuna, M., d'Amato, M. (2013). Appraising building area's index numbers using repeat values model. A case study in Paternò (CT). In: *Dinamics of land values and agricultural policies* (pp. 63–71), Bologna: Medimond International Proceedings, Editografica, ISBN: 978-88-7587-690-6, Palermo, 22–23 November 2012.
Salvo, F., Ciuna, M., d'Amato, M. (2013a). The appraisal smoothing in the real estate indices. In: *(a cura di): Maria Crescimanno, Leonardo Casini and Antonino Galati, Dynamics of land values agricultural policies* (pp. 99–111). Bologna: Medimond Monduzzi Editore International Proceeding Division, ISBN: 978-88-7587-690-6, Palermo, 22–23 November 2012.
Simonotti, M. (1997). *La stima immobiliare*. Torino: Utet Universitaria.
Soderberg, B., & Janssan, C. (2001). Estimating distance gradients for apartment properties. *Urban Studies, 38*, 61–79.
Weaver, W., & Michelson, S. A. (2004). A pedagogical tool to assist in teaching real estate investment risk analysis. *Journal of Real Estate Practice and Education, 7*(1), 45–52.
Wofford, L. E. (1978). A simulation approach to the appraisal of income producing real e American Real Estate and Urban Economics Association, 6(4), 370–394.

Automated Procedures Based on Market Comparison Approach in Italy

Marina Ciuna, Manuela De Ruggiero, Francesca Salvo and Marco Simonotti

Abstract The question about objective valuation is linked on the one hand to the identification of a large number of reliable sale data, on the other hand to the ability to control the critical evaluation protocol. Automating the valuation process, ensured by computerization of real estate data properly detected, provides a concrete answer to the problem of objectivity, allowing a significant control of the subjective component which characterizes the more traditional appraisals. Automatic methods traditionally proposed at international level are often set on multiple regression models, designed to build prediction functions that are valid throughout the study area, calibrating the coefficients on the basis of the real estate data contained on the supporting database. In accordance with the International Valuation Standards, this paper proposes a different approach to automated appraising that, far from the idea of defining appraisal equations generally valid on the basis of regression models, proposes the implementation of an automatic procedure based on the Market Comparison Approach, with the aim to define equations related to the peculiarities of a marketplace in a very circumscribed area. The proposed method has been implemented through the Model Builder tool of ArcGIS and has been tested on a pilot GIS for residential real estate properties.

Keywords Automated valuation models · Market comparison approach · GIS · Model builder

M. Ciuna · M. Simonotti (✉)
University of Palermo, Palermo, Italy
e-mail: marco.simonotti@unipa.it

M. De Ruggiero · F. Salvo
University of Calabria, Rende, Italy

M. d'Amato and T. Kauko (eds.), *Advances in Automated Valuation Modeling*,
Studies in Systems, Decision and Control 86, DOI 10.1007/978-3-319-49746-4_21

1 Introduction

The appraisal approach is linked, on the one hand, to the detection of an adequate number of reliable data, on the other hand to the possibility to critically control the evaluation protocol. Automating the appraising process, made possible by computerization of real estate data appropriately recognized, provides a concrete solution to the problem of objectivity, allowing the control of the information used for the estimate. The implementation of automated valuation methods is based on the establishment of a computerized database of real estate and also on the set of mathematical algorithms representative of the appraisal procedures. Automatic methods traditionally proposed at international level are often set on multiple regression models, designed to build prediction functions that are valid throughout the study area, calibrating the coefficients on the basis of the real estate data contained on the supporting database. The risk involved in this type of analysis is often linked to an excessive generalization of the function itself, which is likely to ignore the specifics of the property and the locational context, with relevant approximations and economically significant practical implications. In theory, an approach based on regression models is not easily applicable in some real estate markets, such as the Italian one, considering the number of real estate data necessary for a statistical modeling which is particularly high, usually not corresponding to the number of real estate data that can actually be found for the above mentioned real estate markets. In accordance with the International Valuation Standards, this paper proposes a different approach to automated appraising that, far from the idea of defining appraisal equations generally valid on the basis of regression models, proposes the implementation of an automatic procedure based on the Market Comparison Approach, with the aim to define equations related to the peculiarities of a marketplace in a very circumscribed area. The approach is aimed at identifying a spatial selection of comparable properties that fall within a limited geographical area, discretionally wide, resulting in a very high level of detail that does not require extrapolation. Expected that the selection of comparable is in a market segment limited in spatial terms, the biasing effect related to the location of the property has been essentially canceled. The procedure implemented by the tool Model Builder of ArcGIS is characterized by its flexibility of use: the algorithm, while using data contained in a specific GIS estate, is still separate from it, and then characterized with universality of application. The method proposed in this work involves the implementation of equations for appraising the value of the property that refers to Market Comparison Approach's modus operandi. The automatic model includes: (1) the selection of property in the ordinary market segment investigated, (2) the determination of the hedonic prices based on the whole sample, according to the information required by the Market Comparison Approach, (3) the determination of a sales adjustment vector constituted by the average price of the ordinary trades selected and the average of the hedonic prices of the drivers characteristics, (4) the determination of the appraising equation. The approach can easily be used in contexts other than the pilot one, always providing that the real estate databases have the same structure as the original one.

2 Literature Review

Some academicians address the study of the sales comparison approach in order to offer methods that could reduce or eliminate the appraisers' subjective judgments. A seminal work focusing on the sales comparison approach was written by Colwell et al. (1983); the study demonstrates how to derive adjustment factors using the ordinary least squares (OLS) method rather than the appraiser judgment. Some interesting academic studies which focused on the sales comparison approach were carried out by Isakson (1986, 1988) who presented a technique called the Nearest Neighbours Appraisal Technique (NNAT) according to which the final estimate of value is calculated as a weighted average of the actual selling prices of the comparable properties. Manaster (1988, 1991) reported that when the sales comparison approach was followed correctly, different appraisers using slightly different methods calculated final estimates of value that were very similar. Vandell (1991) presented a minimum variance (among the adjusted values of the comparable sales) approach for selecting and weighting comparable properties, while Gau et al. (1992, 1994) proposed a variation of Vandell's techniques where the coefficient of variation replaced Vandell's variation as the measure to be minimized. The idea of automating the sales comparison approach has appeared sporadically in the real estate literature over time and has been used in some CAMA (Computer-Assisted Mass Appraisal) methods (Dilmore 1974). Graaskamp and Robbins (1987) developed an automated sales comparison system that they referred to as Market-Comp based on choosing small samples with very similar properties—an approach similar to sales comparison adjustment grids used in conventional individual property appraisals. Detweiler and Radigan (1996, 1999) published an article describing their Computer-Assisted Real Estate Appraisal System (CAREAS). Their work described a statistically derived dissimilarity index used to select comparables and a regression model to create adjustment factors. The use of sophisticated fitting techniques can account for more diverse functional forms or, in some cases, complete lack of functional relationships. Spline regressions, nonparametric regressions, and autoregressive techniques are some examples. Incorporating spatial information in pricing models through the use of direct spatial modeling with Cartesian coordinates (Fik et al. 2003), geostatistical models (Dubin 1998), or response surfaces (O'Connor 2008) has improved the precision of price estimates. Other studies (Goodman and Thibodeau 1998, 2003, 2007; Bourassa et al. 2007; Borst and McCluskey 2008) have focused on improving sample selection by delineating submarkets of homes in which the marginal price contributions of independent variables are more likely to be similar. Predicted residuals from nearby sales (spatial errors) have been used in two separate but related ways in the literature. Case et al. (2004), in particular, developed a two-stage method in which errors from a single-stage ordinary least squares (OLS) model are used as predictors in the two stage model; conversely, Pace and Gilley (1997, 1998), among others, used a simultaneous autoregressive (SAR) model to account for nearby residuals in a single stage model. Bourassa et al. (2010, 2013) recently have dealt with uses of autoregressive models. In Italy, the MCA has been introduced with substantial changes as regards the methodology used in the U.S., particularly in relation to the approach used to estimate the hedonic prices of real estate features (Simonotti 2006). Essentially, the methodology developed aimed at quantifying the marginal prices of real estate properties without resorting to approaches like ADAM, APAM, and MPAM, which are difficult to apply in Italy (d'Amato 2008). Instead, the technique proposed was based on indications of a mathematical purpose, designed and formulated by applying economic criteria of the market value, the cost value, the value of transformation, the complementary value, and the value of subrogation. Later studies conducted at a national level have allowed the simplification and improvement of the methodology proposed by Simonotti as regards the duplication of characteristics and the reliability of real estate sales data (Salvo and De Ruggiero 2011, 2013).

3 Methods

According to International Valuation Standards, appraisal methodologies can be delineated according to well-defined operational protocols enabling them to be well implemented and automated. The valuation analysis is certainly a spatial analysis, which involves the application of criteria and calculations, most of all based on the location of the property, which is a key element in the mechanism of price composition. The appraisal procedure requires, on one the hand, the identification of comparable properties on the basis of logical criteria (belonging to the same market segment) and geographic ones (proximity between properties), on the other hand, the application of simple mathematical operations on data. Geographic Information Systems have the peculiarity to provide answers to these kinds of question, thanks to the selection tools based on attributes and spatial relationships and on basic computing functions applicable to the tables associated with the different thematic layers. For these reasons, the appraisal analysis can be fully represented by the appropriateness models implemented in Model Builder of ArcGIS, applying a standard, unique and fully defined methodology on different layers. It is a new investigative opportunity which has its strength in the reusability of the work done. This workproposes an approach which, far from the idea of defining appraising equations generally valid built through regression models, proposes the implementation of an automatic procedure based on the Market Comparison Approach using a GIS aimed at defining appraisal equations related to the peculiarities of a very circumscribed market area. In order to exemplify the methodology implemented, an illustration of the guidelines for the application of the AVM and of the Market Comparison Approach applied to the AVM will follow. The MCA illustrated and applied in the present work is a methodological evolution of the Sales Comparison Approach indicated in the Standard on Mass Appraisal of Real Property (d'Amato 2004, 2010; d'Amato and Kauko 2008), with the intention of overcoming the limitations of application of the Sales Approach in the Italian real estate market.

4 Automated Valuation Methods

An AVM is a calculation software with mathematical foundations able to appraise the market value of the property through the analysis of the real estate market of the place, its parameters and the characteristics of the property, using information gathered previously and separately, available on a computer database. The distinctive feature of an AVM is the use of mathematical models; this distinguishes AVM from traditional methods in which the evaluator physically inspects the property and relies mostly on experience and judgment to analyze the data and develop the appraisals. Compared to traditional methods, automated methods have the advantage of objectivity, efficiency, reduced costs and faster delivery times, provided that the database contains accurate and reliable data, the analysis is consistent with the appraisal theory accepted by international standards (International Valuation Standards, Uniform Standards of Professional Appraisal Practice, European Valuation Standards), and that modeling is properly tested before application. If these requirements are met, values determined with the AVM may be certified, and can be used in the private sector to estimate the market value of the properties, such as in the public sector, for assessment and taxation purposes (d'Amato and Kauko 2012). In practical terms, the automated valuation methods are built implementing one or more of the three valuation approaches (market approach, cost approach and income approach) through an appropriate programming language. It is clear that the reliability of the results produced with an AVM is related to the accuracy of the real estate data and transactions available on the supporting database, which must be continually monitored and updated in order to verify its integrity. In principle, automated valuation methods are based on statistical analysis and resort to simple and multiple regression models, applying principles and techniques on a large scale in which a sample of properties is analyzed in order to develop a model properly calibrated, or more specifically an appraisal equation, which can be applied to all similar properties falling in the same market area (d'Amato and Siniak 2008). Regardless of the chosen model and the specific formulation, modeling must be scientific and rigorous, as indicated by the international valuation standards. The market values produced by an AVM that uses international valuation standards and processes reliable data available on a computer database can be considered certified appraisals.

5 Market Comparison Approach

The Market Comparison Approach (MCA) is based on the fundamental principle which states that the market determines the price of a property in the same way as it fixes the prices of other properties (similia similibus aestimentur). The procedure is based upon the survey of data of recent trading of immovable properties similar to the technique used to appraise sites in the same area. The comparison between the immovable property to be appraised and the similar immovable properties allows determining the value through a complex of monetary adjustments to the market prices, based on the different formalities introduced by the common characteristics. The market value of the subject is appraised through a series of adjustments to the comparables' market prices according to the different formalities introduced by these in comparison to the property to esteem; particularly, the price of market of every comparable is corrected through a process of unitary adjustments (or hedonic prices) multiplied by the differences in the amount of the quantitative characteristics between the comparables and the subject. The correct prices represent the prices of the market that the compared immovable properties would introduce in the appraisal process if they possessed the same formalities of the characteristics of the subject. Thus, they represent different estimates of the value of the subject. The adjustment for the property *j*th (j = 1, 2, …, m) and for the *i*th feature (i = 1, 2, …, n) is equal to the product of the marginal price p_i (defined as the change in the total price corresponding to a unit change in the amount of real estate property) and the difference between the characteristics of the property being valued x_{j0} and those of the building to be compared x_{ij}:

$$p_{ij} \cdot (x_{0i} - x_{ij} \cdot (1 + r_j)) \tag{1}$$

where r_j is;

$$r_j = \left(\frac{x_{SUR_0} - x_{SUR_j}}{x_{SUR_j}} \right) \tag{2}$$

In the previous formula x_{SUR0} and x_{SURj} are respectively the surface amount of the subject and of the *j*th comparable. The correct price Pcj of the generic purchase *j*th is obtained by adding algebraically to the purchase price adjustments Pj, which correspond to the n estate characteristics in question (Salvo and De Ruggiero 2013):

$$P_{cj} = P_j + \sum_{i=1}^{n} p_{ij} \cdot (x_{0i} - x_{ij} \cdot (1 + r_j)) \tag{3}$$

6 Methodology Algorithm Implementation

This paper proposes the implementation of the Market Comparison Approach using the tool Model Builder available in ArcGIS. It is an extremely flexible tool, able to build automated procedures easily reusable, and for the first time used in the real estate sector. If the tool used for the implementation of the automated methods is absolutely innovative, modeling must comply with the requirements of the international standards in the field of automated valuation methods, even with appropriate adaptations to the specific aims of the project. In this regard, it should be emphasized that the idea of this work is to use an Assisted Automated Valuation Model (AVMAA) rather than a simple one (AVM). In this direction, some of the implementation steps are left to the user (information acquisition about the subject, assumptions and limiting conditions, analysis of data quality), because no automated tool could ever replace the critical capacity of an expert. The implementation of the proposed methodology involves a series of steps that includes: (a) property identification, that is the acquisition of information which characterize the property to be valued. The user notes this information with the aid of survey schemes, and subsequently transfers these data into the system in an input screen specially crafted. The compilation of the input screen is driven by the model. Every time you put the cursor in a cell to be filled, it opens a window on the sidelines that describes the real estate features and indicates the measurement mode (Fig. 1).

For the characteristics which have been defined with a coded values domain, integration is limited to acceptable values (shown in a drop-down menu), avoiding errors which could affect the procedure. Once inserted the subject characteristics, including the address, the model identifies the location on the map by geocoding operation; (b) the key assumption is the intended use. The sale price usually reflects the highest and best use, potentially different from the actual use at the time of the contract, with variations that may also be significant. The real estate data in the pilot database relates only to residential properties, and the model works in the estimation of properties having the same destination. In the area under investigation, the most frequent destination is undoubtedly the housing one, and therefore there should not be inconsistencies. It is clear that those who use the tool should still check every possibility, taking duly into account any differences between the current use and the most profitable one. (c) Data management and quality analysis: once the subject's location has been defined, the system searches for properties in

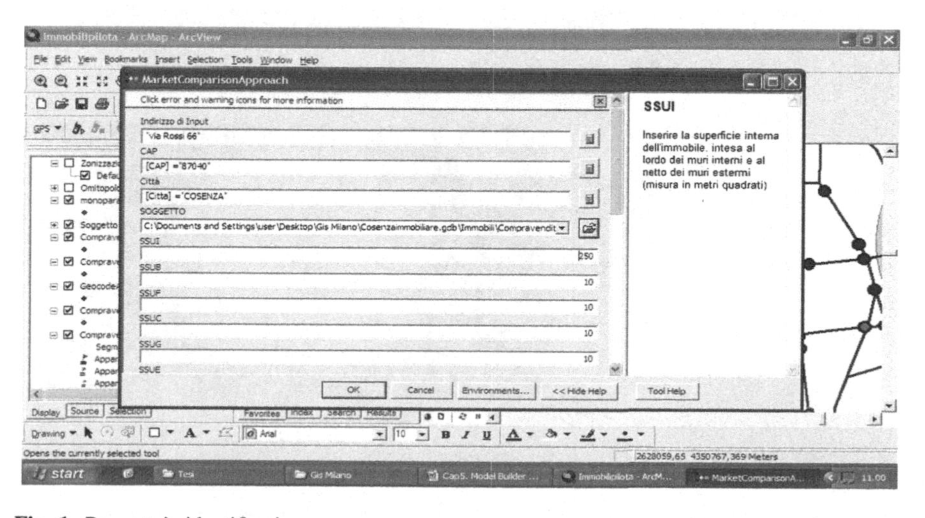

Fig. 1 Property's identification screen

comparison in a small circle around the subject, the magnitude of which can be managed directly from user (Kauko and d'Amato 2011). This selection is implemented through a buffer operation (which defines a circular area around a point, once defined the radius) and a subsequent operation of intersection (which extrapolates from a set of point data those which fall within a specific area) (Fig. 2).

The selection of comparable is then automatic, being the buffer distance the only one variable of this phase of the process (in each case the radius of the surroundings should not be more than a few hundred meters (even if without a limit value), in order to limit distortions related to the locational factor). The quality of the comparison data, which, however, has been extensively tested in the preliminary phase of the creation of the computerized database, should be tested in relation to the specificity of the subject. In fact, it may occur that in the buffer area there is a particular group of properties which are not suitable for comparison (e.g. positioned beyond the railway line), or that do not fall in the area (not covered area). The user has the task of analyzing these circumstances, eliminating the comparable singular or vice versa increasing the radius of the buffer; (d) stratification is the process of grouping properties for modeling and analysis. The application of an appraisal procedure requires the identification of a specific market segment common to subject and comparables. The comparison must relate to properties with similar characteristics (type of contract, type of property, intended use, size) and for which the locational factor is constant. All real estate data contained in the database fall within the same market segment (for residential condominiums located in multi-storey buildings and falling in the segment of the used market), except for the size and location. The size factor is managed by subtypes, which limit the search of comparable to those that fall in a specific segment. The incidence associated with the location is through the operations buffer and intersection. The selection of the comparables is carried out within a little round of the subject, in which we may be assumed that all the parameters related to the location (accessibility, services,

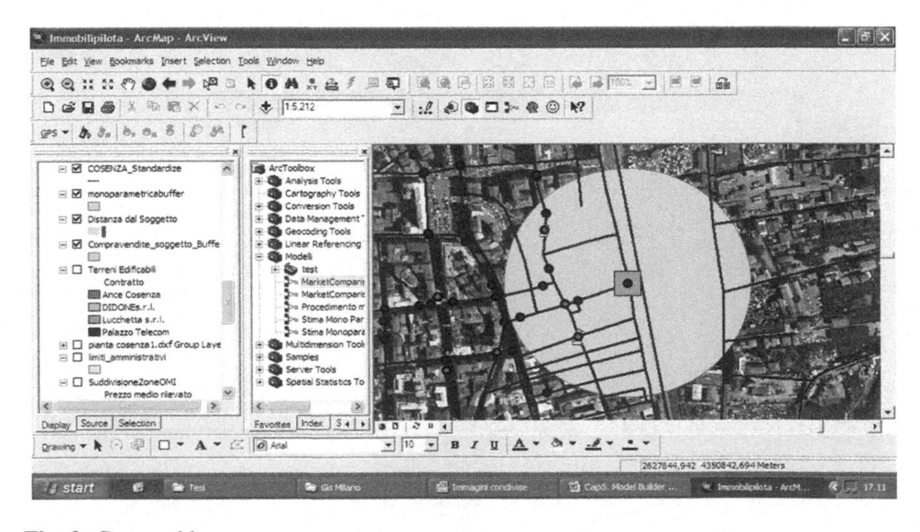

Fig. 2 Comparable screen

infrastructure, etc.) are uniform; (e) the specification is to identify, describe and implement the model configuration and define the variables to be included in the system. In this work the variables are the real estate features, below described in paragraph 4.1, while the model configuration is the hedonic prices analysis which is done according to the appraisals criteria described in literature (Simonotti 2006); (f) calibration is the process of determining the adjustments or the coefficients of the variables used in the AVM through market analysis. If most of the existing AVM uses statistical tools such as multiple linear regression and nonlinear regression to calibrate the model coefficients, it is also possible to use different calibration methods, based on the use of comparison functions and valuation criteria. In this project, oriented to the implementation of automatic procedures that lead to the identification of appraising equations generally valid in a circumscribed market context, the calibration is carried out through the use of postulates cost and comparison functions. The calibration of coefficients in MCA is represented by the valuation of the hedonic prices, carried out in accordance with the International Valuation Standards in relation to the Italian real estate market. The hedonic prices are therefore indirectly appraised, using the formulas derived from the valuation criteria (Simonotti 2006; d'Amato 2015a, b), using specially crafted fields, added in the execution phase in the Model Builder procedure. The market data required for the calculation of the hedonic prices (commercial ratios, annual revaluation rate, restrooms' installing costs, building land unit quotation, maintenance costs, etc.) are reported in the attribute table of the Land Registry and combined with the data of the comparable properties, in order to appraise the corresponding hedonic price for each comparable and for each real estate feature. It can be demonstrated that if the subject's characteristics are averages of the surveyed sample, their incidence is nothing in the evaluation, assuming the same value of the ceteris paribus features. In light of this observation, the Market Comparison Approach procedure can be significantly simplified by reducing the adjustment table (sales adjustment grid) into a column vector (sales adjustments vector), in which each element is calculated as:

$$\bar{p}_i(x_{0i} - \bar{x}_i) \tag{4}$$

where $\bar{p}_i$ is the average hedonic price of the *i*th feature, x_{0i} is the *i*th feature of the subject, while $\bar{x}_i$ the average of the *i*th feature in the sample. The appraisal equation is obtained by adding algebraically the elements of the vector, such as:

$$V = \bar{P} + \sum_{i=1}^{n} \bar{p}_i \cdot (x_{0i} - \bar{x}_i \cdot (1 + \bar{r})) \tag{5}$$

where r is the average value of those calculated with Formula (2). Once coefficients have been calibrated, the next step concerns Sales Adjustment Vector. In the data table, each line represents a real estate feature, whose amount is reported for every comparable (column). In order to build the Sales Adjustment Vector a new record referable to an average property whose features are the average of that of the comparables and whose hedonic prices are average of the corresponding hedonic ones of the comparables. The adjustment vector is then built through a record in which in each field the adjustment is calculated by multiplying the average hedonic price and the difference between the corresponding amount of real estate features for the subject and that of the average comparable. In the latest field, finally, the market value is calculated adding every adjustment to the average sale price (Table 1).

The ability of the method of weighing the incidence of multiple quantitative characteristics reflects in a certain complexity of the block diagram created in Model Builder which, however, shows the algorithm faithfully, graphically organized in a logical and consequential procedure.

7 Case Study

The methodology proposed in this paper has been tested on a pilot real estate information system built for the city of Cosenza, Italy (Ciuna et al. 2014a).

The real estate data were acquired from the Observatory of Real Estate Market, a regional real estate market observatory instituted by University of Calabria, one of the rare examples of databases containing information on the real market prices, although on a local scale, initially used for teaching purposes. A sample of approximately 866 data relating to real estate sale transactions has been surveyed, all data related to segment of the used market for condominiums in multistory buildings, in a period between 1993 and 2013. Properties are physically, technically, economically and functionally similar, the use is exclusively the housing one, and motivations of purchase and sale are usually attributable to transfer. The shape of the market is restricted due to monopolistic competition, in which franchise agents are the most common form of intermediation. From the survey forms associated with each real estate datum, information about real estate features deemed relevant in the mechanism of price formation in the study area were acquired. More specifically, the variables considered in the sample are: Address (ADD): indicates where the apartment is located—street name and number;

Age of building (AB): indicates the period when the property was built; Sale contract Date (SCD): measured with a cardinal scale, indicates the year when the property was sold (Ciuna et al. 2014b, c; Salvo et al. 2013); Rent contract Date (RCD): measured with a cardinal scale, indicates the year when the rent contact was done; Total Surface Area (TSA): measured in square meters, is the sum of the principal surface and the secondary ones, annexed (balconies) and associates (attic, basement, garage), each in respect of its commercial ratio (Ciuna and Simonotti 2014); Car Box (BOX): measured with a cardinal scale, indicates the number of exclusive property's car box; Parking place (PAP): measured with a cardinal scale, indicates the number of assigned external car parking spaces (each one means a parking space of 12.5 m^2); External surface condominium in portions (ESCP) measured in square meters, indicating the portion of the external communal surface attributable to property; External surface in exclusive property (ESEP): measured in square meters, indicates the exclusive external surface; Restrooms (RES): n°

Table 1 Adjustment vector (€)

Average sale price	Feature 1	Feature 2	…	Feature n − 1	Feature n	Value
$\bar{p}$	$\bar{p}_1(x_{01} - \bar{x}_1)$	$\bar{p}_2(x_{02} - \bar{x}_2)$			$\bar{p}_n(x_{0n} - \bar{x}_n)$	$\bar{P} + \sum_{i=1}^{n} \bar{p}_i \cdot (x_{0i} - \bar{x}_i \cdot (1 + \bar{r}))$

measured, indicates the number of toilets; Floor Level (LEV): indicates the apartment floor level, measured with an ordinal scale; Elevator (ELE): expressed as a dichotomous variable, with 0 indicating the absence of the lift, with 1 its presence; Heating (HEA): expressed as a dichotomous variable, with 0 indicating the absence of heating, with 1 its presence; Views (VIE): measured with a cardinal scale, indicate the number of sides on which there are windows; Maintenance and conservation status (MSC): measured with a scale score, was attributed 0 to indicate a state of poor maintenance, 1 for mediocre maintenance, 2 for adequate maintenance, 3 for a discrete state, 4 if the maintenance status is good, 5 if the maintenance condition is excellent; Orientation (ORI): measured with a scale score, indicates the predominant exposure of the property. The value is 2 if the exposure is east-west, 1 if the exposure is north-east or south-east, 0 otherwise.

Sale Price (SPR): measured in €, indicates the true price paid in the real estate transaction; Sale Average Unit Price (SAUP): measured in €/sqm, is the average unit price on the retail space. Sample statistics are reported in Fig. 3.

The construction of the real estate GIS can be summarized as follows:

Logical design of the system; definition of the cartographic base; creation of thematic layers; insertion of sale data through address geocoding.

The creation and the development of a GIS system must be preceded by a careful and detailed design of its constituent parts, depending on the purpose and use of the computer database.

The clarification and specification of the types of data, the format that represents them, the attributes that serve to characterize and identify possible relationships between classes, and the selection of the appropriate reference system are essential elements of the design, to ensure consistency and functionality of the database.

It is clear that the system can be changed during the operation, but some changes may be burdensome in terms of cost and time, as well as undermining the integrity and uniformity of the whole project.

Once you have specified purpose and use of the information system, the essential intention of the conceptual phase is to identify the constituents logical drives and to organize schematic diagrams in order to provide a guided visual reading of the draft.

The final result of the analysis is the creation of a comprehensive and detailed document which fully describes: • feature classes and subtypes; • attributes of the classes, data format and rules attributes (domains and default values); • classes relationships; • spatial relations and topological rules.

Performing spatial analysis using GIS means to query different information layers in an integrated way, by analyzing the spatial relationships between the elements or their attributes. Prerequisite to apply all GIS functions is that all the layers are perfectly stacked into a single and unique geographic location. For this project, we chose to use a Gauss-Boaga coordinates system, with Datum Rome 40. It refers to meters map units, and uses as reference ellipsoid the international Hayford one, oriented in Rome Monte Mario with astronomical data referable to 1940.

With this coordinate system, all the thematic layers have been screened, including the plan of the city in dxf format and the orthophotos of the urban area.

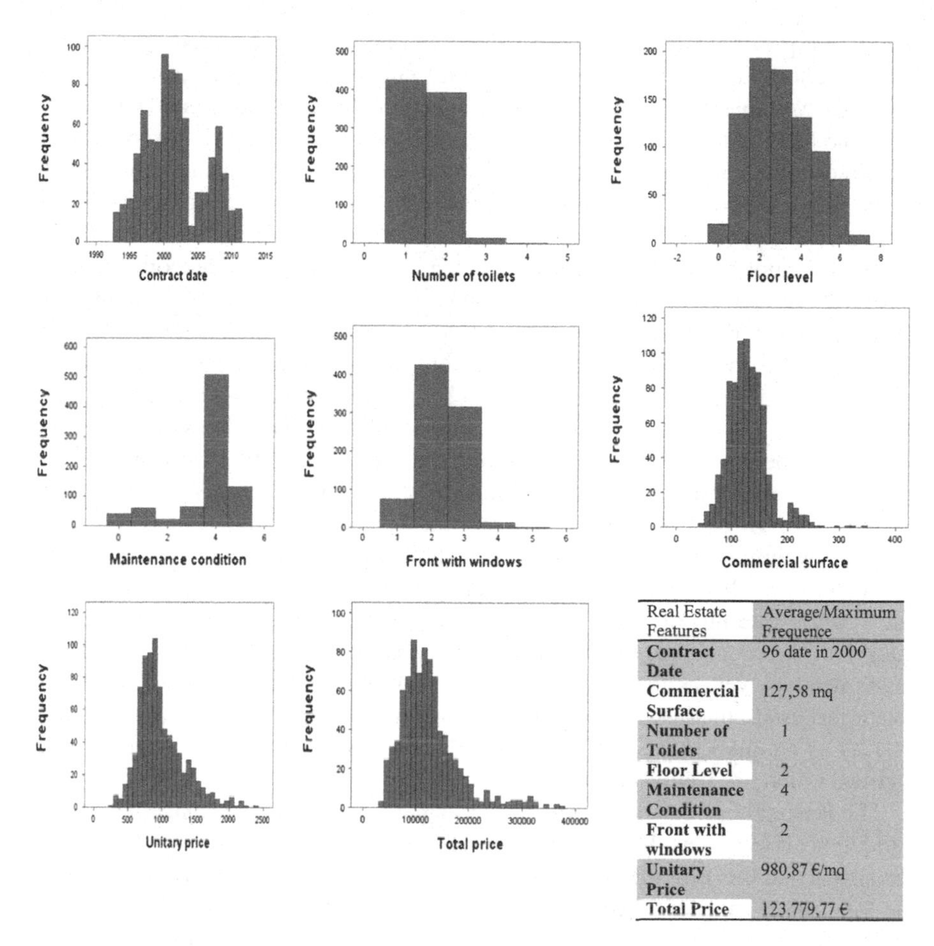

Real Estate Features	Average/Maximum Frequence
Contract Date	96 date in 2000
Commercial Surface	127,58 mq
Number of Toilets	1
Floor Level	2
Maintenance Condition	4
Front with windows	2
Unitary Price	980,87 €/mq
Total Price	123.779,77 €

Fig. 3 Sample statistics

The central part of the construction of the information system consists of the definition of the thematic layers. In this work, we decided to use the data format Personal Geodatabase.

Once the structure of the database has been created, all layers were imported inside it, each with its own peculiarities, but continuously related to the others. In particular, within the Geodatabase Cosenzaimmobiliare.gdb the following have been uploaded: a linear feature class intended to represent the main and secondary roads of the city. It is a geographical feature where the normalized attribute table contains some standard fields which identify each arc road, stating the name of the street but also the number of intervals on the left and right sides of the street. The peculiarities of the feature class road links make it possible to implement procedures for geocoding; A feature dataset, containing: (a) a point layer, intended to store real estate sales data; (b) a point layer, aimed at cataloging information about rents; (c) a polygon feature class containing the information reported by the Land Registry about average unitary prices, average unitary monthly rents, capitalization rates; (d) two tables in Access format data relating to the sale and rental data derived from the Observatory Real Estate Market and already available in Excel format; (e) orthophotos and the plan of the city in dxf format. Supplementary documents (floor plans, photos, survey forms) are associated to each sale (or rent) record through the Hyperlink function. The sales and rent data, available in excel and access, have not been manually placed on the map, but stored in the correspondent layers through a semi-automatic procedure known as address geocoding. The geocoding (or address matching), is the process of creating point elements on a map through the description of the address locations. It is a procedure able to connect an address to a geographical location through the transformation of a list of route information in points with coordinates x, y: in practice, to any address is assigned a location in the real world. The procedure is really useful in the real estate sector, and particularly in automated procedures. First, geocoding allows for immediate placement of all data by simply applying the operations required to access table format containing all the real estate information, with a clear gain in terms of time and efficiency. Geocoding also significantly limits the possibility of making mistakes when entering the data properties, resulting in a quality control and data integrity. The procedure, finally, enables a simple and immediate updating of real estate data, as well as the extension of the pilot information system to other real estate real estate market, in a fast, flexible and automated way. The data structure, despite its simplicity, has the ability to add additional information with different themes, which are useful for economic analysis or estimation of a different nature.

The Personal Geodatabase format is particularly useful in the management of the validation classes rules. In the geodatabase, you can assign two types of validation: the spatial and the attributes ones. Spatial validation is assigned by the definition of the topological rules of adjacency, proximity, connectivity, provided that there is a feature dataset where topologically related feature classes exist. Attribute validation defines behaviors on the attributes of the geographical elements, through the provision of domains, subtypes and relationships between classes. These types of validation are used with tabular data to ensure data integrity and efficiency in their management and visualization.

The spatial and attribute validation rules are therefore designed to ensure the quality control of information, as well as enabling to automate the behavior of the attributes and the streamlining of the introduction of the data. The spatial validation aims to control the behavior of the geographic elements from a geometrical point of view, ensuring that these have interrelationships in conformity with reality. In this work, two different topology rules have been defined to improve the behavior of the Land Registry polygon layer: polygon must not overlap; polygon must not have gaps (Table 2).

The definition of the aforementioned topological rules is essential to the proper functioning of the polygon layer, because they ensure that there are not portions of the territory which fall into two different areas, or there are not areas without a specific classification (not covered area). Besides the spatial validation rules, it is

Table 2 Subdivision of the territory as indicated by the Land Registry (OMI)

OMI homogeneous area
B1/Central/CENTRAL—VIA ROMA, MONTESANTO, XXIV MAGGIO, PARISIO, C.SO D ITALIA, V.LE TRIESTE, P.ZZA LORETO
B2/Central/CENTRAL—C.SO MAZZINI, V.LE ALIMENA, P.ZZA FERA
C1/Semi-central/SEMI-CENTRAL—V.LE DELLA REPUBBLICA, VIA ARNONE, C.SO UMBERTO, VIA VENETO, P.ZZA EUROPA, VIA RIVOCATI
C2/Semi-central/SEMI-CENTRAL—V.LE DELLA REPUBBLICA, VIA ARNONE, C.SO UMBERTO, VIA VENETO, P.ZZA EUROPA, VIA RIVOCATI
D1/Peripheral/PERIPHERAL EST—VIA POPILIA, REGGIO CALABRIA
D2/Peripheral/PERIPHERAL NORTH—CITTA 2000, VIA PANEBIANCO, V.LE COSMAI
D3/Peripheral/PERIPEHRAL (HISTORICAL CENTER)—VIA PAPARELLE, C.SO TELESIO, COLLE TRIGLIO, PORTAPIANA, P.ZZA SPIRITO SANTO
D4/Peripheral/PERIPHERAL (STADIUM AREA)—SAN VITO, SERRA SPIGA

useful to define rules of behavior for the attributes of the geographic features, including domains and subtypes. For the real estate database, it was necessary to define domains for the characteristics measured with nominal or ordinal scales, including qualitative ones, with the aim to "drive" the data entry and reduce errors. In particular, we have chosen to use the domains of the coded values type. (Number of views: measured with a cardinal scale from 1 to 4; Heating: the value 0 is assigned when the heating is absent, 1 when there is central heating and 2 when the heating is autonomous; Elevator: the value 0 is assigned when the lift is not present and the value 1 if it is present; Maintenance: 0 was attributed to indicate a state of poor maintenance, 1 for mediocre maintenance, 2 for adequate maintenance, 3 for a discrete state, 4 if the maintenance status is good, 5 if the maintenance condition is excellent (the maintenance value has to be based on a critical subjective judgment); car box: the value 0 is assigned when it is not present, and instead a value of 1 when the car box is present; parking: 0 is assigned when the parking place is not present, the value 12.5 when there is a parking space and the value 25 when there are two; Water Technical systems: 0 is assigned a value when the system is absent, 1 when it is present Orientation: the value is 2 if the exposure is east–west, 1 if the exposure is north–east or south–east, 0 otherwise.). A further validation attribute system is represented by subtypes, which have the function of sorting the elements of a layer in function of the values assumed by a specific field. It is an extremely useful tool for splitting real estate data because of the market segment associated with the property's surface consistency. In particular, the subtypes were used to classify the sales data considering the surface, identifying four distinct market segments: small apartments (45 sqm ÷ 90 sqm), medium apartments (90 sqm ÷ 130 sqm) medium-large apartments (130 sqm ÷ 180 sqm), large apartments (area exceeding 180 sqm). The possibility to define subtypes and then split the data into multiple real estate market segments increases the system's performance. The selection of the comparable properties, in fact, took place only on the segment corresponding to that of the subject and not on the entire sample of real estate data, with appraisal results more accurate and reliable.

8 Results

The method presented in this work, while sharing with those produced at internationally level many operational phases, is distinguished from these by the originality of the instruments used, the reference methodology and the characteristics of the results. The procedure is characterized by an interpretive purpose on the one hand, predictive on the other. As for the interpretative purpose, the method is able to query and to analyze data in order to understand and predict the dynamics of the real estate sector for the investigated geographical area, ensuring the required transparency to the real estate market. Regarding the predictive purpose, it is important to note that automatic methods traditionally offered at international level are generally set on multiple regression models and designed to build a final prediction function generally valid in quite large homogeneous areas, calibrating the coefficients on the basis of the real estate data contained in the support database. The risk involved in this type of analysis is often linked to excessive generalization of the function itself, which is likely to ignore the specifics of the property and the locational context, with significant approximations and economic practical implications. If in theory an approach based on regression models may involve underestimates and overestimates in many concrete cases, in practical terms it is not even applicable to the Italian real estate market (or similar ones), because the large number of reliable data required for the application of multiple regression is normally difficult to survey (Salvo et al. in print).

The automated method described in this study, based on the use of blocks tools programming tools (Model Builder), is proposed to implement analytical procedures designed to return appraisal equations referred to a limited geographical area, amplitude discretionary, resulting in a very high level of detail which does not require extrapolations. The appraisal of the market value of each property, therefore, restarts every time the procedure, never referring to an ordinary function, but according to the specificity of the property and, even more, to the locational characteristics of a small neighborhood of the subject, inside which a prediction function is defined in order to appraise all properties within the range selected by the GIS system:

$$V = \bar{P} + \sum_{i=1}^{n} \bar{p}_j \cdot (x_{0i} - \bar{x}_i \cdot (1 + \bar{r})) \tag{6}$$

The values obtained through the proposed method have to be verified using a Sales ratio analysis, a type of statistical study based on comparisons between an estimated value and market value as indicated by sales prices (Borst et al. 2008). For AVM use, the numerator would be the estimated value generated from the model, while the denominator would be the sale price. The ratios thus calculated are subjected to statistical analysis aimed to determine variability statistics able to provide information about the degree to which model-determined values for individual properties are similar with respect to the market value. In the proposed work, the sales based ratio studies have been implemented on the basis of the Standard on Ratio Studies at experimental level only on a limited sample of data in a semiautomatic way, which, however, have provided acceptable levels of variability. Further details of the research are intended to improve the automation of the test ratio through appropriate programming languages.

9 Conclusions

The question about objective valuation is linked on the one hand to the identification of a large number of reliable sale data, on the other hand to the ability to control the critical evaluation protocol. Automating the valuation process, ensured by computerization of real estate data properly detected, provides a concrete answer to the problem of objectivity, allowing a significant control of the subjective component which characterizes the more traditional appraisals (Renigier-Biłozor et al. 2014a, b).

The method proposed in this work is based on the use of the Market Comparison Approach and the selection of comparable takes place in a market segment very limited in spatial terms, so that the biasing effect related to the location of the property is essentially canceled, a variable which, more than any other real estate feature, affects in price formation and which a regression coefficient can hardly represent. The most interesting aspect of the work done is represented by the extreme flexibility of use of the proposed method. The procedure implemented in the Model Builder, while using data contained in the specific GIS, is still separate from it, and then has universality of application. The method proposed in this work, in fact, not providing the generalization inherent in regression models, can be easily used in contexts other than real estate pilot one, provided that the real estate database has the same structure as the original one. The results hint at the possibility of further investigation, the type WebGIS applications, the extension to new contexts and different types of real estate properties, the implementation of other valuation methods (d'Amato 2015), with significant changes in the aspects of detail, but with integrity of the methodological approach. It is important to emphasize the fact that the automated evaluation should not be in lieu of the role of the professional evaluator, but only to help, simplifying and speeding up the steps of the selection of comparable and application of procedural algorithms, and leaving to the critical ability of the professionals considerations and value judgments. Given the relative ease of implementation of automated procedures and the exportability of the same, the operation of the procedures is subject to the acquisition systematic, codified and consistent data property essential to the creation of a computerized database support. The real estate data are prerequisite to the application of appraisal procedures, not only in the construction phase of the methods themselves, but in the continuous supervision of the support database, which must be systematically monitored and updated.

References

Borst, R., & McCluskey, W. (2008). The modified comparable sales method as the basis for a property tax valuations system and its relationship and comparison to spatially autoregressive valuation models. In T. Kauko, & M. d'Amato (Eds.), *Mass appraisal methods: An international perspective for property valuers* (pp. 49–69). Chicester, UK: Wiley Blackwell.

Borst, R. A., Des Rosiers, F., Renigier, M., Kauko, T., & d'Amato, M. (2008). Technical comparison of the methods including formal testing accuracy and other modelling performance using own data sets and multiple regression analysis. In T. Kauko, & M. d'Amato (Eds.), *Mass appraisal an international perspective for property valuers.* Wiley Blackwell.

Bourassa, S., Cantoni, E., & Hoesli, M. (2007). Spatial dependence, housing submarkets, and house price predictions. *Journal of Real Estate Finance and Economics, 35*, 143–160.

Bourassa, S., Cantoni, E., & Hoesli, M. (2010). Predicting house prices with spatial dependence: A comparison of alternative methods. *Journal of Real Estate Research, 32*, 139–159.

Bourassa, S., Cantoni, E., & Hoesli, M. (2013). Robust repeat sales indexes. *Real Estate Economics, 41*(3), 517–541.

Case, B., Clapp, J., Dubin, R., & Rodriguez, M. (2004). Modeling spatial and temporal house price patterns: A comparison of four models. *Journal of Real Estate Finance and Economics, 29*, 167–191.

Ciuna, M., & Simonotti, M. (2014). Real estate surfaces appraisal. *AESTIMUM, 64*, 1–13 (Giugno).

Ciuna, M., Salvo, F., & Simonotti, M. (2014a). The expertise in the real estate appraisal in Italy. In *Proceedings of the 5th European Conference of Civil Engineering (ECCIE '14). Mathematics and Computers in Science and Engineering Series* (pp. 120–129), North Atlantic University Union. ISSN:2227-4588, Firenze, 22/24 November 2014.

Ciuna, M., Salvo, F., & De Ruggiero, M. (2014b). Property prices index numbers and derived indices. *Property Management, 32* (2), 139–153. DOI(Permanent URL):10.1108/PM-03-2013-0021

Ciuna, M., Salvo, F., & Simonotti, M. (2014c). Multilevel methodology approach for the construction of real estate monthly index numbers. *Journal of Real Estate Literature: 2014, 22* (2), 281–302.

Colwell, P. F., Cannaday, R. E., & Wu, C. (1983). The analytical foundations of adjustment grid methods. *Journal of the American Real Estate and Urban Economics Association, 11*(1), 11–29.

d'Amato M. (2004). A Comparison Between RST and MRA for Mass Appraisal Purposes. A Case in Bari, International Journal of Strategic Property Management, Vol.8, pp. 205–217.

d'Amato, M. (2008). Rough set theory as property valuation methodology: The whole story. In T. Kauko, & M. d'Amato (Eds.), *Mass appraisal an international perspective for property valuers* (Chap. 11, pp. 220–258). Wiley Blackwell .

d'Amato, M. (2010). A location value response surface model for mass appraising: An "iterative" location adjustment factor in Bari, Italy. *International Journal of Strategic Property Management, 14*, 231–244.

d'Amato, M. (2015). Income approach and property market cycle. *International Journal of Strategic Property Management, 29*(3), 207–219.

d'Amato, M. (2015a). Stima del valore di trasformazione utilizzando la funzione di stima. Il MCA a tabella dei dati ridotta, Territorio Italia, pp. 1–12.

d'Amato, M. (2015b). MCA a Tabella dei Dati Ridotta e Sistema Integrativo di Stima. Un Secondo Caso a Bari, Territorio Italia, pp. 49–60.

d'Amato, M., & Kauko, T. (2008). Property market classification and mass appraisal methodology. In T. Kauko, & M. d'Amato (Eds.), *Mass appraisal an international perspective for property valuers* (Chap. 13, pp. 280–303). Wiley Blackwell.

d'Amato, M., & Kauko, T. (2012). Sustainability and risk premium estimation in property valuation and assessment of worth. *Building Research and Information, 40*(2), 174–185 (March–April 2012).

d'Amato, M., & Siniak, N. (2008). Using fuzzy numbers in mass appraisal: The case of belorussian property market. In T. Kauko, & M. d'Amato (Eds.), *Mass appraisal an international perspective for property valuers* (Chap. 5, pp. 91–107). Wiley Blackwell.

Detweiler, J., & R. Radigan. (1996). Computer-assisted real estate appraisal. *Appraisal Journal,* January, 91–101.

Detweiler, J., & Radigan, R. (1999). Computer-assisted real estate appraisal: A tool for the practicing appraiser. *Appraisal Journal,* July, 280–286.

Dilmore, G. (1974). Appraising houses. *Real Estate Appraiser* July–August: 21–32.

Dubin, R. (1998). Estimation of regression coefficients in the presence of spatially autocorrelated error terms. *Review of Economics and Statistics, 70*(3), 466–474.

Fik, T., Ling, D., & Mulligan, G. (2003). Modeling spatial variation in housing prices: A variable interaction approach. *Real Estate Economics, 31*(4), 623–646.

Gau, G. W., Lai, T.-Y., & Wang, K. (1992). Optimal comparable selection and weighting in real property valuation: An extension. *Journal of the American Real Estate and Urban Economics Association, 20*(1), 107–124.

Gau, G. W., Lai, T.-Y., & Wang, K. (1994). A further discussion of optimal comparable selection and weighting, and a response to green. *Journal of the American Real Estate and Urban EconomicsAssociation, 22*(4), 655–663.

Goodman, A., & Thibodeau, T. (1998). Housing market segmentation. *Journal of Housing Economics, 7*, 121–143.

Goodman, A., & Thibodeau, T. (2003). Housing market segmentation and hedonic prediction accuracy. *Journal of Housing Economics, 12*(3), 181–201.

Goodman, A., & Thibodeau, T. (2007). The spatial proximity of metropolitan area housing submarkets. *Real Estate Economics, 35*, 209–232.

Graaskamp, J., & Robbins, M. (1987). *Business 868 lecture notes*. University of Wisconsin–Madison.

Isakson, H. R. (1986). The nearest neighbors appraisal technique: An alternative to the adjustment grid methods. *Journal of the American Real Estate and Urban Economics Association, 14*(2), 274–286.

Isakson, H. R. (1988). Valuation analysis of commercial real estate using the nearest neighbors appraisal technique. *Growth and Change, 19*(2), 11–24.

Kauko, T., & d'Amato, M. (2011). *Neighbourhood effect, international encyclopedia of housing and home.* Elsevier Publisher.

Manaster, M. S. (1988). Sales comparison approach: A comparative analysis of three appraisal technique. *Growth and Change, 19*(2), 11–24.

Manaster, M. S. (1991). Sales comparison approach: A comparative analysis of three appraisal reports on the same property. *The Real Estate Appraiser*, May, 12–26.

O'Connor, P. (2008). Automated valuation models by model-building practitioners: Testing hybrid model structure and GIS location adjustments. *Journal of Property Tax Assessment and Administration, 5*(2), 5–24.

Pace, R., & Gilley, O. (1997). Using the spatial configuration of data to improve estimation. *Journal of Real Estate Finance and Economics, 14*, 333–340.

Pace, R., & Gilley, O. (1998). Generalizing the OLS and grid estimators. *Real Estate Economics, 1*, 331–346.
Renigier-Biłozor, M., Dawidowicz, A., & Radzewicz, A. (2014a). An algorithm for the purposes of determining the real estate markets efficiency in land administration system. *Public Survey Review, 46*(336), 189–204.
Renigier-Biłozor, M., Wiśniewski, R., Biłozor, A., & Kaklauskas, A. (2014b). Rating methodology for real estate markets—Poland case study. *Pub. International Journal of Strategic Property Management*, 18(2), 198–212. ISNN:1648-715X.
Salvo, F., & De Ruggiero, M. (2011). Misure di similarità negli Adjustment Grid Methods. *AESTIMUM, 58*, 47–58. ISSN:1592-6117.
Salvo, F., & De Ruggiero, M. (2013). Market comparison approach between tradition and innovation. A simplifying approach. *AESTIMUM, 62*, 585–594. ISSN:1592-6117.
Salvo, F., Ciuna, M., & d'Amato, M. (2013). The appraisal smoothing in the real estate idices. In (a cura di): *Maria Crescimanno, Leonardo Casini and Antonino Galati, Dynamisc of land values agricultural policies* (pp. 99–111), Bologna: Medimond Monduzzi Editore International Proceeding Division. ISBN:978-88-7587-690-6, Palermo, 22–23/11/2012.
Salvo, F., Simonotti, M., Ciuna, M., & De Ruggiero, M. (in print). Measurements of rationality for a scientific approach to the Market Oriented Methods. *Journal of Real Estate Literature*. ISSN:0927–7544.
Simonotti, M. (2006). *Metodi di stima immobiliare* (pp. 177–246). Palermo: Dario Flaccovio Editore.
Vandell, K. D. (1991). Optimal comparable selection and weighting in real property valuation. *Journal of the American Real Estate and Urban Economics Association, 19*(2), 213–239.

Short Tab Market Comparison Approach. An Application to the Residential Real Estate Market in Bari

Maurizio d'Amato, Vladimir Cvorovich and Paola Amoruso

Abstract The Italian real estate market is characterised by the general absence of an organized set of data. Real estate market information is rarely available because is not repeatable in time and space, as in the case of exact sciences. Consequently, appraisal methods, which work on small samples, may be useful. The present work proposes a new methodology to calculate appraisal functions based on a reduced number of comparables (d'Amato, Stima del valore di trasformazione utilizzando la funzione di stima. Il MCA a tabella dei dati ridotta, Territorio Italia 2015a; d'Amato, MCA a Tabella dei Dati Ridotta e Sistema Integrativo di Stima. Un SecondoCaso a Bari, Territorio Italia b), named "Short Tab Market Comparison Approach". With reference to this method, MCA is used to define an appraisal function (Simonotti 1997) whose dependent variable is the price while independent variables are the characteristics of the analysed property. Specifically, this method has been applied to the residential market segment in Bari.

1 Introduction

The forecast of prices arising from possible transformations is one of the recurrent problems typical of investment property analysis. The value assumed by real estate properties after their placement on the market is a relevant factor in the forecasting worth assessment. In practice, this forecasting process typically ends with the

---

The paper has been written in strict cooperation among the three authors. Therefore the credit of the article should be equally divided among them.

M. d'Amato (✉)
DICATECh Technical University Politecnico di Bari, Bari, Italy
e-mail: madamato@fastwebnet.it

V. Cvorovich
Montenegro Business School, Podgorica, Montenegro

P. Amoruso
Lum Jean Monnet University, Casamassima, BA, Italy

M. d'Amato and T. Kauko (eds.), *Advances in Automated Valuation Modeling*,
Studies in Systems, Decision and Control 86, DOI 10.1007/978-3-319-49746-4_22

definition of an average price of the property per square meter, obtained as result of the transformation process. The aforementioned price frequently disregards available information, following the developed project. Existing literature has stressed the need to use the Sales Comparison Approach in case of inadequate market information, as a possible alternative to linear regression (Prizzon 2001). This study proposes a method for determining the appraisal function to forecast the market value, following transformation processes (Ciuna 2010; Ciuna 2011; Ciuna et al. 2014a; Ciuna et al. 2014b; Ciuna et al. 2015; Ciuna and Simonotti 2014). Typically, the application of MCA is a comparison between two properties, the former with known price and characteristics and the latter with unknown price and known characteristics. With reference to the investigated context, the MCA has been applied to contingent valuation of a property (Saltari 2011). In this regard, the comparison with a real property is not a significant element as it is necessary to evaluate and approximate actors' behaviour in a specific market segment. The present work was organized as follows: in the first instance a comparison between MCA and Short Tab was conducted. Secondly, the investigated method was applied to a sample of ten real estate residential properties under construction in Santo Spirito, a suburb of Bari, in order to determine their market value. In view of the performed analysis, future research directions are explained in the conclusion.

2 Short Tab MCA

The Market Comparison Approach is a comparative evaluation method used to determine the value of real estate properties. It is a method of appraising property by analyzing the prices of similar properties sold in the recent past and then making adjustments based on differences among the properties and the relative age of the other sale. According to this method, the determination of subject value is obtained as a result of member to member subtraction of two linear functions and additive estimation (Simonotti 2003), as explained by the following formula:

$$S - P_j = L/ - L/ + p'_{1,j'} + (x_{S,1} - x_{J,1}) + \cdots + p'_{n,j'}(x_{S,n} - x_{J,n}) \quad (1)$$

Since all observations are referred to the same market segment, as a result the formula can be approximated as follows:

$$S = P_j + p'_{1,j'}(x_{S,1} - x_{J,1}) + \cdots + p'_{n,j'}(x_{S,n} - x_{J,n}) \quad (2)$$

The analysed method is structured in three different steps. As a first stage the data table is organized, with characteristics in a row and comparables in a column. The marginal cost analysis of selected variables is then carried out, representing total price variation following the variation of one unit, with reference to one of the selected characteristics. Finally the valuation table is organized, with adjusted prices which define the price that each comparable would have if it had been in the same conditions as the subject. The process is formally synthesized in Formula (1). Short Tab MCA has the ultimate aim of defining an appraisal function to evaluate contingent properties. According to a financial interpretation contingent properties would exist upon the occurrence of certain events or conditions. Real estate properties under construction fall into this category, since they depend on the implementation of investment decisions. VBT, the value obtained as result of the transformation process, is calculated as follows:

$$V_{BT} = L_0 + V_{fj} \quad (3)$$

In Formula (3) V_{BT} refers to the future value of the residential property built on the land, L_0 is the location variable while the last term is the vector of characteristics of the contingent good to be built. Specifically the last term can be described as follows:

$$Vfj = p_{1,j'}' x_{J,1} + \cdots + p_{n,j'}' x_{J,n} \quad (4)$$

This rapport is obtained by using a data table which does not refer to a specific subject to estimate. Marginal costs are defined at a later stage, by applying a traditional MCA, while values of the appraisal function Vfj are determined in the third phase. These are particularly relevant since they allow for an initial quantification of the function. Locational variables as well as other qualitative variables are isolated by determining the difference between Vfj and the transaction price. The final evaluation table, which follows the quantification of all variables involved, includes all marginal costs with related acronyms. After the initial application on the real estate market segment of second homes in the province of Bari, the analysed methodology will focus on a residential construction project in a suburb of Bari, called Santo Spirito.

3 Case Study: Bari

The investigated case study is the third application of Short Tab MCA. As mentioned above, the first focused on the real estate market segment of second homes in Bari (d'Amato 2015a; Kaklauskas et al. 2012; d'Amato and Kauko 2008; d'Amato and Siniak 2008; Kauko and d'Amato 2008; Kauko and d'Amato 2011), while the second was in the sector of residential property (d'Amato 2015b).

The present case study has the final aim of forecasting the placing value of real estate properties in the suburb of Santo Spirito. Needless to say, evaluation may also be affected by the real estate market cycle (d'Amato 2015; d'Amato 2010; d'Amato and Kauko 2012; d'Amato 2004; d'Amato 2008). The ground cubage allows for the construction of 10 real estate units and a building project is now under construction for which it is necessary to determine a possible placing value. In the present work Short Tab MCA is integrated with a supplementary appraisal method; the latter allows for the definition of the marginal cost of an *inaestimabilis* variable along with a locational variable (Table 1).

Table 1 Short Tab MCA

	A	B	C
PRZ	€150,000.00	€120,000.00	€100,000.00
DAT	5	2	4
SUB	90	93	95
SUP	10	15	13
SUBX	12	12	15
PROX	1	1	0

In the Short Table, defined as such since the analysed subject is not a part of it, comparables A, B and C are highlighted in the column, while characteristics considered significant are identified in the row. DAT variable is related to months and is retrospectively counted; SUP represents the main area in square meters, determined through cardinal measurement while SUB variable, cardinal measured as well, is expressed in square meters. SUBX variable represents the area of the relative pit, with cardinal measured in square meters. Selected comparables are placed near to the area under consideration and have similar characteristics. Specifically, they are residential buildings, with a rigid frame. Commercial ratios collected in the market segment are as Table 2.

Marginal cost of the main area is determined via the ratio between prices in relative areas multiplied by commercial ratios:

$$
\begin{aligned}
p_A{}'(SUP) &= \frac{150{,}000\,€}{90 + 10 \cdot 0,3 + 12 \cdot 0,4} = 1{,}533.74\,€/mq \\
p_A{}'(SUP) &= \frac{120,000\,€}{90 + 12 \cdot 0,3 + 15 \cdot 0,4} = 1{,}169.59\,€/mq \\
p_A{}'(SUP) &= \frac{100{,}000\,€}{95 + 13 \cdot 0,3 + 15 \cdot 0,4} = 953.29\,€/mq
\end{aligned}
\tag{5}
$$

Marginal cost, expressed as the minor among calculate average, is 953.29 €/m^2. Marginal cost of the balcony area can be easily obtained, after calculating the marginal cost of the main area

$$p\,'(SUB) = 953.29\,€/mq \times 0,3 = 285.99\,€ \tag{6}$$

and that of pits

$$p\,'(SUBX) = 953.29\,€/mq \times 0,4 = 381.32\,€ \tag{7}$$

Table 2 Commercial ratio and price index

Π balcony	0.3
Π box	0.4
S riv	0.01

The data is retrospectively calculated in months and, consequently, assumes a negative value.

$$p\,'(DAT) = -\frac{0.01}{12} \cdot = -0.00083 \qquad (8)$$

This is a percentage adjustment, which will be applied to a comparable price. At this stage, a value from appraisal function shall be determined. Specifically, the value from appraisal function, with reference to comparable A is determined as the sum of the below mentioned amounts, as shown in Table 3, by using estimated marginal costs.

VfA the value from appraisal function, with reference to comparable A, is 92,606.65 €. A similar process is applied to comparable B which is shown in Table 4.

VfB, the value from appraisal function, is 97,607.44 € with reference to comparable B. The same argument is proposed, in regard to comparable C, whose value is obtained as the sum of thc following marginal costs as shown in Table 5.

VfC, the value from appraisal function, is 99,633.33 € with reference to comparable C. A check previously expressed, which would need experimental verification was introduced at this stage of the process, in order to test whether divergence between calculated values was not excessive. In this regard, an upper limit for percentage divergence of 0,1 has been previously proposed (Simonotti

Table 3 VfA determination

DAT	€−625.00	5*(−0.00083)*€150,000 = −€ 625.00
SUP	€85,796.00	90*€1,048.62 = €94,375.60
SUB	€2,859.87	10*€314.59 = €3,145.85
SUBX	€4,575.79	12*€419.45 = €5,033.37
PROX	1	–

Table 4 VfB determination

DAT	€−200.00	2*(−0.00083)*€120,000 = €199.20
SUP	€88,655.86	93*€1,048.62 = €97,521.00
SUB	€3,431.84	12*€314.59 = €3,775.00
SUBX	€5,719.73	15*€419.45 = €6,291.70
PROX	1	–

Table 5 VfC determination

DAT	€−366.67	4*(−0.00083)*€110,000 = −€199.20
SUP	€90,562.44	95*€1,048.62 = €99,618.00
SUB	€3,717.83	13*€314.59 = €4,089.60
SUBX	€5,719.73	15*€419.45 = €6,291.70
PROX	0	–

2006). In the present case, the aforementioned limit is maintained, with a value of 0.0758. Determined prices are exclusive of two variables: the locational variable (LOC), whose existence distinguishes between average and marginal costs and the variable related to shore to a particular context (PROSP). These two variables are determined by integrating Short Tab MCA with a valuation system. The proposed process differs from the general valuation system based on the determination of marginal costs, obtained as the difference between the two appraisal functions, one referred to the subject and the other to the selected comparable, as described below:

$$P_j - S = p'_{1,j}(x_{1,j} - x_{s,j}) + \cdots \tag{9}$$

Likewise, it differs from the distribution system, whose aim is to spread prices for several variables, in order to obtain an average cost:

$$P_j = \overline{p}_{1,j} \cdot x_{1,j} + \overline{p}_{2,j} \cdot x_{2,j} \cdots \tag{10}$$

The proposed process is an integrated appraisal system, which links calculated values by using appraisal function (VfA; VfB; VfC) to the locational variable and additional *inaestimabilis* others. Specifically it is described as follows:

$$\begin{aligned} P_A - V_{fA} &= LOC + p_{PROSP'}X_{PROSP\_A} \\ P_B - V_{fB} &= LOC + p_{PROSP'}X_{PROSP\_B} \\ P_C - V_{fC} &= LOC + p_{PROSP'}X_{PROSP\_C} \end{aligned} \tag{11}$$

From which readily:

$$\begin{bmatrix} P_A - V_{fA} \\ P_B - V_{fB} \\ P_C - V_{fC} \end{bmatrix} = \begin{bmatrix} 1 & 1 \\ 1 & 1 \\ 1 & 0 \end{bmatrix} \begin{bmatrix} LOC \\ p'_{PROSP} \end{bmatrix} \tag{12}$$

In the present case it is:

$$\begin{bmatrix} 130{,}000€ - 92{,}606.65€ \\ 120{,}000€ - 97{,}607.44€ \\ 110{,}000€ - 99{,}633.33€ \end{bmatrix} = \begin{bmatrix} 1 & 1 \\ 1 & 1 \\ 1 & 0 \end{bmatrix} \begin{bmatrix} LOC \\ p'_{PROSP} \end{bmatrix} \tag{13}$$

With the Ordinary Least Squares, the system is solved as shown in Table 6.

The result of the integrated system determines the completion of the appraisal function, which can be described in the evaluation table (Table 7).

Table 6 Quantifying locational variable and proximity

€10,366.67	Locational variable
€19,526.29	Proximity variable

Table 7 Valuation short tab

Acronym	Variable	Amount
LOC	**LOCATION**	**€10,366.67**
SUP	MAIN AREA	€953.29
SUB	BALCONY AREA	€285.99
SUBX	PIT AREA	€381.32
DAT	DATA	−0.00083*P
PROX	**PROXIMITY**	**€19,526.29**

Variables with grey underlay were determined through integrated valuation system, while other variables were obtained by applying marginal cost theory, at the basis of the MCA method. The variable presented is a percentage adjustment of placing value, present in both the first and second sides of the balance. The problem is easily solved through a mathematical process, by means of which final determination is obtained. Indeed, in the light of that data table we can state that:

$$\begin{aligned} V_{BT} &= 10{,}366.67\,€ - 0.00083 \cdot V_{BT} \cdot \mathrm{DAT} + 953.29\,€ \cdot \mathrm{SUP} + 285.99 \cdot \mathrm{SUB}\ + \\ &+ 381.32\,€ \cdot \mathrm{SUBX} + 19{,}526.29\,€ \cdot \mathrm{PROSP} \end{aligned} \tag{14}$$

It follows that:

$$V_{BT} = \frac{10{,}366.67\,€ + 953.29\,€ \cdot \mathrm{SUP} + 285.99 \cdot \mathrm{SUB} + 381.32\,€ \cdot \mathrm{SUBX} + 19{,}526.29\,€ \cdot \mathrm{PROSP}}{(1 - 0.00083) \cdot \mathrm{DAT}} \tag{15}$$

The mathematical expression allows the value of property under transformation to be determined, by including locational and other qualitative variables which is shown in Table 8.

Table 8 Marginal cost application to characteristics of real estate properties to estimate

	Loc	Sup	Sub	Subx	Prox	Dat	Estimated value
1	1	90	9	10	1	6	**€122,686.97**
2	1	92	8	10	1	6	**€124,315.67**
3	1	90	10	12	1	6	**€123,740.84**
4	1	95	10	10	1	12	**€128,407.36**
5	1	102	8	12	0	6	**€115,038.70**
6	1	102	8	10	0	6	**€114,272.25**
7	1	90	10	15	0	12	**€105,795.99**
8	1	95	8	12	0	12	**€108,877.20**
9	1	97	8	12	1	15	**€130,854.80**
10	1	102	9	12	1	15	**€135,970.92**

As the calculation shows, each individual variable has been considered. Each real estate property has an intrinsic placing value, depending on its specific characteristics.

It is also possible to provide an interval measure of the value. These ranges can be determined through statistical methods. In probability theory, Central Limit Theorem states that if a population is normally distributed with reference to a phenomenon, all extracted samples will do the same. This claim remains valid even if population is not normally distributed, but is sufficiently numerous. Generally, if the population is normally distributed, it is possible to determine the statistical confidence within which its average is included by referring to the standard normal curve or, more precisely, by means of the following formula:

$$\bar{x} - z_{\alpha/2} \cdot \frac{\sigma}{\sqrt{n}} \leq \mu \leq \bar{x} + z_{\alpha/2} \cdot \frac{\sigma}{\sqrt{n}} \tag{16}$$

where X is the sample average, μ the population average and σ standard population deviation; $z_{\alpha/2}$ is the normal standard variable and n represents the size of the sample. The same method may be used, even if the data sample is not normally distributed, as long as standard population deviation is known. If information about standard deviation is not available, reference can be made to the *Student's t*-distribution variable:

$$\bar{x} - t_{(\alpha/2;n-1)} \cdot \frac{s}{\sqrt{n-1}} \leq \mu \leq \bar{x} + t_{(\alpha/2;n-1)} \cdot \frac{s}{\sqrt{n-1}} \tag{17}$$

where x is the sample average, μ is the population average and s is sample standard deviation; $t_{(\alpha/2;\ n-1)}$ is the student t-distribution for n−1 degrees of freedom, n represents the size of the sample, and consequently, n−1 are degrees of freedom. Nevertheless, results show some approximations. Further measures, which allow for the determination of statistical confidence in the case of a gap in information regarding character distribution, are the domain of definition and Tchebysheff's inequality. The latter is represented by the following formula, with knowledge of average and sample standard deviation:

$$(\bar{x} - k\sigma \leq \mu \leq \bar{x} + k\sigma) \geq 1 - \frac{1}{k^2} \tag{18}$$

where x is the sample average, μ is population average and σ is sample standard deviation; k is a symmetric range depending on which probability theory is applied. In the present context Formula (17) is applicable as in Table 9:

Table 9 Interval estimation of Short Tab MCA final results

	Observations (€)	Average (€)	Stand. Dev. (€)	n	$t^{(\alpha/2;\ n-1)}$:
A	92,606.65	99,615.81	3,616.78	2.00	4.30
B	97,607.44				
C	99,633.33				

Table 10 The interval valuation

	Value	Percentage
Minimum	€85,618.78	0.1284
Media	€96,615.81	
Maximum	€107,612.83	0.1138

Hence it is possible to write:

$$96{,}615.81\,€ - 4{,}30\frac{3{,}616.78\,€}{\sqrt{2}} < 96{,}615.81\,€ < 96{,}615.81\,€ + 4{,}30\frac{3{,}616.78\,€}{\sqrt{2}} \quad (19)$$

Consequently the variation of the price is described in the final Table 10.

The values may be 13 % less than the mean or may increase by up to 11 % on the mean. The same consideration may be applied to the selling price of the property to be built.

4 Conclusions

The second Short Tab MCA application was combined with an integrated appraisal system, in order to determine the role and influence of *inaestimabilis* variables. The investigated process showed a potential implementation, even though it is needed to develop additional researches about divergence upper limit among all values, obtained by using the appraisal function Vfj, which in the present case is 0.1. The aforementioned limit should be tested on broader range of case studies in order to be confirmed. A further interesting comparison may concern the determination of locational variables, by using Short Tab MCA and a linear regression model. Such comparisons could lead to indirect validation of the proposed model. In conclusion, the possible implementation of nonlinear models could be considered, in order to determine the appraisal function. Furthermore, the importance of the analysed process is related to its use as a professional tool.

References

Ciuna, M. (2010). L'Allocation method per la stima delle aree edificabili (Vol. 57, pp. 171–184). AESTIMUM.

Ciuna, M. (2011). The valuation error in the compound values. AESTIMUM [S.l.] (pp. 569–583), August 2013. ISSN 1724-2118.

Ciuna, M., Salvo, F., & De Ruggiero, M. (2014a). Property prices index numbers and derived indices. *Property Management*, *32*(2), 139–153. DOI (Permanent URL). 10.1108/PM-03-2013-0021.

Ciuna, M., Salvo, F., & Simonotti, M. (2014b). Multilevel methodology approach for the construction of real estate monthly index numbers. *Journal of Real Estate Literature, 22*(2), 281–302.

Ciuna, M., Salvo, F., & Simonotti, M. (2015). Parametric measurement of partial damage in building. *Proceedings of XLIV INCONTRO DI STUDI Ce.S.E.T. Il danno. Elementi giuridici, urbanistici e economico-estimativi* (pp. 171–188). Bologna, Italy. November 27–28, 2014. ISBN: 978-88-99459-21-5.

Ciuna, M., & Simonotti, M. (2014) Real estate surfaces appraisal. *AESTIMUM 64 Giugno*, 1–13.

d'Amato (2010). A location value response surface model for mass appraising: An "Iterative" location adjustment factor in Bari, Italy. *International Journal of Strategic Property Management, 14*, 231–244.

d'Amato, M. (2015). Income approach and property market cycle. *International Journal of Strategic Property Management, 29*(3), 207–219.

d'Amato, M., & Kauko, T. (2012). Sustainability and risk premium estimation in property valuation and assessment of worth. *Building Research and Information, 40*(2), 174–185.

d'Amato, M. (2004). A comparison between RST and MRA for mass appraisal purposes. A Case in Bari. *International Journal of Strategic Property Management, 8*, 205–217.

d'Amato, M. (2008). Rough set theory as property valuation methodology: The whole story. Chapter 11, In T. Kauko & M. d'Amato (Eds.), *Appraisal an international perspective for property valuers* (pp. 220–258). Wiley Blackwell.

d'Amato, M. (2015a). Stima del valore di trasformazione utilizzando la funzione di stima. Il MCA a tabella dei dati ridotta, Territorio Italia, pp. 1–12.

d'Amato, M. (2015b). MCA a Tabella dei Dati Ridotta e Sistema Integrativo di Stima. Un Secondo Caso a Bari, Territorio Italia, pp. 49–60.

d'Amato, M., & Kauko, T. (2008). Property market classification and mass appraisal methodology. Chapter 13, In T. Kauko & M. d'Amato (Eds.), *Mass appraisal an international perspective for property valuers* (pp. 280–303). Wiley Blackwell.

d'Amato, M., & Siniak, N. (2008). Using fuzzy numbers in mass appraisal: The case of belorussian property market. Chapter 5, In T. Kauko & M. d'Amato (Eds.), *Mass appraisal an international perspective for property valuers* (pp. 91–107). Wiley Blackwell.

Kaklauskas, A., Daniūnas, A., Dilanthi, A., Vilius, U., Lill Irene, Gudauskas, R., D'Amato, M. et al. (2012). Life cycle process model of a market-oriented and student centered higher education. *International Journal of Strategic Property Management, 16*(4), 414–430.

Kauko, T., & d'Amato, M. (2011). *Neighbourhood effect, international encyclopedia of housing and home.* Edited by Elsevier Publisher.

Kauko, T., & d'Amato, M. (2008). Introduction: Suitability issues in mass appraisal methodology. In T. Kauko & M. d'Amato (Eds.), *Mass appraisal an international perspective for property valuers*, (pp. 1–24). Wiley Blackwell.

Prizzon, F. (2001). *Gli Investimenti Immobiliari Analisi di Mercato e Valutazione Economico-Finanziaria degli Interventi*. Torino: Celid.

Renigier-Biłozor, M., Dawidowicz, A., & Radzewicz, A. (2014a). An algorithm for the purposes of determining the real estate markets efficiency in land administration system. *Pub. Survey Review., 46*(336), 189–204.

Renigier-Biłozor, M., Wiśniewski, R., Biłozor, A., & Kaklauskas, A. (2014). Rating methodology for real estate markets—Poland case study. *Public International Journal of Strategic Property Management, 18*(2), 198–212 ISNN. 1648-715X.

Saltari, E. (2011). *Appunti di economia finanziaria*. Bologna: Esculapio Economia.

Simonotti (2006) Metodi di Stima Immobiliare, Ed. Flaccovio, Palermo.

Simonotti M. (1997) La Stima Immobiliare, Utetlibreria, Torino.

Simonotti, M. (2003). L'analisi estimativa standard dei dati immobiliari, Genio Rurale, n.10, pp. 26–35.

Conclusions

Maurizio d'Amato and Tom Kauko

In this book five connected themes were brought up. The first one is the approach to the theoretical foundations of the AVM, the second one is an analysis of several experiences of using AVMs from various circumstances institutional contexts. In the third part. Based on these experiences—ranging from diagnosed system failures to best practices as well as more theoretically oriented literature—the question is as to what kind of challenges lie ahead in promoting this approach. Further challenges are the use of spatio and spatio temporal modelling and how to address the problem of Automated Valuation inputs, what the suggested pathways to tackle these challenges would be. At the end for each part some questions can be raisen.

The first part identifies some emerging problems. It starts with an introductive contribution on the non-agency crisis. Although the crisis can be seen from different points of view, the contributors to the book decided to interrogate the current methodology, either in academia or in practice. For this reason, we (i.e. the editors of this book) began by examining the root of the valuation process: the relationship between property valuation and mathematical modelling in AVM. After that another view was presented by Mooya: linking AVMs to NCE theory. The main question raised from this part is the role of the AVM in the non agency mortgage crisis and the relationship between the sophisticated automatic valuation method and the valuation in persons.

The second part comprised contributions of the application of AVM valuation in Germany and Italy and the relationship between real estate and financial sector in emerging markets. Eilers and Kunert examine the specific legal and regulatory standards for the valuation of property used as collateral in Germany and conclude that, without high quality, reliable and up-to date property data, to produce reliable market values and mortgage lending values would be impossible. Ciuna and Simonotti present an estimative model in Italy. Arslanh and Pekdemir explain Emerging market of Turkish REIT's performance, concerning finance sector reforms and REIT's with respect to Basel III requirements. The cross-national team of d'Amato, Cvorovich and Amoruso try to create an appraisal "function" in emerging real estate market based on few comparables in the city of Podgorica (Montenegro), based on a Short Tab Market Comparison Approach originally proposed to provide a forecast for the price of residential properties to be built on

M. d'Amato and T. Kauko (eds.), *Advances in Automated Valuation Modeling*,
Studies in Systems, Decision and Control 86, DOI 10.1007/978-3-319-49746-4

underdeveloped land. Ciuna, Salvo and Simonotti implement an automatic procedure based on the Market Comparison Approach and implemented through the Model Builder tool of ArcGis, with the aim to define equations related to the peculiarities of a marketplace in a very circumscribed area (again in Italy, where a large share of the empirical evidence for this compilation comes from). In the last paper of this part, Marina and colleagues document an experimental study in Cosenza (Italy), in the context of the Income Approach: an automated valuation model of the capitalization rate based on a real estate database built on a computerized geocoding automatic procedure. The question raised in this part are related to the procedure itself and how the procedures proposed address the results in several contexts even with limited number of data.

The third part highlighted some methodological challenges of AVM. This discussion concerns theoretical and empirical trends on the topics of spatial variables, the use of multilevel modelling to improve deterministic modelling, the role of non-deterministic reasoning, and finally valuation models and their inputs. Del Giudice and De Paola find that, in the central urban area of Naples (Italy), some of the observed variables can have non-linear relationships with the response variable, and therefore kriging techniques are combined with additive models to obtain the geoadditive models. Based on Italian evidence, Curto, Fregonara and Semeraro demonstrate that the liquidity of the market, proxied by the time on the market and the discount ratio, is not associated with geographical submarkets. The income approach is revisited by d'Amato, Coskun and Amoruso in discussions on risk premium factor in the regressed DCF application. After that, LVRS Modelling for AVM has prospects for dealing with spatial autocorrelation, d'Amato and Amoruso suggest. Bidanset and colleagues use geographically weighted regression (GWR) to improve upon accuracy of ordinary least squares (OLS)-based AVMs; here the impacts of various kernel and bandwidth combinations employed in GWR models are crucial for optimal results. Keskin, Watkins and Dunning demonstrate the advantages of using multi-level modelling when the appropriateness of the individual model is contingent upon the characteristics of the housing market in question; in circumstances where the housing market is spatially granulated and data is sparse, distinguishing impacts of environmental event or externalities on different spatial segments of the market as we move away from the source requires that we extend the standard model; in this particular case the impact of a natural disaster on the housing market in Istanbul (Turkey). Then again Ciuna, Salvo and Simonotti design an experimental mass appraisal model by collecting a data sample of sales of apartments in the city of Palermo (Italy), in the five years 2008–2012, and by formulating and testing a multivariate statistical model (multilevel model) within in a scheme of online real estate valuations. In this part the fundamental question to address is how location deals with the AVM.

The fourth part described and evaluates different AVM approaches. Stumpf González designed a two level—fuzzy system with multiple regression hedonic price models. This scheme is based on two main characteristics of real estate market: location and segmentation in sub-markets; here fuzzy logic permits to take in account continuity of variables, and in this way, is well-placed to offer improvement of AVM systems. Elsewhere, according to d'Amato and Renigier-Biłozor lack or unavailability of data in same cases affords limited possibilities for using statistical methods; therefore, rough set theory was applied on a small sample of commercial properties in the city of Bari (Italy), by taking into account the specific nature of information referring to this market. The question raised in this part are related to the relationship between modelling and observations in specific contexts where data are not abundant or may present a relationship between dependent and independent variable that may be modelled using different approaches.

The fifth part discussed inputs and models. First McCluskey and Borst pointed out that valuation—in particular the sales comparison approach—is often described as being the application of 'art' or a 'science' or a combination of them both. The distinction is more readily applicable to the techniques employed to calculate value and not necessarily to the underlying concept of valuation; in reality advances in research can minimize the subjectivity associated with comparable selection and determination of variable weights, and therefore, the sales comparison approach can be viewed as a scientific approach. Lausberg and Dust, in turn, demonstrated how the anchoring effect can be reduced with a tool helping the valuer to make better decisions. In this experiment an office building was appraised with the help of a self-written valuation software. The question raised in this part are related to the input of the models and the relationship between models and variable.

The book demonstrated the strategic role of valuation automata and the wide range of possible uses that can be done using various analytical and computing methods. From the five parts of the book it is quite clear that the final word on the important question raised in the book is far from declared. The difficulty here is that, although AVM techniques become more and more sophisticated, we must admit that our solution are sometimes slower than the change we observe in the real estate market. Although this it is quite evident how the stability of our economic systems depends on these valuation system combined with in person valuation in different ways further contributions debate and analysis are more and more required to improve and deal with the unknown as we did in the previous book on mass appraisal.

Glossary

Maurizio d'Amato and Tom Kauko

Accessibility The number of opportunities available within a certain distance or travel time. An important attribute in the property basket, but due to its difficulty to measure is often approximated as CBD distance

The analytic hierarchy process (AHP) A technique for eliciting relative weights for competing elements (attributes or alternatives). Based on pair-wise comparison of elements it produces a comparison matrix in which the relative importance of each element is determined (the Saaty method of elicitation)

Artificial intelligence (AI) The study and design of 'intelligent agents', where an intelligent agent is a system that perceives its environment and takes actions which maximizes its chances of success

Artificial neural network (ANN) A powerful data modelling tool that is able to capture and represent complex input/output relationships. The motivation for the development of neural network technology stemmed from the desire to develop an artificial system that could perform 'intelligent' tasks similar to those performed by the human brain

Automated valuation method (AVM) Automated, often computerised procedures for carrying out the task of valuing one or more properties

Comparable sales The most comparable is the least dissimilar

Computer assisted mass appraisal (CAMA) Systems for particular tax assessment, where sophisticated AVMs are utilised

Deterministic A certain relationship between two variables that does not involve probability (Opposite to stochastic.)

Error term The explained variation of the model; the residuals.

Externality Cost or benefit which accrue to bodies other than the one sponsoring the project. Side-effect of economic activity that may result in nuisance for individuals

M. d'Amato and T. Kauko (eds.), *Advances in Automated Valuation Modeling*,
Studies in Systems, Decision and Control 86, DOI 10.1007/978-3-319-49746-4

Fuzzy logic Derived from fuzzy set theory dealing with reasoning that is approximate rather than precise

Genetic algorithm (GA) A search technique used in computing to find solutions to optimization and search problems. A particular class of evolutionary algorithms (evolutionary computation) that use techniques inspired by evolutionary biology such as inheritance, mutation, selection and crossover (recombination)

Geostatistical model A model based on statistical methods developed for and applied to geographic data. These statistical methods are required because geographical data does not usually conform to the requirements of standard statistical procedures, due to spatial autocorrelation and other problems associated with spatial data

Geographical information system (GIS) A computer system for capturing, storing, checking, integrating, manipulating, analysing and displaying data related to positions on the Earth's surface. Typically, a Geographical Information System (or Spatial Information System) is used for handling maps of one kind or another. These might be represented as several different layers where each layer holds data about a particular kind of feature. Each feature is linked to a position on the graphical image of a map

Geographically weighted regression (GWR) An estimation and modelling technique for analysis of spatially varying, local relationships in multivariate datasets, based on a MRA framework

Hedonic price modelling A theoretical-methodological framework for estimating the implicit prices of attributes pertaining to heterogeneous goods. Original contributions were in price index research. Today widely used in empirical property price studies and automated property valuation

Hierarchical trend model In the hierarchical trend model a general trend, local and house type time components and specific housing characteristics play a role. It is a hierarchical model in which both a general trend and cluster level aspects as deviations from the general trend are specified

Implicit price The monetary worth of one attribute on the property basket; the part price of the property; the market adjustment factor; hedonic price. In a hedonic regression model the implicit prices and the error term sum in the total price estimate

Location adjustment factor It is a component of a hedonic model calculated on the distance between an observation and a Value Influence Center. It may be related both to the error term and to the mathematical relationship between the distance and the observation

Location value response surface models A multiple regression model integrated with a Location Adjustment Factor or LAF in order to take into account proximity effect of central place of an urban area

Market segment Clusters that are combinations of location (in administrative or geographic terms) and house type

Market value The most probable price that should be paid for in competitive and transparent market setting

Mass appraisal A systematic appraisal of groups of properties using standardized procedures

Multiple regression analysis A statistical technique that enables explanation of a phenomenon based on known independent variables. It is based on identifying correlations between two variables, when controlling for as much other influences as possible. It provides a solid basis for identifying drivers and forecasting

Pattern recognition To classify dat (patterns) based on either *a priori* knowledge or on statistical information extracted from the patterns. The patterns to be classified are usually groups of measurements or observations, defining points in an appropriate multidimensional space. A subtopic of machine learning

Residuals of the model The error terms. The unexplained part of the variation

Rough set theory (RST) A technique for defining a causal relationship between attributes of an object through if then rules (Boolean algebra)

Self-organising map (SOM) A type of supervised competitive neural network, and is a type of flexible regression technique (the Kohonen Map). Mapping from a high-dimensional data space onto, a (usually) two-dimensional lattice of points

Spatial autocorrelation The degree of relationship that exists between two or more spatial variables, such that when one changes, the other(s) also change. This change can either be in the same direction, which is a positive autocorrelation, or in the opposite direction, which is a negative autocorrelation

Spatial autoregression (SAR) The spatial autoregression model is an extension of the linear regression model designed at accounting for spatial autocorrelation. In hedonic modelling, two types of SAR models are used, namely the spatial lag model (SLM), also referred to as 'mixed autoregressive model', and the spatial error model (SEM)

Spatial dependence Also referred to as spatial autocorrelation (SA), spatial dependence is concerned with the degree to which objects or activities located in space are similar (positive autocorrelation) or dissimilar (negative autocorrelation) to other objects or activities located nearby. According to Tobler's first law of geography, SA is assumed to decrease with distance. Several statistics do exist for measuring spatial dependence, with the Moran's I being the most widely used. Analogous to the conventional correlation coefficient, Moran's I usually ranges from +1 (strong positive SA) to –1 (strong negative SA)—although it is not formally constrained within these limits—, with 0 indicating the absence of spatial dependence

Spatial error model (SEM) The SAR-SEM model differs from the former one in that the autoregressive term accounts for the spatial influence of neighbouring model residuals. The SAR-SEM equation thus writes: $Y = \beta X + \lambda W\varepsilon + u$, where W is a distance-weighted spatial matrix of the residuals and λ an endogenous parameter accounting for the spatial influence of neighbouring residuals

Spatial lag model (SLM) With the SAR-SLM model, a dependent variable Y is regressed on a set of independent variables X to which is added a spatial autocorrelation term designed at modelling the strength of the spatial dependencies among the elements of the dependent variable (*i.e.* the neighbouring values of Y). The SAR-SLM equation thus writes: $Y = \beta X + \rho Wy + \varepsilon$, where W is a distance-weighted spatial matrix of the Y elements while ρ is an endogenous parameter accounting for the additional influence of neighbouring observed values on the dependent variable

Spatial weight matrix A spatial weight matrix provides the structure of assumed spatial relationships among a set of n observations (*e.g.* house prices) and contains a distance 'd' term for each of the $n \times n$ combinations of observations in the data set, with zero terms on the matrix diagonal. More often than not, 'd' terms are the inverse distance between observations or are either set to 0 or 1 if observations are connected, *e.g.* if they share a border (0) or a vertex (1)

Stochastic A relationship between two variables that involves a casual relationship, opposite to deterministic that requires a causal relationship

Structural time series model Time series models in which the observations are described in terms of its components of interest, such as trend, seasonal and regression components plus error

Valued tolerance relation (VTR) A procedure to deal with the indiscernibility relation in RST and make the technique more flexible

Value influence center Area or Location that influence property and value and prices in the surroundings in positive or negative ways

Printed in the United States
By Bookmasters